RECONSTRUCTING PAST MONASTIC LIFE VOLUME 1: BIOARCHAEOLOGY, LIFE AND DEATH

NEW TRENDS FROM ARCHAEOLOGICAL, BIOANTHROPOLOGICAL AND DOCUMENTARY PERSPECTIVES

Edited by

LLUÍS LLOVERAS, CARME RISSECH,
JORDI NADAL AND PHILIP BANKS

OXBOW | books
Oxford & Philadelphia

Published in the United Kingdom in 2025 by
OXBOW BOOKS
81 St Clements, Oxford OX4 1AW

and in the United States by
OXBOW BOOKS
1950 Lawrence Road, Havertown, PA 19083

Paperback Edition: ISBN 979-8-88857-173-6
Digital Edition: ISBN 979-8-88857-174-3 (epub)

A CIP record for this book is available from the British Library

Library of Congress Control Number: 2025934513

Printed in the United Kingdom by Short Run Press

Typeset in India by Lapiz Digital Services, Chennai.

For a complete list of Oxbow titles, please contact:

UNITED KINGDOM
Oxbow Books
Telephone (0)1226 734350
Email: oxbow@oxbowbooks.com
www.oxbowbooks.com

UNITED STATES OF AMERICA
Oxbow Books
Telephone (610) 853-9131, Fax (610) 853-9146
Email: queries@casemateacademic.com
www.casemateacademic.com/oxbow

Oxbow Books is part of the Casemate Group

Front cover: Top (clockwise): Manuscript from the Notre-Dame hospital in Tournai (15th century); Giotto di Bondone – The Death of St Francis from the Bardi Chapel; Medieval manuscript (https://commons.wikimedia.org/wiki/File:Medieval_manuscript_%288079764946%29.jpg); Human remains from Santa Caterina friary (Barcelona, Spain). Photo: Antony Cevallos. *Middle:* Santa Caterina cloister (Barcelona, Spain). Auguste Blanchard 1826, Arxiu Històric de la Ciutat de Barcelona. *Bottom (left to right):* Santa Caterina cloister (Barcelona, Spain). M. Redon, Arxiu Històric de la Ciutat de Barcelona; Livre de la vie active. Illuminated manuscript of the late 15th century, written by Jehan Henry (https://commons.wikimedia.org/wiki/File:Livre_de_la_vie_active.jpg).
Back cover: Human remains from Sant Agustí friary (Barcelona, Spain). Photo: Carme Rissech.

The Publisher's authorised representative in the EU for product safety is Authorised Rep Compliance Ltd., Ground Floor, 71 Lower Baggot Street, Dublin D02 P593, Ireland.
www.arccompliance.com

Contents

List of contributors

Walter Alegría Tejedor
Universitat de Barcelona (Spain)
walteralegria@ub.edu

Izaskun Ambrosio Arcas
Iltirta Arqueologia SL (Spain)
i.ambrosio.a@gmail.com

Philip Banks
Universitat de Barcelona (Spain)
pjbanks52@gmail.com

David Bennett-Jones
University of Reading (UK)
d.n.bennett-jones@pgr.reading.ac.uk

Annalisa Biselli
R.T. Restauro Tessile (Italy)
erreti@restaurotessile.it

Marc Bouzas
Universitat de Girona (Spain)
marc.bouzas@udg.edu

Francesco Breglia
Università di Padova (Italy)
francesco.breglia@unipd.it

Nuno Carolino
Instituto Nacional de Investigação Agrária e Veterinária (Portugal)
nuno.carolino@iniav.pt

Giuseppina Carta
Università degli Studi di Genova (Italy)
giuseppina.carta@edu.unige.it

Anna Castellano-Tresserra
Museu Monestir de Pedralbes (Spain)
acastellano@bcn.cat

Lídia Catarino
Universidade de Coimbra (Portugal)
lidiagil@dct.uc.pt

Araceli Coll Sanabra
Universitat de Barcelona (Spain)
araceli.coll@hotmail.com

Neus Coromina
Universitat de Girona (Spain)
neuscoromina@gmail.com

Ruy Dantas
ACERG (Portugal)
rui.dantas@cachena.pt

Mathilde Daumas
Université Libre de Bruxelles (Belgium)
Mathilde.Daumas@ulb.be

Elena Dellú
Soprintendenza Archeologia, Belle Arti e Paesaggio (Italy)
elena.dellu@pec.it

Andreu Falcó
Universitat de Barcelona (Spain)
andreufalcosanchiz@gmail.com

Marina Fernández-Liarte
Universitat de Barcelona (Spain)
Marina_ferli@hotmail.com

Carlos Fernández-Rodríguez
Universidad de León (Spain)
c.fernandez@unileon.es

Francesco Maria Galassi
University of Lodz (Poland)
francesco.galassi@biol.uni.lodz.pl

Gianni Gallello
Universitat de València (Spain)
gianni.gallello@uv.es

Marina Gascà
Universitat Rovira i Virgili (Spain)
marina.gasca@estudiants.urv.cat

Laura González-Garrido
Universidad de León (Spain)
bioantropologica@gmail.com

Eduardo González Gómez De Agüero
Universidad de León (Spain)
gomezaguero_@hotmail.com

Jesús Herrerín López
Universidad de Alcalá de Henares (Spain)
jesus.herrerin@uam.es

José V. Leite
ACERG (Portugal)
jose.leite@amiba.pt

Anna Llauradó
Universitat Rovira i Virgili (Spain)
llauradolopez1414@gmail.com

Lluís Lloveras
Universitat de Barcelona (Spain)
lluislloveras@ub.edu

Louise Loe
Oxford Archaeology (UK)
louise.loe@oxfordarchaeology.com

Susana Lopes
ACERG (Portugal)
acerg@mail.telepac.pt

Stephane Louryan
Université Libre de Bruxelles (Belgium)
stephane.louryan@ulb.be

Sílvia Marín Ortega
ESCRBCC (Spain)
smarin35@xtec.cat

María Martín-Seijo
Instituto de Ciencias del Patrimonio CSIC (Spain)
maria.martin-seijo@incipit.csic.es

Germán Martínez Correal
ASOCRIOLLANOS (Colombia)
germanmartinez@ganadocriollo-colombiano.com

Marta Merino
Universitat Rovira i Virgili (Spain)
martamerinop87@gmail.com

Álvaro Monge Calleja
Universidade de Coimbra (Portugal)
alvaromonge23@gmail.com

Jordi Nadal
Universitat de Barcelona (Spain)
jordinadal@ub.edu

Benjamin Neil
Cambridge University (UK)
bjn23@cam.ac.uk

Lluís Palahí
Universitat de Girona (Spain)
lluis.palahi@udg.edu

Ginevra Panzarino
Universitat de València (Spain)
gipan@alumni.uv.es

Pere M. Parés-Casanova
Universitat Oberta de Catalunya (Spain)
ppares@uoc.edu

Agustín Pastor
Universitat de València (Spain)
agustin.pastor@uv.es

António Pereira Coutinho
Universidade de Coimbra (Portugal)
cafe@bot.uc.pt

Donatella Pian
Soprintendenza Archeologia, Belle Arti e Paesaggio (Italy)
donatella.pian@cultura.gov.it

Israel Picón Platas
Citania Arqueoloxía S.L (Spain)
chartinez@gmail.com

Caroline Polet
Royal Belgian Institute of Natural Sciences (Belgium)
caroline.polet@sciencesnaturelles.be

David Eduardo Rangel
Universidad Cooperativa de Colombia (Colombia)
David.rangel@campusucc.edu.co

Santiago Riera
Universitat de Barcelona (Spain)
rieram@ub.edu

Carme Rissech
Universitat Rovira i Virgili (Spain)
carme.rissech@urv.cat

Rafael M. Rodríguez Martínez
Deputación de Pontevedra (Spain)
rafael.rodriguez@depo.es

Ana Luisa Santos
Universidade de Coimbra (Portugal)
alsantos@antrop.uc.pt

Andrea Sanz Boscà
Universitat de Barcelona (Spain)
sanzbosca@gmail.com

Arcesio Salamanca-Carreño
Universidad Cooperativa de Colombia (Colombia)
asaca_65@yahoo.es

Natasa Sarkic
AITA Bioarchaeology (Spain)
nsarkic@gmail.com

Maria Soler Sala
Universitat de Barcelona (Spain)
mariasoler@ub.edu

Abu B. Siddiq
Mardin Artuklu University (Turkey)
siddiq@artuklu.edu.tr

Stefano Vanin
Università degli Studi di Genova (Italy)
stefano.vanin@unige.it

Elena Varotto
Flinders University (Australia)
elena.varotto@flinders.edu.au

Mauricio Vélez-Terranova
Universidad Nacional de Colombia (Colombia)
ovelez@unal.edu.co

Jordi Vivo
Universitat de Girona (Spain)
jordi.vivo@outlook.com

Sofía Zdral
CSIC-SODERCAN-Universidad de Cantabria (Spain)
sofia.zdral@unican.es

List of reviewers

Carmen Abad Zardoya
Área de Historia del Arte, Universidad de Zaragoza, Spain

Bibiana Agustí
In situ SCP, Centelles, Spain

Silvia Albizuri
Departament d'Història i Arqueologia, Universitat de Barcelona, Spain

Núria Armentano
Laboratori d'Osteoarqueologia i Paleopatalogia, Museu d'Arqueologia de Catalunya, Spain

Guy Bar-Oz
Department of Archaeology, University of Haifa, Israel

Lucie Biehler-Gomez
Laboratory of Forensic Anthropology and Odontology (LABANOF), University of Milan, Italy

Jordi Bolós
Departament d'Història, Universitat de Lleida, Spain

Jo Buckberry
School of Archaeological & Forensic Sciences, University of Bradford, UK

Simon Davis
Laboratório de Arqueociências, DGPC, Portugal

Pablo C. Díaz
Departamento de Prehistoria, Historia Antigua y Arqueología, Universidad de Salamanca, Spain

Anton Ervynck
Erfgoedonderzoeker Natuurwetenschappen, Agentschap Onroerend Erfgoed, Belgium

Héctor R. Francisco
IMHICIHU-CONICET, Argentina

Erika Gal
Institute of Archaeology of the Research Centre for the Humanities, Hungarian Academy of Sciences, Hungary

Rebeca García Gonzalez
Departamento de Historia, Geografía y Communication, Universidad de Burgos, Spain

Roberta Gilchrist
Department of Archaeology, University of Reading, UK

Alexandra Ion
Human-Computer Interaction Institute, Carnegie Mellon University, USA

Andrew Jotischky
Department of History, Royal Holloway, University of London, UK

Nicholas Marquez-Grant
Cranfield Forensic Institute, Cranfield University, UK

Lluís Palahí
Càtedra Roses d'Arqueologia i Patrimoni Arqueògic, Universitat de Girona, Spain

Nicholas Passalacqua
Anthropology Department, The Ohio State University, Columbus, USA

Patxi Perez Ramallo
Department of Archaeology and Cultural History, Norwegian University of Science and Technology, Norway

Juan Antonio Quirós Castillo
Departamento de Geografía, Prehistoria y Arqueología, Universidad del Pais Vasco, Spain

Mateu Riera Rullán
Departament de Ciències de l'Antiguitat i l'Edat Mitjana, Universitat Autònoma de Barcelona, Spain

Santiago Riera
Departament d'Història i Arqueologia, Universitat de Barcelona, Spain

Felipe Serrano
Departamento de Patrimonio Histórico, Universidad de Jaén, Spain

Richard Thomas
School of Archaeology and Ancient History, University of Leicester, UK

Introduction

Lluís Lloveras, Carme Rissech, Jordi Nadal and Philip Banks

This book showcases recent developments in studies about monasticism and monastic life in the past. It represents the culmination of the MonBones research project, which concluded with the organisation of the conference *MonBones24: Reconstructing Past Monastic Life. Inferences from Archaeological, Bioanthropological and Documentary Perspectives*, which took place at the University of Barcelona and the Monastery of Santa Maria de Pedralbes in Barcelona, Spain, on 25 and 26 January 2024.

Monasticism is a religious way of life in which individuals devote themselves to spiritual practice, often characterised by asceticism, communal living or, alternatively, solitude. The monastic individual separates himself or herself from society either by living as a hermit or anchorite (religious recluse) or by joining a community (coenobium) of others who profess similar intentions (Johnston 2000).

Monasticism is observed in various religious traditions, including Christianity (*e.g.* Benedictines, Trappists), Buddhism (*e.g.* Theravāda monks, Tibetan lamas), Hinduism (*e.g.* sannyasis), and Jainism. Common features of monastic life include vows of poverty, chastity and obedience, as well as adherence to specific rules or codes of conduct that guide their daily lives and spiritual practices.

The Christian monastic movement began in eastern Mediterranean communities in Egypt, Palestine and Syria in the late 3rd and early 4th centuries (de Hoz 2009; Brooks Hedstrom 2018). During the 6th century, monastic communities began to spread to Europe and flourished during the early medieval period, particularly with the spread of Benedictine monasticism. The Rule of Saint Benedict, written around AD 530, became the foundation for Western monastic life, emphasising prayer (*ora*), work (*labora*) and community living. As monasticism developed, different orders emerged, which addressed what they considered the most pressing concerns of their time.

From their earliest years, monastic communities had a range of important functions, which included their role as economic, technological, cultural and spiritual centres (Gilchrist 2014; Lawrence 2015). Life in medieval monastic houses was highly regimented. Nuns and monks followed a strict schedule centred around the Divine Office, a series of prayers recited at specific times of the day and night. The day typically began with Matins before dawn and concluded with Compline in the evening (Knowles 1962). Aside from prayer, monks engaged in manual labour, such as farming and tending livestock, as well as intellectual pursuits like copying manuscripts, studying theology, and teaching. Monks also produced illuminated manuscripts, which exemplify the artistic achievements of medieval monastic communities (Brown 2003).

Monasteries were not only spiritual centres but also played crucial economic and social roles. They owned vast tracts of land, making them significant agricultural producers. The innovations in farming techniques developed by monks, including crop rotation and water mills, contributed to agricultural productivity (White 1962). Monasteries also served as centres of charity, offering food, shelter and medical care to the poor and pilgrims. The hospitality extended by monasteries reinforced their position as vital institutions within medieval society (Lawrence 2015).

Due to the importance that monastic houses achieved in the medieval and post-medieval periods, an in-depth study of monasteries is essential to help us to understand biological, social, economic, legal, religious, cultural and many other aspects of our past. In fact, the study of monastic life in the past has a long tradition that dates back to the end of the 19th century and has been based mainly on two types of sources, documentary and archaeological evidence, which have enabled us to gain deeper knowledge of the implementation, organisation and ways of life within monasteries. The most frequently used have been documentary sources, mainly texts. First of all, monastic rules have been widely used as an essential source for establishing what monastic life was like, but other books and manuals (accounts, briefs, midwives, etc.), letters, chancery records, etc. have also provided significant information (Davis 2018).

On the other hand, monastic and church archaeology has traditionally focused on the study of individual monuments and their buildings and landscapes (Gilchrist 2014). It is only in the last decades that the emphasis has shifted away from reconstructing the layouts of monastic buildings to a better understanding of the wider monastic environment. Monastic archaeology has become a distinctive branch of medieval archaeology that encompasses the study of religious buildings, landscapes, burials and archaeological material remains (Alvarez Palenzuela 1993; Gilchrist 2014). These archaeological remains include metal, pottery, glass, fabrics and other artifacts, and also bioarchaeological vestiges, such as human, animal, plant and other organic remains. All of them play a key role in the study of past human societies and in recent years, studies related to monastic contexts and based on these materials have proved to be essential to understand monastic life in the past. Special interest has been shown by studies based on human remains buried in monasteries and animal bones recovered from monastery rubbish dumps (Clavel 2001; Ervynck 2004; Gilchrist and Sloane 2005; Mays 2009; Pluskowski 2010; Serjeantson *et al.* 2018; Sarkic *et al.* 2019; among many others). These works are an essential source of a different type of data, providing direct evidence of the population under study and revealing information about very diverse questions such as diet, health, economy, culture and society.

It was with the aim of disseminating knowledge and the latest results on monastic studies derived from all the different topics of study mentioned above that we organised the MonBones24 conference. In total, 70 researchers from Australia, Belgium, Canada, China, Colombia, Croatia, Egypt, France, Germany, Iceland, Italy, the Netherlands, Norway, Portugal, Serbia, South Korea, Spain, the United Kingdom and the United States of America participated in the conference sharing emerging trends and discoveries, and engaging in discussions about studies focusing on palaeopathology, diet, health, economy, society, lifestyle, gender, architecture, monuments and landscapes within historical monastic contexts.

The multidisciplinarity of the contributions presented at MonBones24 supposes an unprecedented sharing of recent developments in studies about past monasticism coming from different points of view that resulted in the publication of the conference proceedings in two volumes. The chapters included in the two books are authored by many contributors to the MonBones24 conference, bringing together specialists from different disciplines that contribute new trends and findings in research on different topics and fields of study.

The first volume, *Reconstructing Past Monastic Life. Volume 1. Bioarchaeology, Life and Death* is dedicated to discussing health status and ways of life in monastic environments from data provided by human and animal remains, as well as other archaeological materials. **The second volume**, *Diet, Landscape and Monastic Space*, presents new findings about diet, buildings and landscapes, mostly based on animal bones, documentary sources and architectural analysis.

This volume **(Volume 1)** starts with three contributions that deal with a topic of study of great interest, the state of health and the study of the pathological conditions of people living in monastic houses. **Laura González-Garrido**'s chapter employs a multi-method approach to diagnose the possible aetiology for an unusual dental calculus growth displayed by a 40–50-year-old nun (18th–19th century) from the Dominican Convent of Santa Catalina de Siena (Cuenca, Spain). The macroscopic, radiological and chemical analyses point to facial paralysis as the most plausible diagnosis, increasing the palaeopathological record of this type of neuromechanical problems. This research shows how rigorous archaeological excavation techniques and refining dental calculus conservation methodologies are crucial for enhancing diagnostic accuracy and deepening our understanding of past populations. The study of the state of health is also the subject of **Mathilde Daumas**'s paper discussing, from a palaeopathological perspective, notable cases of disabled individuals displaying serious disorders, such as disproportionate dwarfism, bilateral calcaneus varus with evidence of crutch use, and cerebral palsy, in a Belgian medieval Cistercian monastic community. The prevalence of disabilities within this monastic community raises important questions about the social consequences of visible congenital anomalies in the Middle Ages. A third contribution on the health status of members of religious orders can be found in **David Bennett-Jones**'s chapter. Four segments of calcified artery from an elderly male with Diffuse Idiopathic Skeletal Hyperostosis, who was buried in a stone coffin in a 13th-century Lichfield Cathedral chapel, are analysed using focus-stacking photography. This is the first published archaeological example of calcified arteries found in association with the skeletonised remains of a named individual, providing crucial data about vascular disease and diet in medieval England.

Past monastic life pathways, explored in different contexts and from different perspectives, are the focus of interest of nine contributions. By means of bioanthropological analysis, **Andrea Sanz**'s paper provides new information on the life and health conditions of a sample of individuals buried in the friary of Santa Caterina (Barcelona) in the 13th to 15th centuries. Crucial data about both friars and lay people buried in different areas of the convent are discussed. **Marta Merino**'s contribution elucidates the lifestyle and health status of the individuals buried at the 9th–10th-century necropolis discovered at the Antic Jutjats Municipals site (Barcelona). A short life expectancy, a high infant mortality rate, and a wide range of skeletal indicators and postcranial pathologies point to a physically demanding lifestyle and to the likelihood of physical violence having taken place. **Carme Rissech**'s chapter analyses the human remains buried in Funerary Unit 228 at the friary of Santa Caterina in Barcelona. A total of 18 individuals are examined. The biological profile and pathological conditions together with the location of this funerary unit in the friary indicate that they correspond to a moderately prosperous family. **Marc Bouzas**'s paper presents the results of the work carried out in recent years in the cloister area of the monastery of Santa Maria de Roses (Girona, Spain), a space intensely used in the early modern period for burials. This research delves deeply into a type of population context that has been little studied. The military, monastic and funerary uses of this space are also analysed. **Ginevra Panzarino**'s study explores the life and death of the Poor Clare nuns of the Holy Trinity in Monte Sant'Angelo (Puglia, Italy). The archaeological, anthropological, pathological, botanical, entomological, textile, chemical and documentary data assembled provide considerable insight into the community's lifestyle, health, clothing and rituals as well as the composition of the community and the social relations between the nuns and the external world. **Maria Soler**'s contribution presents the results of the interdisciplinary study of the tomb of Prioress Jerònima de Gort (1586–1601), excavated at the female Hospitaller Monastery of Santa Maria d'Alguaire (Lleida, Spain). The approach takes into account archaeology, physical anthropology, written documentation and restoration techniques, rendering this tomb a particularly compelling case study for understanding the living conditions and funerary practices of a female community within the Hospitaller Order. **Benjamin Neil**'s chapter investigates the osteobiographical and osteological analysis of 11 individuals from the Cambridge Austin Friary, spanning the years 1302 to 1538. The research contextualises skeletal remains within broader historical, ethical and sociocultural frameworks. Integrating osteological evidence with historical and clinical knowledge, the study provides insights into pain, suffering and behaviour, enriching the understanding of past human experiences in religious and urban medieval contexts by proposing an 'Avatar Model' that allows for imaginative reconstructions of lived experiences in religious and urban medieval contexts. **Ginevra Panzarino** presents a second chapter on the reconstruction of the monastic community of the Teutonic monastery of San Leonardo di Siponto in Manfredonia (Puglia, Italy). The anthropological and palaeopathological data offer new insights into the lifestyle and health of the monastic community. **María Martín-Seijo**'s paper explores the everyday life of the female religious community at the convent of Santa Clara (Pontevedra, Spain) also from a multidisciplinary approach. The archaeological data, together with archaeobotanical and zooarchaeological analyses, provide an unprecedented amount of information on the various activities associated with life and death in this enclosed convent.

In the last section of this volume, three contributions focus on the study of animal evidence in the monastic environment. **Jordi Nadal**'s chapter analyses the paw prints on cloister tiles in the Monastery of Santa Maria de Pedralbes (Barcelona, Spain). The main species represented are dogs, cats and fowl, which were not food animals in medieval Europe. This study provides information about the origin of the tiles and other clay building materials in this monastery. **Pere Miquel Parés-Casanova**, through geometric morphometrics, evaluates the presence of directional and fluctuating asymmetries in horses. These observations can help us to establish the horses' rusticity and infer their use in different human contexts, such as, for example, in monasteries, friaries and nunneries, where horses, donkeys and mules were kept for different purposes. The last chapter in this first volume is by **Arcesio Salamanca-Carreño** and five co-authors. In their study, they evaluate cattle cephalic measurements in order to estimate withers height and body length through a linear equation. This formula can be useful in zooarchaeological studies, given the rarity of complete skeletal elements in archaeological cattle assemblages.

To conclude, this book presents a multitude of approaches to the study of monastic contexts demonstrating the current vitality and interest of this area of study as well as the increasing importance of the application of interdisciplinary methodologies to monastic studies.

References

Álvarez Palenzuela, V.A. (1993) Expansión de las órdenes monásticas en España durante la Edad Media. In J. I. de la Iglesia Duarte (coord.) *III Semana de Estudios Medievales (Nájera 3 al 7 de agosto de 1992)*, 161–78. Nájera, Instituto de Estudios Riojanos.

Brooks Hedstrom, D.L. (2018) The archaeology of early monastic communities. In D.K. Pettegrew, W.R. Caraher and T.W. Davis (eds) *The Oxford Handbook of Early Christian Archaeology*, 147–66. Oxford, Oxford University Press.

Brown, M. (2003) *The Lindisfarne Gospels: Society, Spirituality and the Scribe*. London, The British Library.

Clavel, B. (2001) L'animal dans l'alimentation médievale et moderne en France du nord (XIIè-XVIIè siècles). *Revue Archéologique de Picardie* 19, 1–204.

Davis, S.J. (2018) *Monasticism: A Very Short Introduction*. Oxford, Oxford Academic.

De Hoz J. (2009) Orígenes del monacato cristiano y de su arquitectura. *AXA* (Universidad Alfonso X el Sabio) 2, 1–19.

Ervynck, A. (2004) Orant, pugnant, laborant. The diet of the three orders in the feudal society of medieval north-western Europe. In S. O'Day, W. van Neer and A. Ervynck (eds), *Behaviour Behind Bones. The Zooarchaeology of Ritual, Religion, Status and Identity*, 215–23.Oxford, Oxbow Books.

Gilchrist, R. (2014) Monastic and church archaeology. *Annual Review of Anthropology* 43, 235–50.

Gilchrist, R. and Sloane, B. (2005) *Requiem: The Medieval Monastic Cemetery in Britain*. London, Museum of London Archaeology Service.

Johnston, W.M. (ed.) (2000) *Encyclopedia of Monasticism*, 2 vol. Abingdon, Routledge.

Knowles, D. (1962) *The Monastic Order in England: From the Times of St. Dunstan to the Fourth Lateran Council, 940–1215*. Cambridge, Cambridge University Press.

Lawrence, C.H. (2015) *Medieval Monasticism: Forms of Religious Life in Western Europe in the Middle Ages*. New York, Routledge.

Mays, S. (2009) The osteology of monasticism in medieval England. In R. Gowland and C. Knusel (eds) *The Social Archaeology of Funerary Remains*, 179–89. Oxford, Oxbow Books.

Pluskowski, A. (2010) The zooarchaeology of medieval 'Christendom': ideology, the treatment of animals and the making of medieval Europe. *World Archaeology* 42, 201–14.

Sarkic, N., Herrerín López, J., López-Costas, O. and Grandal-d'Anglade, A. (2019) Eating in silence: isotopic approaches to nuns' diet at the convent of Santa Catalina de Siena (Belmonte, Spain) from the sixteenth to the twentieth century. *Archaeological and Anthropological Sciences* 11, 3895–911.

Serjeantson, D., Crabtree, P., Mulville, J., Ayres, K., Ingrem, C. and Locker, A. (2018) How pious? How wealthy? The status of Eynsham and St Albans Abbeys between the 8th to the 12th centuries re-examined in the light of their food consumption. In B. Jervis (ed.) *The Middle Ages Revisited. Studies in the Archaeology and History of Medieval Southern England presented to Professor David A. Hinton*, 115–40. Oxford, Archaeopress.

White, L. (1962) *Medieval Technology and Social Change*. Oxford, Oxford University Press.

Part 1

State of health and human remains

1

Unusual and exuberant dental calculus in an 18th- to 19th-century nun from Santa Catalina de Siena, Belmonte (Cuenca, Spain): oral/joint dysfunction or facial paralysis?

Laura González-Garrido, Álvaro M. Monge Calleja, Natasa Sarkic, Sofía Zdral, António Pereira Coutinho, Lídia Catarino, Jesús Herrerín López and Ana Luisa Santos

*This study employs a multi-method approach to diagnose the possible aetiology for an unusual dental calculus growth in individual Sk.44, a 40- to 50-year-old nun (18th–19th-century) from the Dominican Convent of Santa Catalina de Siena (Cuenca, Spain). This individual preserved 25 teeth, five lost antemortem and both upper lateral incisors missing due to agenesis. The right maxillomandibular premolars and molars are covered with ~15mm of 'fern-like' greenish dental calculus, including on the occlusal surfaces. The remaining dentition shows minor beige calculus deposits, caries, periodontitis and linear hypoplasia. Scanning electron microscopy of the greenish calculus reveals filaments with small (<1 μm) ovoid fungal sporophytes. X-ray fluorescence shows a higher proportion of strontium in teeth compared to other samples. Contamination from funerary goods (rosary, crucifix and medals) was ruled out due to their location away from the oral cavity, but the use of antimicrobial remedies to relieve pain or inflammation cannot be excluded. Isotopic analysis indicates no significant dietary differences between this individual and the other nuns. The identification of plant starches indicates the consumption of wheat (*Triticum *sp.), possibly contaminated by the fungus *Ustilago tritici*. Although rarely documented in palaeopathological literature, anomalous dental calculus growths have been linked to unilateral mastication and inadequate oral hygiene. In individual Sk.44, there are no indications of malocclusion or missing teeth in the affected area. Differential diagnoses include oral dysfunction, temporomandibular joint disorders, trauma or buccal immobilisation due to partial facial paralysis. The macroscopic, radiological and chemical features found in individual Sk.44 point to facial paralysis as the most plausible diagnosis, increasing the palaeopathological record of this type of neuromechanical problems.*

Keywords: Dental anthropology, palaeopathology, SEM, pXRF, plant starch, facial paralysis.

Introduction

Bioarchaeological research on dental calculus has increased exponentially in a short period, having been described as a biographical, historical and behavioural reservoir from birth to death (Warinner *et al.* 2017; Akcali and Lang 2017; Radini and Nikita 2017; Radini *et al.* 2023). The origin of these deposits comes from the interaction between the complex microbial biofilm (Warinner *et al.* 2017) and the salivary and gingival crevicular fluid (Adler *et al.* 2013; Akcali and Lang 2017). Dental calculus comprises both organic (~15–25%) and inorganic (~75–85%) components (Greene *et al.* 2005). The inorganic portion is primarily composed of calcium phosphates precipitated from saliva and gingival crevicular fluid (Akcali and Lang 2017). The crystalline structure of the calculus passes through several stages during its formation in the mouth (Abraham *et al.* 2007). In mature deposits (more than six months), the most abundant crystalline form is hydroxyapatite ($Ca_{10}[PO_4]_6,OH_2$), with minor amounts of octacalcium phosphate ($Ca_8[HPO_4]_4$), whitlockite ($Ca_3[PO_4]_2$) and brushite ($Ca[HPO_4]\cdot 2H_2O$). A longitudinal study of the growth of calculi shows that brushite appears in the first stages, then octacalcium phosphate develops, and as they develop, whitlockite and hydroxyapatite arise (Abraham *et al.* 2007). The organic component is mainly composed of bacterial cells and their secreted matrix but may also include traces of food remains (*e.g.* starch grains, pollen, proteins), host biomolecules (*e.g.* DNA, proteins) and various environmental debris (*e.g.* smoke particles, pigments, textile fibres) (Singh *et al.* 2010; Warinner *et al.* 2017). Starch grains are increasingly used in archaeobotanical research as key indicators of diet, plant domestication, tool usage and site organisation (Aceituno-Bocanegra and López-Sáez 2012).

The accumulation of dental calculus on human teeth is a continuous and irreversible process (Hillson 2005), influenced by a complex array of factors such as diet, genetic variation, salivary composition, age, gender, ancestry and certain pathologies (Lieverse 1999). This multifaceted interplay renders the intra- and interpopulation variability of calculus a highly intricate and not yet fully understood phenomenon (Akcali and Lang 2017). Additionally, unusual locations, abnormal sizes and distinct pigmentations of calculus can provide insights into pathological conditions in past populations (Hansen and Alt 2012), though such findings are rare in the bioarchaeological record (Hanihara *et al.* 1994).

This investigation aims to analyse an asymmetrical and substantial deposition of supragingival dental calculus found in a nun exhumed from the cemetery of the Convent of Santa Catalina de Siena (Belmonte, Cuenca), and to explore potential aetiologies using an osteoarchaeological approach.

Individual and methods

Archaeological background and individual description

The fortress of Infante Don Juan Manuel in Belmonte (14th century; Cuenca, Spain, Fig. 1.1A) was converted into the Dominican Convent of Santa Catalina de Siena (Fig. 1.1B) in the 16th century and remained in use as a cloistered nunnery until 1960, when the last surviving nuns abandoned it (Caballero García and Sánchez Peláez 2013).

In 1836, the Spanish government conducted an inventory of this religious congregation's assets in accordance with the confiscation laws.[1] In this document the simplicity of life in the convent is evident.[2] The existence of two infirmaries located in the cloister could indicate the need for more space for the care of the sick or that the nuns were housed in infirmaries as in other Dominican convents in Spain (Sanz Fuentes 1993; Moreno González and Rubio Masa 2007).

During the refurbishment of the convent into a spa hotel in 2009, 116 individuals were exhumed (87.9% were female individuals) from two burial areas belonging to members of the monastic population, one located in the cloister and dating from the 16th to 17th century (27 tombs, 85 individuals), and the other in the choir area from the 19th to 20th century (45 tombs, 31 individuals) (Sarkic 2017). Individuals were buried in simple graves or in wooden coffins, some of them with religious objects (Caballero García and Sánchez Peláez 2013; Sarkic 2017).

According to the inscriptions engraved on some tombstones from the more recent burial area, the individual Sk.44 under study was inhumed between the years 1850 and 1921 (Caballero García and Sánchez Peláez 2013). Individual Sk.44 was buried in supine position within a wooden coffin, oriented west–east, according to the canons of Catholicism (Fig. 1.1C), with 15 rosary beads (Fig. 1.1D), a crucifix on top of the sternum (Fig. 1.1E), and a belt with a metal buckle around the waist (Fig. 1.1F). The location of the tomb and the grave goods allows it to be assumed that she was a nun from the convent (Caballero García and Sánchez Peláez 2013).

Methods

The human remains of individual Sk. 44 were macroscopically analysed in the laboratory using a lens (three dioptres) with good lighting. Sex and age at death were estimated according to standard methods (Lovejoy *et al.* 1985; Brooks and Suchey 1990; Buikstra and Ubelaker 1994; Bruzek 2002). The palaeopathological analysis followed the recommendations of Ortner (2003) and Neville *et al.* (2023). The macroscopic study of the cranium followed Tesorieri's (2016) description of limb asymmetry. The inclination of the articular eminence was then measured on dry skulls using the top-roof line method (Jasinevicius *et al.* 2006). The mandible was radiographed with a Siemens Vertix U device.

The criteria of Hillson (2001) were used to determine if any enamel hypoplasia or caries were present. Dental wear and calculus were recorded according to Smith (1984). Periodontal disease was evaluated using Kerr's method (1988).

The second maxillary right molar was selected for scanning electron microscope (SEM) analysis using a Tescan Vega 3 SBH device, without prior manipulation of the sample (Monge Calleja 2015). A small piece of calculus from this molar (Sk.44) was then subjected to the extraction and isolation of micro-remains, following the protocol established by Monge Calleja *et al.* (2021). Light microscope images were captured at 600× magnification using a Motic™ BA310 LED microscope with a digital camera.

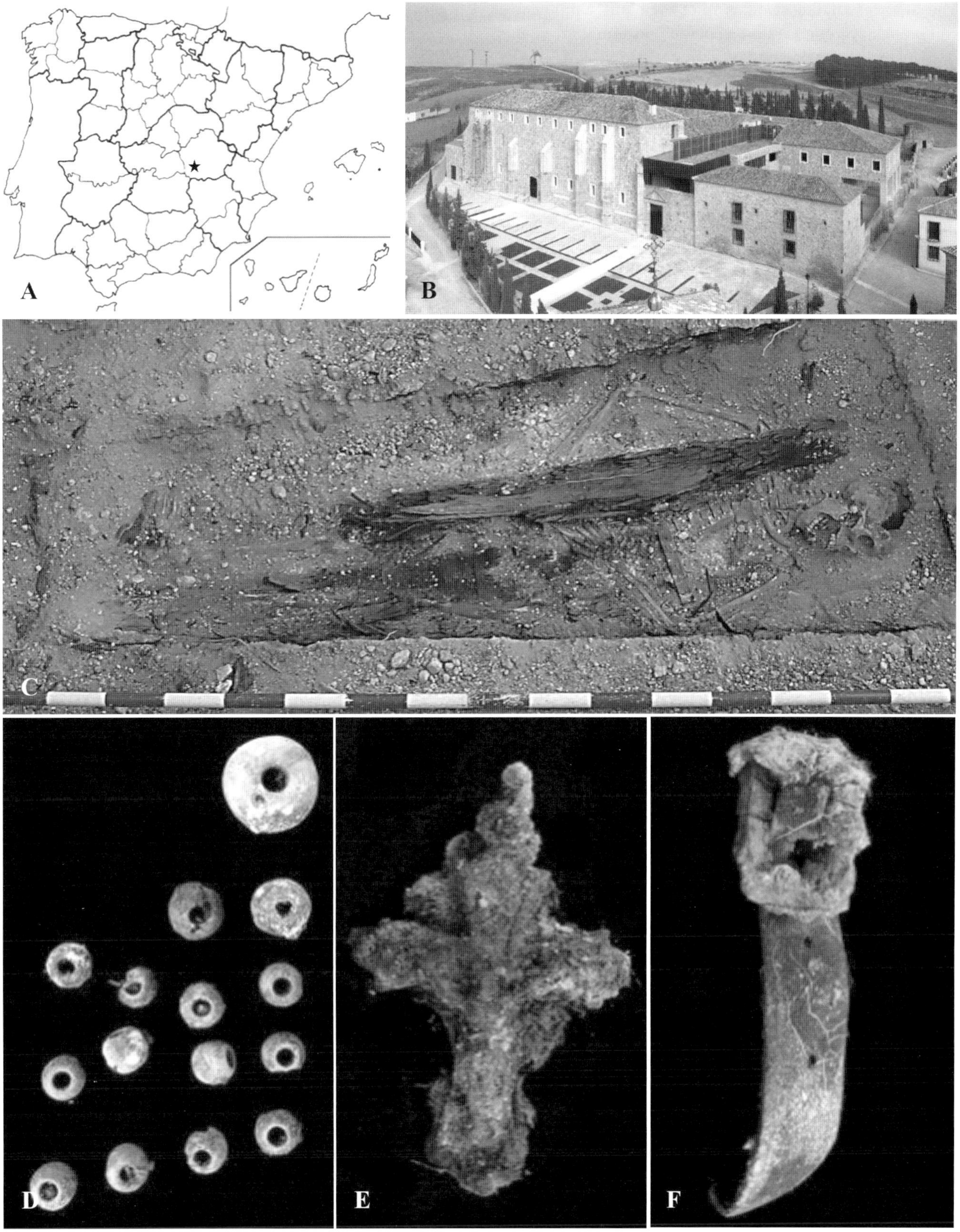

Figure 1.1. A: Geographical location of Belmonte (star) (Cuenca, central Spain). B: Dominican Convent of Santa Catalina de Siena currently operating as a hotel. C: Sk.44 during the exhumation. Sk.44 funerary grave goods: different sizes of rosary beads (D), a crucifix (E), and a belt with a buckle (F).

The analysis of the calculus micro-remains employed a comprehensive comparative approach, examining remains from plants, animals and crafts (*e.g.* Piperno *et al.* 2004; Aceituno-Bocanegra and López-Sáez 2012; Kruse *et al.* 2018; Watt 2023; Plant Parasites of Europe 2024). To avoid contamination, sterile Eppendorf tubes and sterile steel filters (mesh = 1/10 mm), disposable gloves and double-distilled water were used.

The elemental composition of the dental calculus of this tooth was quantified using a portable X-ray fluorescence (pXRF) Thermo Scientific™ Niton™ XL3t GOLDD XRF analyser, which measures up to a total of 43 chemical elements emitting a beam of 6–50kV and 0–200μA. The precision and the accuracy of those measurements were tested by the relative Technical Error of Measurement (Perini *et al.* 2005) and a Principal Component Analysis (PCA) was applied in order to investigate the dispersion of the elemental proportion both between individuals and teeth. Statistical data were calculated by Excel XLSTAT-Base v.2018 software.

For comparative purposes, analyses also included the examination of other individuals from the same contemporary religious cemetery: a canine (Sk.4, female, 40–50 y.o.), an incisor (Sk.42, female, undetermined age); and an incisor from an adult male from Arraiolos, Portugal (E31). The molar ratios of Ca/P were subsequently employed to estimate the primary crystallographic structure of the dental calculus by comparing them with the values of pure calcium phosphates (Abraham *et al.* 2005).

Results

Macroscopic and radiological examination

The skeletal remains of individual Sk.44 represent a female aged 40–50 years whose skeleton was poorly preserved (Fig. 1.2A). Macroscopic observation revealed lipping on both occipital condyles, the second, fourth and fifth cervical vertebrae, on the seventh and ninth thoracic vertebrae, the distal epiphyses of the right humerus and femur, and the right patella (Fig. 1.2A). Schmörl's nodes were noted on the inferior surface of the seventh and ninth thoracic vertebrae.

Twenty-five out of 32 teeth were *in situ*. The first maxillary right molar, first maxillary left premolar, second and third mandibular left molars, and third mandibular right molar were lost ante-mortem. Both maxillary lateral incisors were absent (Fig. 1.2B). Examination revealed linear enamel hypoplasia in the central incisors, canines, and in the mandible's first left premolar, right canine and first premolar (Fig. 1.2B and C).

Dental wear ranges from grades 1 to 5, progressively increasing from the posterior molars to the incisal midline. A large area of dentine (grade 5) was observed in both mandibular central incisors. The left mandible exhibits more wear than the right, with no significant wear differences on the right due to the absence of maxillary molars (Fig. 1.2C). Maxillary central incisors show paramasticatory wear on the incisal edge, with a buccolingual groove near the midline exposing secondary dentine (Fig. 1.2B).

Sk.44 has occlusal caries in the upper left third molar. The septal morphology in the maxillary and mandibular regions is disorganised and riddled, classified as Kerr's third category of periodontal disease, except for the lower right second molar and lower left first molar, which are in Kerr's fifth category (Fig. 1.2B and C, Fig. 1.3A).

Buccal exostosis was localised on the maxilla along the alveolar surfaces of canine, premolar and molar areas (Fig. 1.2B and C).

The Sk.44 skull was one of the best-preserved skeletal portions of the individual, except for the fragmented calvaria and the left mandibular ramus which was not recovered (Figs 1.2C, 1.3A). During analysis, it was observed that the insertion of the musculus digastricus was symmetrical and there was no asymmetrical tension in the insertion of the nuchal musculature (Fig. 1.3A). The mandibular fossae of Sk.44 display asymmetry, with the left fossa maintaining a closed shape (Fig. 1.3B), while the right fossa exhibits a slight burr in the posterior area near the articular tubercle and a more open, eroded surface (Fig. 1.3C) compared to the left. What is reflected in the articular eminence inclination was measured at 70° on the right side (Fig. 1.3B) and 46° on the left side (Fig. 1.3C).

The right mandibular condyle is disposed slightly oblique towards the oral cavity and has some porosity in its articular surface. The radiograph did not reveal any signs of mandible fracture.

Dental calculus macroscopy

The dental calculus was present on all the preserved teeth of individual Sk.44, primarily manifesting as supragingival deposits with significant variability in terms of location, volume and pigmentation. These deposits exhibit asymmetry between the left and right sides (Fig. 1.4), as well as between the maxilla and mandible, with the mandible generally showing slightly more extensive accumulations on the lingual surface (Fig. 1.4C). On the left side, thin beige layers of calculus can be observed near the cementoenamel junction on the lingual and buccal surfaces of the molars and premolars. The calculus increases towards the left canines and incisors (Fig. 1.4A). This pattern is consistent on the right side's incisors and canines, with a marked increase in calculus from the mesial surface of the first premolar to the retromolar region. Consequently, the posterior right dentition exhibits the most substantial and voluminous calculus deposits, adhering to all surfaces of the tooth crowns (Fig. 1.4A). No indications of malocclusion or ante-mortem tooth loss were observed in the areas affected by dental calculus.

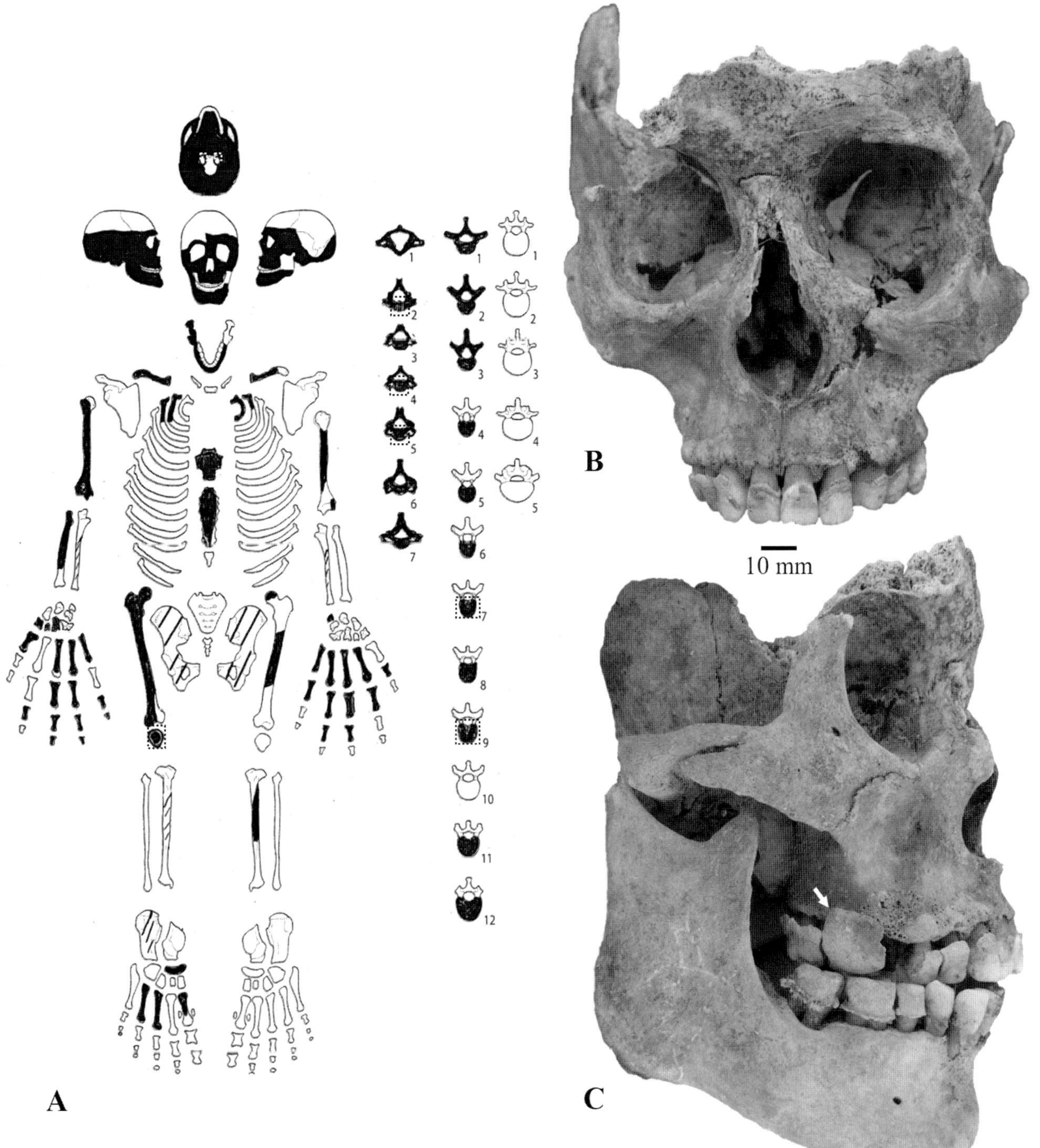

*Figure 1.2. A: Preserved elements of the skeleton (black), bones with post-mortem damage (striped) and lipping on several anatomic regions (dotted square). B: Anterior view of the cranium of individual Sk.44, showing clear evidence of notching on the incisal edges of both maxillary first incisors. C: Right buccal aspect of the individual Sr.44 maxilla and mandible. A white arrow indicates the sampled calculus (*i.e.* from the upper right permanent second molar).*

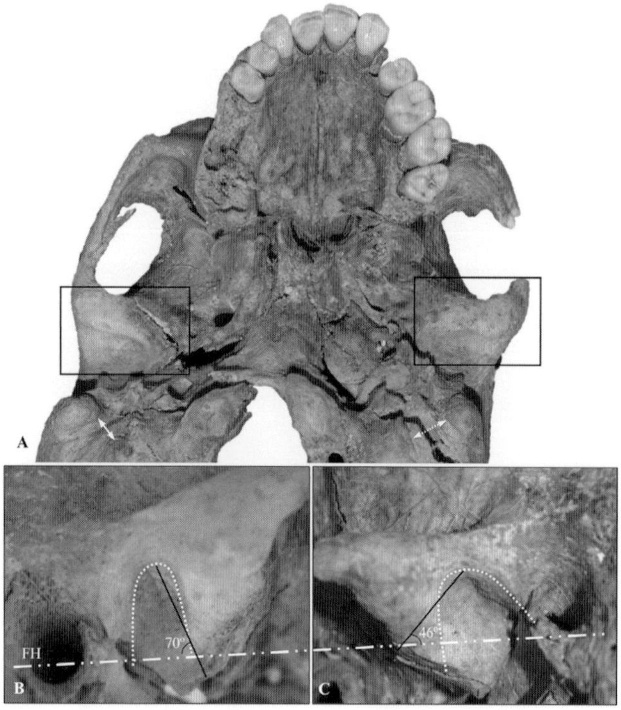

Figure 1.3. A: Inferior view of the skull showing symmetry of the digastric notches (arrows) and asymmetrical mandibular fossa. Note that the left side was broken post mortem. B: Right mandibular fossa (immobile) and articular eminence inclination (70°). C: Left mandibular fossa with wide mobility and articular eminence inclination (46°) (FH: Frankfort horizontal plane).

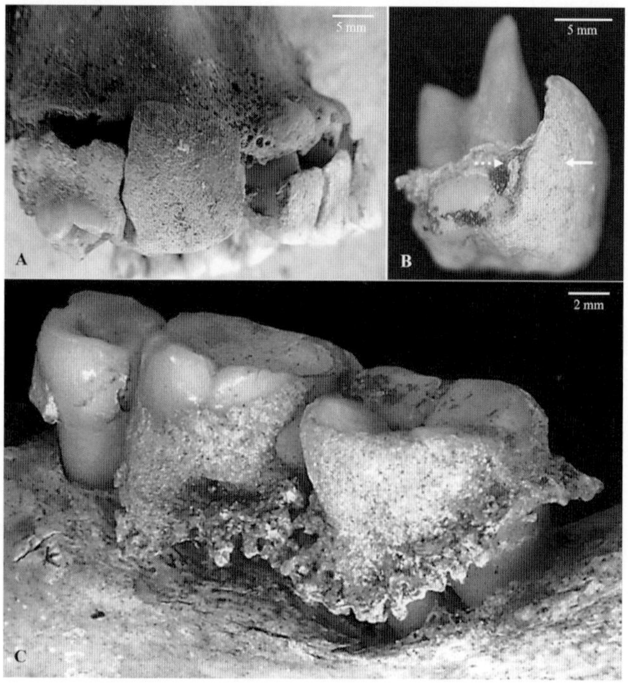

Figure 1.4. A: Close up of the buccal surface of superior maxilla showing the second and third right molar covered by exuberant dental calculus. B: Distal surface of the second upper right molar with exuberant calculus. Note the difference in colour: beige on the buccal surface (arrow) and green in inner area (dotted arrow). C: Lingual surface of the mandibular first and second right molars with calculus. Note the 'fern-like' shape of the calculus deposit.

The upper right second molar shows a pronounced calculus deposit on the buccal surface (13.67 × 13.63 × 6.14 mm) (Fig. 1.4A and B). On other teeth surfaces, two distinct layers are identifiable: a smoother, more porous greyish-white layer, and a granular greenish-white layer. The calculus adheres to the buccal surfaces and extends upwards, creating a gap that would be filled by the gum in a living individual (Fig. 1.4B). In the right posterior maxillary dentition, a wedge-shaped calculus area (7 × 16 mm) is noted on the buccal surface extending towards the alveolar bone, associated with severe periodontitis (Fig. 1.4A). Within this deposit and along the contour of the cementoenamel junction, a greenish calculus with darker spots is also present (Fig. 1.4B).

The right posterior mandibular dentition shows an irregular, 'fern-like' morphology on its lingual surface (5 × 12 mm) (Fig. 1.4C). Additionally, there is evidence of periodontitis, and calculus on the occlusal surface (Fig. 1.4C). These mandibular deposits appear greener and more exposed than those observed in the maxilla (Fig. 1.4).

Dental calculus microscopy: microorganisms and vegetal starches

Within the calculus matrix of the green-pigmented regions in the maxillary molar, fungal debris with a vermiform morphology and variable size was observed (30 × 4 μm). One end was flagellated, while the other was rounded and open to the exterior (Fig. 1.5A). In some areas of its surface, nanometric depressions were visible. Additionally, filaments with small (<1 μm) ovoid fungal sporophytes were identified (Fig. 1.5B).

During analysis of the micro-remains, three distinct structures were identified: 1) some long, thick-walled sclerenchyma fibres (Fig. 1.5C); 2) fungal spores likely related to *Ustilago* sp., possibly associated with *U. tritici* (Pers.) C.N. Jensen, Kellerm and Swingle (Fig. 1.5D); and 3) starch grains with a bimodal distribution. Two types of starch grains were identified: large lenticular grains (10–31 μm) with a centric hilum (Fig. 1.5E and F) and small circular grains (5–7 μm). The morphometric characteristics suggest they are compatible with the Triticeae tribe, likely *Triticum* sp. (wheat), and less probably *Hordeum vulgare* L. (barley) or *Secale cereale* L. (rye).

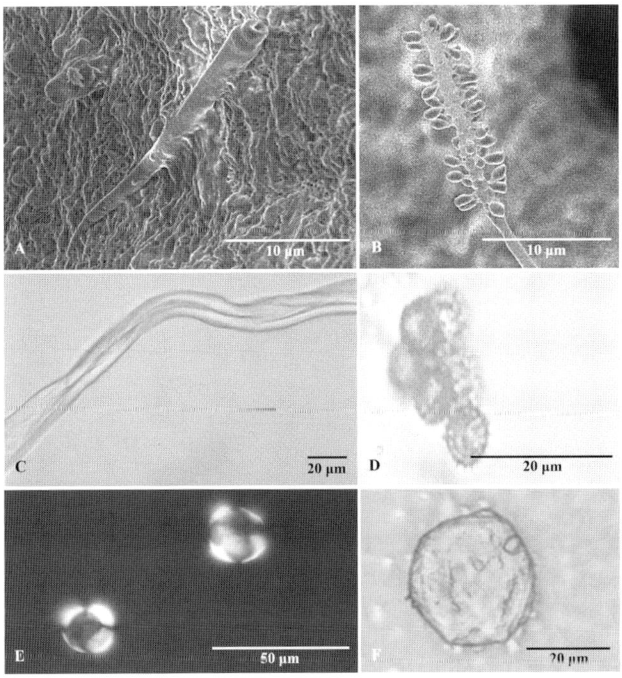

Figure 1.5. Microscopic analysis of the green dental calculus. A: Unidentified bipolar vermiform structure embedded in the calculus matrix. B: Fungal sporophytes. C: Thick-walled sclerenchyma fibre. D: Spores of Ustilago *sp. E: Triticeae starches exhibit a Maltese cross of their own. F: Triticeae starch grain (probably* Triticum *sp.).*

Dental calculus elemental composition

The pXRF analytical results are summarised in Table 1.1, where the concentrations of all the elements analysed are reported. The highest calcium (Ca) concentration was found in the dental calculus of individual Sk.44, specifically in the beige samples (Table 1.1), in which the molar ratio Ca/P (calcium/phosphorus) was 1.64, indicating mature calculus (hydroxyapatite) (Abraham *et al.* 2005). In contrast, lower ratios suggest that these areas are composed of hydroxyapatite-phase precursors, possibly a mixture of octacalcium phosphate (with a Ca/P ratio close to 1.33) or whitlockite (with a Ca/P ratio close to 1.43) (Abraham *et al.* 2005). The Arraiolos (E31-IB) sample, with a ratio of 1.89, corresponds to a more compact and stable stage of hydroxyapatite (with a Ca/P ratio between 1.7 and 2.0) mineralisation (Abraham *et al.* 2005). In conclusion, as the Ca/P molar ratio increases and the colour changes to white/beige, the stability of the hydroxyapatite phase becomes higher.

Higher concentrations of zinc (Zn) and copper (Cu) were found on the greenish surface when compared to the beige area and with the premolar calculus, while higher concentrations of iron (Fe), sulphur (S), strontium (Sr) and silicon (Si) were present in the beige calculus. The lower lead (Pb) concentrations were in the outer/beige area of Sk.44 molar while the Arraiolos (E31-IB) sample had a higher concentration.

The PCA (Fig. 1.6A) reveals that the dental calculus samples form three distinct clusters based on their colour: 1) Arraiolos E31 sample (E31-IB); 2) Belmonte molar Sk.44 samples (44-MG, 44-MB, 44-MF: molar fragment); and 3) Other Belmonte samples: Sk.4 (4-CB), Sk.42 (42-IB); and premolar of Sk.44 (44-PB, 44-PG). The correlation circle (Fig. 1.6B) highlights the unique composition of the calculus on Sk.44's molar, which shows significantly higher concentrations of Ca, S, Zn, and an exceptionally high Sr content in both the beige and green sections. Notably, the beige calculus on Sk.44's molar contains Sr levels approximated as being five times higher than the Arraiolos sample (E31-IB) and more than the premolar of the same individual. The Arraiolos sample exhibits a higher Ca/P ratio and elevated levels of manganese (Mn), aluminium (Al), Fe, Si, potassium (K) and Sc compared to the Belmonte samples. The other calculus samples (premolar of Sk.44, Sk.4, Sk.42) show higher concentrations of chlorine (Cl), caesium (Cs), tellurium (Te), barium (Ba), and tin (Sn) compared to both the Sk.44 and Arraiolos samples. The concentration of arsenic (As) remains relatively constant across all samples, while Sn is more prevalent in the Belmonte samples and Cu in the Arraiolos sample (E31-IB).

Discussion

Individual Sk.44 was a nun over 40 years old, exhibiting disperse osteoarthritis and Schmörl's nodes in some thoracic vertebrae, both probably related to her routine monastic activities. As Neves *et al.* (2022) suggest, the 'incisor-dominant' dental wear indicates that her teeth were used as a 'third hand'. The notching likely resulted from monastic chores, such as fibre processing or holding objects in her mouth.

The formation of dental calculus is influenced by factors such as age, misaligned teeth, salivary composition, diet, dental hygiene and some pathologies (Lieverse 1999). In addition, rapid calculus formation appears to be linked to patients with more severe periodontal diagnoses (Fons-Badal *et al.* 2020). Occlusal calculus has been related to the absence of an isomer tooth piece that makes chewing difficult, dental malocclusion or even oral immobilisation (Carvalho *et al.* 1989; Hansen and Alt 2012). Since dental malocclusion was discarded in individual Sk.44, and the antemortem loss of the first maxillary right molar does not explain the huge calculus deposit, only oral immobilisation can be considered a viable option for this case study.

Huge dental calculus deposits have been highlighted in palaeopathological cases of Down's syndrome (Rivollat *et al.* 2014) and leprosy (Roffey *et al.* 2017), yet Sk.44 did not display any bone alteration related to those conditions. In both diabetic and chronic renal failure patients, altered urea salivary composition alkalinises the oral environment, contributing to higher concentrations of sodium and potassium

Table 1.1. Chemical compositions of supra-gingival dental calculus, measured parts per million (ppm)

Element (ppm)	*Belmonte (Sk.44)*				*Belmonte (Sk.4)*	*Belmonte (Sk.42)*	*Arraiolos (E31)*
	Upper second molar		*Upper premolar*		*Upper canine*	*Lower canine*	*Incisive buccal side*
	MB	*MG*	*PB*	*PG*	*CB*	*CB*	*IB*
	mean±SE	*mean±SE*	*mean±SE*	*mean±SE*	*mean±SE*	*mean±SE*	*mean±SE*
Molar Ca/P	1.64	1.52	1.38	1.33	1.51	1.42	1.89
Ca	319327.78±2250.15	315499.50±1901.08	354808.69±2099.54	313190.50±1914.92	293451.34±1672.48	265224.41±1855.05	283192.84±1764.75
P	150654.23±1121.82	160807.64±1173.6	199032.25±1298.42	182158.48±1222.63	159608.58±1056.76	136074.45±1156.89	115968.76±957.51
S	20457.11±358.49	7195.31±252.15	5574.52±248.05	5687.73±230.33	10266.91±285.6	11192.62±285.99	1676.18±161
Si	6059.65±620.34	3309.61±614.58	4739.65±674.02	6574.18±645.32	9570.18±693.71	11096.39±701.3	79747.88±1155.83
Al	na	na	na	3413.73±1916.96	3659.25±1646.32	3195.13±1870.78	26398.29±2424.5
Sr	1154.87±13.75	877.94±9.62	935.23±10.33	897.53±9.92	705.77±8.85	777.11±8.84	202.39±4.58
Fe	928.40±63.94	554.09±42.94	814.39±51.21	708.95±47.59	1059.75±43.93	659.20±55.1	10016.67±176.52
Cl	877.00±107.65	1856.00±122.55	3244.56±142.53	1610.25±120.81	1369.02±106.9	1048.71±118.64	290.13±94.39
K	868.95±128.72	1091.01±179.01	628.31±125.51	720.05±119.00	678.44±114.15	775.89±120.24	3321.13±165.31
Sc	504.55±131.79	228.17±119.82	266.45±146.29	457.10±120.11	260.61±89.7	379.42±99.71	825.62±137.09
Zn	504.49±23.48	695.50±21.82	777.68±23.88	766.67±23.25	1035.54±16.47	391.93±26.85	387.45±16.21
Ba	409.84±40.14	537.31±38.22	802.65±41.98	632.36±43.99	585.29±55.7	na	463.77±37.32
Mn	158.21±56.71	151.72±44.85	282.47±51.74	198.01±47.28	150.98±55.45	na	1265.31±71.43
Te	98.44±20.65	140.24±21.67	216.03±23.57	158.00±22.48	174.86±31.02	na	129.24±20.31
Ni	87.40±33.65	na	159.72±25.17	127.37±26.28	66.85±32.53	na	113.71±22.96
Cs	69.65±7.00	101.16±7.36	151.26±7.99	114.62±7.64	107.83±10.49	na	75.28±6.8
Zr	22.37±5.81	12.84±4.10	12.30±4.37	12.89±4.22	na	11.59±5.56	5.31±2.45
Sn	20.76±6.33	28.09±6.65	47.46±7.28	32.55±6.9	33.35±9.7	na	25.14±6.21
Cu	na	31.47±12.93	24.58±13.46	38.31±13.62	41.70±18.33	na	49.26±12.87
Pb	14.56±5.39	48.74±5.98	57.69±6.63	42.48±5.83	108.60±9.38	147.83±8.70	145.73±9.2
As	18.23±24.46	24.46±18.23	21.98±19.3	19.30±21.98	24.16±17.98	17.98±24.16	18.58±18.58

IB: incisive beige, CB: canine beige, PB: premolar beige, PG: premolar with greenish pigmentation, MB: molar beige, MG: molar with greenish pigmentation, na: data not available

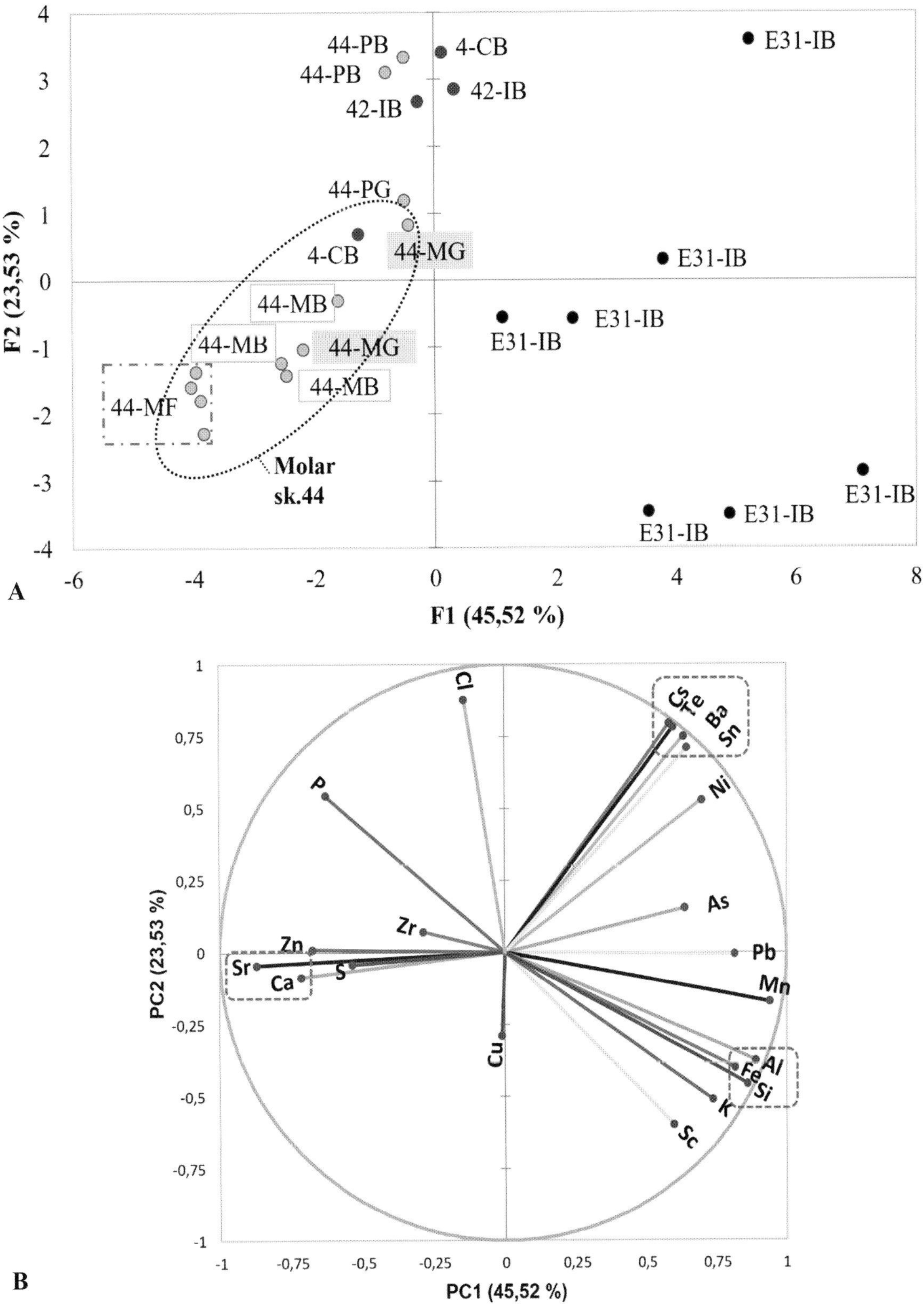

Figure 1.6. A: Correlation values of the 21 chemical elements quantified by pXRF and factor loadings of the PCA. B: Correlation circle showing the distribution of the samples by chemical elements. In bold the loading of the component with the largest value for each element (4-CB: Belmonte, Sk.4, female, 40–50 y.o.; 42-IB: Belmonte, Sk.42, female, undetermined age; E31-IB: Arraiolos, E31, male, adult; 44-M: molar of Sk.44: MB beige and MG greenish pigmentation calculus, and MF a fragment).

linked with excessive calculus formation (Tomás *et al.* 2008; D'Souza *et al.* 2023). The pXRF analysis conducted on the dental calculus of Sk.44 did not indicate any potassium augmentation. Additionally, no lower limb fractures, osteomyelitis, Charcot neuropathic arthropathy, osteochondritis dissecans, and/or vascular calcifications, which could be potentially linked to diabetes (Biehler-Gomez *et al.* 2019), were found.

Based on the macroscopic examination of the cranium, the symmetrical insertion of the musculus digastricus and the nuchal musculature in individual Sk.44 suggests that disproportionate use of a single upper extremity can be excluded (Tesorieri 2016). However, the differing inclinations of the articular eminence can indicate jaw dysfunction. The 60–90° inclination on the right side is associated with a younger mandibular condyle (Katsavrias 2002) and individuals with arthritic changes exhibit shallow angles <60° (Yamada *et al.* 2004). These results suggest that eminence flattening may occur during transitions from erosion and anterior disc displacement. This is likely to represent an adaptation of the condyle, articular disc and articular eminence to changes in loading (Yamada *et al.* 2004).

Jaw dysfunction is usually observed in cases of facial paralysis, a condition that causes orofacial disruptions and may increase the risk of excessive calculus on one (unilateral facial paralysis) or both sides (bilateral facial paralysis) due to compromised oral hygiene (Strobelt *et al.* 2022). Conditions leading to long-term facial paralysis with significant impairment of mouth movement include neurological complications of chronic viral infections such as herpesvirus (Kleinschmidt-DeMasters and Gilden 2001), ischemic strokes (Cardoso *et al.* 2023) and neuromuscular diseases such as multiple sclerosis or dystonia (Tesorieri 2016).

The herpesvirus family, particularly the zoster and varicella-zoster viruses, can lead to oral manifestations when they affect the mandibular and maxillary branches of the trigeminal nerve in chronic viral infections (Kleinschmidt-DeMasters and Gilden 2001). This can result in devitalised teeth, internal resorption, tooth exfoliation and necrosis (Mendieta *et al.* 2005). Ischemic strokes can affect mouth movement and lead to prolonged facial paralysis when the brain regions controlling facial muscles are impacted (Schimmel *et al.* 2017). They often result in dysphagia as well (Cardoso *et al.* 2023). Additionally, a stroke severely affects masticatory efficiency and maximum lip force, but not maximum bite force (Schimmel *et al.* 2011). Therefore, complications arising from herpesviruses and ischemic strokes do not explain the restricted temporomandibular joint movement observed in Sk.44.

Identifying and diagnosing neurological disorders in archaeological contexts poses significant methodological challenges due to the highly variable and often non-specific skeletal manifestations (Brothwell and Browne 2002). Novak *et al.* (2014) described cases of hemiplegia and triplegia from Croatia with ambiguous diagnoses of cerebral palsy or poliomyelitis. Other cases described are likely to have been caused by poliomyelitis (Ciesielska and Stark 2020; Kozakaitė *et al.* 2022; Gómez-González *et al.* 2024). Few definite cases of cerebral palsy have been identified in the archaeological record and most often only include the shortening of one or two long bones (Tesorieri 2016), which does not happen in Sk.44.

A common cause of facial paralysis is idiopathic, known as Bell's palsy, where moderate to severe facial asymmetry may persist in 25% of patients (Zhang *et al.* 2020). Nerve damage can result in either excessive or reduced production of tears and saliva, along with an increased risk of dental caries (Danesh and Ouanounou 2022). Charlier (2007) described several representations of facial paralysis in sculptures and masks, as well as an archaeological case (2nd–3rd century AD in a female over forty, who exhibited significant oral lesions and facial asymmetry, including severe dental calculus deposits (4.2 × 2.5 × 2 cm) on the right mandibular molars. The morphology of the mandibular fossae suggests a wider mouth opening on the left side and a more restricted opening of the right temporomandibular joint.

Individual Sk.44 shows dental and maxillofacial pathology consistent with right-sided facial paralysis, including substantial calculus accumulation on the right side of the mouth, uneven wear and periodontal disease. The presence of fewer carious lesions in Sk.44 may be attributed to the potential protective effects of calculus against caries (Keyes and Rams 2016). It was not possible to determine the precise cause of the facial paralysis in individual Sk.44. Santos-Lasaosa *et al.* (1999) report that facial nerve palsy is the most common cranial nerve lesion, with an incidence of 30 per 100,000. In chronic cases, particularly, there is a risk that insufficient or improper movement in the cheek after eating may lead to food particles becoming trapped, compromising oral hygiene (Strobelt *et al.* 2022), which may have been the case in the individual under analysis.

The starch grains identified in the calculus were predominantly from staple cereals of the season, including those from the Triticeae tribe (wheat, barley and rye). The definition of Maltese crosses in the starch grains varied somewhat, though not significantly, and irregular surfaces were observed in *Triticum* sp. starch grains. These variations suggest mild preparation processes, such as grinding, soaking, and either baking or boiling (Henry *et al.* 2009). The genus *Ustilago*, a common parasitic fungus of Poaceae, which was also detected, is frequently present in agricultural areas (Wunderle *et al.* 2012; Plant Parasites of Europe 2024). *U. tritici* is directly associated with the presence of wheat, barley and rye. However, the morphology of the starch grains in this case is more similar to that of wheat than those of barley (aberrant shapes (Yang and Perry 2013) absent), or rye (radial fissures (Yang and Perry 2013; Gismondi *et al.*

2018) absent). The culinary preparation of the Triticeae species must have been relatively mild since the extinction cross is well defined (Henry *et al.* 2009). Sclerenchyma fibres indicate the proximity of woody plants or the bark of certain Rosaceae fruits, typical of warm-temperate climates (Barclay 2007; Lin *et al.* 2022; D'Agostino *et al.* 2024). Another possibility is that sclerenchyma fibres, which are widespread across the Poaceae family (including the Triticeae tribe), could also be involved. Additionally, unidentified fungi and yeast remains found via electron microscopy of dental calculus suggest influences from food storage, hygiene and environmental conditions (Singh *et al.* 2010; MacKenzie *et al.* 2023).

Usually, the greenish pigmentation on teeth and bones has been associated with the presence of copper materials. However, the funerary goods found with Sk.44 were located far from the oral cavity, ruling out this hypothesis. The only reported instance of green dental calculus in the bioarchaeological literature, not related to metals, comes from Cooper (2017). This study focused on 17th- to 18th-century Benedictine monks from San Agustín in San Millán de la Cogolla (La Rioja, Spain) and on monks and upper-class civilians from the Capuchin Catacombs in Palermo (Sicily, Italy). This author identified dark green shades on the deeper surfaces of dental calculus in these populations, attributing this to the presence of high-density hydroxyapatite containing carbonate. This suggests that during a specific period of calculus formation, an increase in carbonate ions in the oral environment led to the incorporation of carbonate into hydroxyapatite instead of phosphate, or the replacement of phosphate with carbonate, resulting in the green coloration, as also supported by Fleet (2014).

The molar from individual Sk.44 shows higher concentrations of Ca, Zn, S, and an exceptionally high Sr content in both the beige and green areas. The highest concentration of Ca was found in the beige dental calculus of Sk.44. However, the sample with the highest mean Ca/P ratio was from Arraiolos (Ca/P = 1.89). Notably, the Arraiolos individual did not exhibit any green coloration, suggesting the absence of carbonate-containing hydroxyapatite according to Cooper (2017). This could indicate that phosphate groups were likely substituted by another ionic group rather than carbonate, a possibility supported by the low carbonate content of the Arraiolos sample (283192.84±1764.75). The higher presence of Al in the Arraiolos sample might indicate that hydroxyapatite was substituted at the phosphate group with AlO_4^{3-} (LeGeros 1981), which could explain the absence of green-coloured calculus. The elevated concentration of Si in Arraiolos means a hydroxyapatite predominant crystalline phase (Damen and ten Cate 1989). These differences suggest a distinct oral environment and different mineralisation processes in the individuals from Belmonte compared to those from Arraiolos. The presence of carbonate-rich hydroxyapatite in the dental calculus, as might be occurring in the green coloration of individual Sk.44, could indicate an oral environment with high alkalinity, allowing for the accumulation of dental calculus, coupled with periods of high acidity, where the bicarbonate buffer system regulates low pH levels (Fleet 2014). This shift between alkaline and acidic conditions suggests a varied diet that included both proteins and fermentable sugars (Lieverse 1999). In contrast, the absence of carbonate-containing hydroxyapatite in the Arraiolos sample might be attributed to the formation of dental calculus in individuals whose diet was low in fermentable sugars and high in proteins (Lieverse 1999).

In this study, the variation in Zn has been attributed to the initial mineralisation process, which is why Arraiolos shows the lowest concentrations with a higher Ca/P ratio. During this initial process, Zn ions are released by bacteria to inhibit the crystallisation of calcium phosphate, which is supported by other research on calculus formation (Pérez *et al.* 2004).

Sulphur has been detected in higher concentrations in the beige area of the dental calculus from individual Sk.44, and in other samples from Belmonte, and to a much lesser extent in the individual from Arraiolos. The most likely explanation is dietary, although periodontal disease has been ruled out due to its external location. Foods rich in sulphur include animal-based proteins and cruciferous vegetables. This aligns with the isotopic study by Sarkic *et al.* (2019), which reveals that individual Sk.44 had a diet primarily consisting of terrestrial proteins ($\delta^{15}N$ = 12.0, $\delta^{13}C$ = -18.6) from C3 plants, with a lesser contribution from marine resources.

The higher concentrations of Sr in Belmonte may also indicate soil contamination, as Sr can readily exchange with Ca during burial (Lambert *et al.* 1985; Pate and Hutton 1988). Another possible explanation is that the Sr concentration in the green calculus of Sk.44 could be due to diet, with Sr derived from marine fish, shellfish and molluscs. This evidence becomes particularly relevant when associated with the high levels of Zn and Cu (Giorgi *et al.* 2005) present in the Belmonte samples.

It has also been demonstrated that dental calculus may accumulate toxic heavy metals, such as lead and arsenic, from oral exposures (Forshaw 2022). The higher concentration of lead (Pb) found in the dental calculus of the Arraiolos sample and individual Sk.42 might be attributed to the prolonged use of metal containers for food and drink (Sierra 2011). This evidence could suggest a different role within the monastery/convent for these two individuals. Arsenicals, despite their notorious toxicity, have been used since ancient times as a treatment for various disorders (Paul *et al.* 2023). However, there are no differences in this element compared to the other samples, so its use was ruled out. Soil studies of burial sites are required to rule out the taphonomic effects of iron concentrations in the samples.

Conclusions

Macroscopic, radiological and chemical characteristics observed in individual Sk.44, along with similarities to a reported palaeopathological case, suggest a probable diagnosis of facial paralysis. The specific cause remains undetermined, but the asymmetrical and exuberant accumulation of dental calculus underlines its importance as a palaeopathological marker.

Diseases caused by problems in the nervous system, such as paralyses, are difficult or impossible to diagnose in palaeopathology. However, this study raises awareness of facial paralysis and its possible identification in skeletal remains related to a large accumulation of asymmetrical dental calculus, which may help to increase the body of evidence from the past. Therefore, adhering to rigorous archaeological excavation techniques and refining dental calculus conservation methodologies are crucial for enhancing diagnostic accuracy and deepening our understanding of past populations.

Acknowledgements

The authors would like to express their gratitude to Pedro Sidónio (University of Coimbra, Portugal) for his support with the SEM analysis, Juan Carrión Bolaños (Departamento de Odontología Clínica, Universidad Europea de Madrid, Spain) for his assistance with the radiological examination of the mandible, and Sarah Delaney (University of York, United Kingdom) for her invaluable suggestions, and the reviewer for the helpful comments. This work is supported by Portuguese National Funds through the FCT (Foundation for Science and Technology, I.P.), under projects UIDB/00283/2020 (https://doi.org/10.54499/UIDB/00283/2020) and UIDB/00073/2020.

Notes

1 The confiscation law (disentailment) in Spain was a historical, economic and social process that began in the late 18th century and continued throughout the 19th century. With the Royal Decree of 19th February 1836, Juan Álvarez Mendizábal implemented a new disentailment, which involved the forced expropriation and public auctioning by the state of properties and lands belonging to the Catholic Church and its religious orders, which had previously been inalienable.

2 'Inventario de los bienes de las monjas dominicas de Belmonte' (21 de marzo de 1836). Archivo Histórico Provincial de Cuenca. Sección: Hacienda, Signatura: 3/2. This important document, which is held in the Historical Archive of Cuenca, enables the nature of the conventual buildings and the belongings that were in them in 1836 to be known in detail.

References

Abraham, J., Grenón, M., Sánchez, H.J., Pérez, C. and Barrea, R. (2005) A case study of elemental and structural composition of dental calculus during several stages of maturation using SRXRF. *Journal of Biomedical Materials Research Part A* 75(3), 623–8. https://doi.org/10.1002/jbm.a.30484

Abraham, J.A., Grenón, M.S., Sánchez, H.J., Valentinuzzi, M.C. and Perez, C.A. (2007) μX-ray fluorescence analysis of traces and calcium phosphate phases on tooth–tartar interfaces using synchrotron radiation. *Spectrochimica Acta Part B: Atomic Spectroscopy* 62(6-7), 689–94. https://doi.org/10.1016/j.sab.2007.03.008

Aceituno-Bocanegra, F.J. and López-Sáez, J.A. (2012) Caracterización morfológica de almidones de los géneros *Triticum* y *Hordeum* en la Península Ibérica. *Trabajos de Prehistoria* 69(2), 332–48. https://doi.org/10.3989/tp.2012.12095

Adler, C.J., Dobney, K., Weyrich, L.S., Kaidonis, J., Walker, A.W., Haak, W., Bradshaw, C.J.A., Townsend, G., Sltysiak, A., Alt, K.W., Parkhill, J. and Cooper, A. (2013) Sequencing ancient calcified dental plaque shows changes in oral microbiota with dietary shifts of the Neolithic and Industrial revolutions. *Nature Genetics* 45, 450–6. https://doi.org/10.1038/ng.2536

Akcali, A. and Lang, N.P. (2017) Dental calculus: the calcified biofilm and its role in disease development. *Periodontology 2000* 76(1), 109–15. https://doi.org/10.1111/prd.12151.

Barclay, G. (2007) Plant anatomy. In K. Roberts (ed.) *Handbook of Plant Science*, 13–26. Chichester, Wiley.

Biehler-Gomez, L., Castoldi, E., Baldin, E., Cappella, A. and Cattaneo, C. (2019) Diabetic bone lesions: a study on 38 known modern skeletons and the implications for forensic scenarios. *International Journal of Legal Medicine* 133(4), 1225–39. https://doi.org/10.1007/s00414-018-1870-0

Brooks, S. and Suchey, J.M. (1990) Skeletal age determination based on the os pubis: a comparison of the Acsádi-Nemeskéri and Suchey-Brooks methods. *Human Evolution* 5, 227–38. https://doi.org/10.1007/BF02437238

Brothwell, D. and Browne, S. (2002) Skeletal atrophy and the problem of differential diagnosis in conditions causing paralysis. *Antropologia Portuguesa* 19, 5–18.

Bruzek, J. (2002) A method for visual determination of sex, using the human hip bone. *American Journal of Physical Anthropology* 117, 157–68. https://doi.org/10.1002/ajpa.10012

Buikstra, J.E. and Ubelaker, D. (1994) *Standards for Data Collection from Human Skeletal Remains*. Fayetteville, Arkansas, Arkansas Archeological Survey Research Series No. 44.

Caballero García, R. and Sánchez Peláez, E. (2013) *Intervención arqueológica en el antiguo Alcázar del infante D. Juan Manuel. Término municipal: Belmonte (Cuenca).* Memoria final de la actuación arqueológica.

Cardoso, A.F., Ribeiro, L.E., Santos, T., Pinto, M., Rocha, C., Magalhães, J., Augusto, B., Santos, D., Duque, F.M., Fernandes, B.L., Sousa, R.C., Silva, R., Ventura, F., Fernandes, A.M., Cardoso, D. and Rodrigues, R. (2023) Oral hygiene in patients with stroke: a best practice implementation project protocol. *Nursing Reports* [online] 13(1), 148–56. https://doi.org/10.3390/nursrep13010016

Carvalho, J.C., Ekstrand, J.R. and Thylstrup, A. (1989) Dental plaque and caries on occlusal surfaces of first permanent molars in relation to stage of eruption. *Journal of Dental Research* 68(5), 773–79. doi: 10.1177/00220345890680050401

Charlier, P. (2007) Un nouveau cas de paralysie faciale sur une terre cuite smyrniote hellénistique. *Histoire des Sciences Médicales* 41, 49–60.

Ciesielska, J.A. and Stark, R.J. (2020) Possible neurogenic disorder in a female buried in the monastic cemetery at Ghazali (ca. 670–1270 CE), northern Sudan. *International Journal of Osteoarchaeology* 30, 33–42. https://doi.org/10.1002/oa.2828

Cooper, K.A. (2017) *The Physical Characterisation and Composition of Archaeological Dental Calculus*. PhD thesis, Cranfield University, Bedford. http://dspace.lib.cranfield.ac.uk/handle/1826/12817.

D'Agostino, A., Di Marco, G., Rolfo, M.F., Alessandri, L., Marvelli, S., Braglia, R., Congestri, R., Berrilli, F., Fuciarelli, M.F., Ferracci, A., Canini, A. and Gismondi, A. (2024) Microparticles from dental calculus disclose paleoenvironmental and palaeoecological records. *Ecology and Evolution* 14(2), e11053. https://doi.org/10.1002/ece3.11053

Damen, J.J. and ten Cate, J.M. (1989) Calcium phosphate precipitation is promoted by silicon. In J.M. ten Cate (ed.) *Recent Advances in the Study of Dental Calculus*, 105–14. Oxford, IRL Press.

Danesh, A. and Ouanounou, A. (2022) Bell's palsy: etiology, management and dental implications. *Journal of the Canadian Dental Association* 88(m8), ISSN 1488-2159.

D'Souza, L.L., Lawande, S.A., Samuel, J. and Wiseman Pinto, M.J. (2023) Effect of salivary urea, pH and ureolytic microflora on dental calculus formation and its correlation with periodontal status. *Journal of Oral Biology and Craniofacial Research* 13(1), 8–12. https://doi.org/10.1016/j.jobcr.2022.10.004

Fleet, M. (2014) Calcium phosphate precipitation is promoted by silicon. *Carbonated Hydroxyapatite: Materials, Synthesis, and Applications*. Boca Raton, Florida, Pan Stanford Publishing.

Fons-Badal, C., Fons-Font, A., Labaig-Rueda, C., Solá-Ruiz, M. F., Selva-Otaolaurruchi, E. and Rubén Agustín-Panadero, R. (2020) Analysis of predisposing factors for rapid dental calculus formation. *Journal of Clinical Medicine* 9(3), 858. https://doi.org/10.3390/jcm9030858

Forshaw, R. (2022) Dental calculus – oral health, forensic studies and archaeology: a review. *British Dental Journal* 233(12), 961–7. https://doi.org/10.1038/s41415-022-5266-7

Giorgi, F., Bartoli, F., Iacumin, P., Bartolini, G., Basile, G., and Turi, B. (2005) Oligoelements and isotopic geochemistry: a multidisciplinary approach to the reconstruction of the paleodiet. *Human Evolution* 20, 55–81. https://doi.org/10.1007/BF02438898

Gismondi, A., D'Agostino, A., Canuti, L., Di Marco, G., Basoli, F. and Canini, A. (2018) Starch grains: a data collection of 40 food species. *Plant Biosystems – An International Journal Dealing with all Aspects of Plant Biology* 153(2), 273–9. https://doi.org/10.1080/11263504.2018.1473523

Gómez-González, S., González-Garrido, L., Rodríguez, L., Serralvo-González, M., Gonzalo-Orden, J.M. and Wasterlain, S.N. (2024) An identified case of poliomyelitis: contribution to diagnosis in ancient human remains. *Anthropological Science* 132(2), 105–15. https://doi.org/10.1537/ase.231108

Greene, T.R., Kuba, C.L. and Irish, J.D. (2005) Quantifying calculus: a suggested new approach for recording an important indicator of diet and dental health. *Homo* 56(2), 119–32. https://doi.org/10.1016/j.jchb.2005.02.002

Hanihara, T., Ishida, H., Ohshima, N., Kondo, O. and Masuda, T. (1994) Dental calculus and other dental disease in a human skeleton of the Okhotsk Culture unearthed at Hamanaka-2 Site, Rebun-Island, Hokkaido, Japan. *International Journal of Osteoarchaeology* 4(4), 343–51. https://doi.org/10.1002/oa.1390040408

Hansen, J.L. and Alt, K.W. (2012) An exceptional case of dental calculus in a Merovingian skeleton from Mannheim-Seckenheim. *Bulletin of the International Association for Paleodontology* 6(2), 70–6. https://hrcak.srce.hr/95444

Henry, A.G., Hudson, H.F. and Piperno, D.P. (2009) Changes in starch grain morphologies from cooking. *Journal of Archaeological Science* 36, 915–22. https://doi.org/10.1016/j.jas.2008.11.008

Hillson, S. (2001) Recording dental caries in archaeological human remains. *International Journal of Osteoarchaeology* 11(4), 249–89. https://doi.org/10.1002/oa.538

Hillson, S. (2005) *Teeth*. Cambridge, Cambridge University Press.

Jasinevicius, T.R., Pyle, M.A., Nelson, S., Lalumandier, J.A., Kohrs, K.J. and Sawyer, D.R. (2006) Relationship of degenerative changes of the temporomandibular joint (TMJ) with the angle of eminentia. *Journal of Oral Rehabilitation* 33(9), 638–45. https://doi.org/10.1111/j.1365-2842.2006.01618.x

Katsavrias, E.G. (2002) Changes in articular eminence inclination during the craniofacial growth period. *The Angle Orthodontist* 72(3), 258–64. https://doi.org/10.1043/0003-3219(2002)072<0258>2.0.CO;2

Kerr, N.W. (1998) Prevalence and natural history of periodontal disease in prehistoric Scots (pre-900 AD). *Journal of Periodontal Research* 33(2), 131–7. https://doi.org/10.1111/j.1600-0765.1998.tb02303

Keyes, P.H. and Rams, T.E. (2016) Dental calculus arrest of dental caries. *Journal of Oral Biology* 3(1), 4. https://doi.org/10.13188/2377-987x.1000017

Kleinschmidt-DeMasters, B.K. and Gilden, D.H. (2001) The expanding spectrum of herpesvirus infections of the nervous system. *Brain Pathology* 11(4), 440–51. https://doi.org/10.1111/j.1750-3639.2001.tb00413.x

Kozakaitė, J., Jakulis, M., Brindzaitė, R., Miliauskienė, Ž., Piombino-Mascali, D. and Jankauskas, R. (2022) A possible case of paralysis in early modern Vilnius and the implications for social care. *International Journal of Paleopathology* 36, 14–23. https://doi.org/10.1016/j.ijpp.2021.09.002

Kruse, J., Dietrich, W., Zimmermann, H., Klenke, F., Richter, U., Richter, H. and Thines, M. (2018) Ustilago species causing leaf-stripe smut revisited. *IMA Fungus* 9(1), 49–73. https://doi.org/10.5598/imafungus.2018.09.01.05

Lambert, J.B., Simpson, S.V., Weiner, S.G. and. Buikstra, J.E. (1985) Induced metal-ion exchange in excavated human bone. *Journal of Archaeological Science* 12(2), 85–92.

LeGeros, R.Z. (1981) Apatites in biological systems. *Progress in Crystal Growth and Characterization* 4(1–2), 1–45.

Lieverse, A.R. (1999) Diet and the aetiology of dental calculus. *International Journal of Osteoarchaeology* 9(4), 219–32. https://doi.org/10.1002/(SICI)1099-1212(199907/08)9:4<219::AID-OA475>3.0.CO;2-V

Lin, S., Lin, D., Wu, B., Ma, S., Sun, S., Zhang, T., Zhang, W., Bai, Y., Wang, Q. and Wu, J. (2022) Morphological and developmental features of stone cells in *Eriobotrya* fruits. *Frontiers in Plant Science* 13, article 823993.

Lovejoy, C., Meindl, R., Pryzbeck, T. and Mensforth, R. (1985) Chronological metamorphosis of the auricular surface of the ilium: a new method for the determination of adult skeletal age at death. *American Journal of Physical Anthropology* 68(1), 15–28. https://doi.org/10.1002/ajpa.1330680103

MacKenzie, L., Speller, C.F., Holst, M., Keefe, K. and Radini, A. (2023) Dental calculus in the industrial age: Human dental calculus in the Post-Medieval period, a case study from industrial Manchester. *Quaternary International* 653–4, 114–126. https://doi.org/10.1016/j.quaint.2021.09.020

Mendieta, C., Miranda, J., Brunet, L.I., Gargallo, J. and Berini, L. (2005) Alveolar bone necrosis and tooth exfoliation following herpes zoster infection: a review of the literature and case report. *Journal of Periodontology*, 76(1), 148–53. https://doi.org/10.1902/jop.2005.76.1.148

Monge Calleja, A.M. (2015) *Reconstruindo a morte na infância: Resultados preliminares da microscopia de varrimento como ferramenta diagnóstica da porosidade em vestígios osteológicos dos séculos XIX-XX*. Master's thesis, University of Coimbra.

Monge Calleja, A.M., Santos, A.L. and Coutinho, A.P. (2021) Novo método de extração de amidos do cálculo dentário e criação de uma coleção identificada para a reconstrução das dietas do passado. *Antropologia Portuguesa* 37, 99–129. https://doi.org/10.14195/2182-7982,37-5

Moreno González, J.M. and Rubio Masa, J.C. (2007) Las enfermerías conventuales de Zafra (siglos XVI al XVIII). *Revista de Estudios Extremeños* 63(extra 1), 23–60.

Neves, D., Silva, A.M., Simões, F. and Wasterlain, S.N. (2022) More than they could chew: activity-induced dental modifications in a Portuguese medieval–modern rural population. *International Journal of Osteoarchaeology* 32(4), 856–865. https://doi.org/10.1002/oa.3113

Neville, B.W., Damm, D.D., Allen, C.M. and Chi, A.C. (2023) *Oral and Maxillofacial Pathology (E-Book)*. Elsevier Health Sciences.

Novak, M., Čavka, M. and Šlaus, M. (2014) Two cases of neurogenic paralysis in medieval skeletal samples from Croatia. *International Journal of Paleopathology* 7, 25–32. https://doi.org/10.1016/j.ijpp.2014.06.002

Ortner, D.J. (2003) *Identification of Pathological Conditions in Human Skeletal Remains*. 2nd edition. Amsterdam, Academic Press.

Pate, F.D. and Hutton, J.T. (1988) The use of soil chemistry data to address post-mortem diagenesis in bone mineral. *Journal of Archaeological Science* 15(6), 729–39.

Paul, N.P., Galván, A.E., Yoshinaga-Sakurai, K., Rosen, B.P. and Yoshinaga, M. (2023) Arsenic in medicine: past, present and future. *Biometals* 36(2), 283–301. https://doi.org/10.1007/s10534-022-00371-y

Peréz, C.A., Sánchez, H.J., Barrea, R.A., Grenón, M. and Abraham, J. (2004) Microscopic X-ray fluorescence analysis of human dental calculus using synchrotron radiation. *Journal of Analytical Atomic Spectrometry* 19(3), 392–7.

Perini, T.A., Oliveira, G.L.D., Ornellas, J.D.S. and Oliveira, F.P.D. (2005) Technical error of measurement in anthropometry. *Revista Brasileira de Medicina do Esporte* 11, 1–5.

Piperno, D.R., Weiss, E., Holst, I. and Nadel, D. (2004) Processing of wild cereal grains in the Upper Palaeolithic revealed by starch grain analysis. Nature 430(7000), 670–673.https://doi:10.1038/nature02734

Plant Parasites of Europe – leafminers, galls and fungi. (2024) *Ustilago tritici*. https://bladmineerders.nl/parasites/fungi/dikarya/basidiomycota/ustilaginomycotina/ustilaginomycetes/ustilaginales/ustilaginaceae/ustilago/ustilago-tritici/

Radini, A. and Nikita, E. (2023) Beyond dirty teeth: Integrating dental calculus studies with osteoarchaeological parameters. *Quaternary International: The Journal of the International Union for Quaternary Research* 653–654, 3–18. https://doi.org/10.1016/j.quaint.2022.03.003

Radini, A., Nikita, E., Buckley, S., Copeland, L. and Hardy, K. (2017) Beyond food: The multiple pathways for inclusion of materials into ancient dental calculus. *American Journal of Physical Anthropology* 162, 71–83. https://doi.org/10.1002/ajpa.23147

Rivollat, M., Castex, D., Hauret, L. and Tillier, A.M. (2014) Ancient Down syndrome: an osteological case from Saint-Jean-des-Vignes, northeastern France, from the 5–6th century AD. *International Journal of Paleopathology* 7, 8–14. http://dx.doi.org/10.1016/j.ijpp.2014.05.004

Roffey, S., Tucker, K., Filipek-Ogden, K., Montgomery, J., Cameron, J., O'Connell, T., Evans, J., Marter, P. and Taylor, M. (2017) Investigation of a medieval pilgrim burial excavated from the leprosarium of St Mary Magdalen Winchester, UK. *PLOS Neglected Tropical Diseases* 11(1), e0005186. https://doi.org/10.1371/journal.pntd.0005186

Santos-Lasaosa, S., Pascual-Millán, L.F., Tejero.Juste, C. and Morales-Asín, F. (1999) Peripheral facial paralysis: etiology, diagnosis and treatment. *European Neurology* 30(11), 1048–53. https://doi.org/10.1159/000007990

Sanz Fuentes, M.J. (1993) Constituciones de las Orden dominicana femenina en un manuscrito de la Biblioteca Universitaria de Oviedo. In M.I. Viforcos Marinas and J. Paniagua Pérez (coords) *I Congreso Internacional del Monacato Femenino en España, Portugal y América. 1492–1992*. Vol. 2, 141–8. León, Universidad de León.

Sarkic, N. (2017) *A Study of the Lifestyle and Health Parameters of Nuns from Convents on the Iberian Peninsula in Modern Times*. PhD thesis, Universidad Autónoma de Madrid.

Sarkic, N., Herrerín, J., López-Costas, O. and Grandal-d'Anglade, A. (2019) Eating in silence: isotopic approaches to nuns' diet at the convent of Santa Catalina de Siena (Belmonte, Spain) from the sixteenth to the twentieth century. *Archaeological and Anthropological Sciences* 31 (2019), 941–56.

Schimmel, M., Leemann, B., Herrmann, F.R., Kiliaridis, S., Schnider, A. and Müller, F. (2011) Masticatory function and bite force in stroke patients. *Journal of Dental Research* 90(2), 230–4. https://doi.org/10.1177/0022034510383860

Schimmel, M., Ono, T., Lam, O.L.T. and Müller, F. (2017) Oro-facial impairment in stroke patients. *Journal of Oral Rehabilitation* 44, 313–26. https://doi.org/10.1111/joor.12486

Sierra, J.M.V. (2011) Vivir como un cura. El hogar de un clérigo acomodado en Extremadura (España) al final del Antiguo Régimen. *Etnicex: revista de estudios etnográficos* (3), 139–72.

Singh, J., Yu, C.W.F. and Kim, J.T. (2010) Building pathology, investigation of sick buildings–toxic moulds. *Indoor and Built Environment* 19(1), 40–7. https://doi.org/10.1177/1420326x09358808

Smith, B.H. (1984) Patterns of molar wear in hunter–gatherers and agriculturalists. *American Journal of Physical Anthropology* 63(1), 39–56. https://doi.org/10.1002/ajpa.1330630107

Strobelt, L., Kuttenreich, A.M., Volk, G.F., Beurskens, C., Lehmann, T. and Schüler, I.M. (2022) Oral health and oral health-related quality of life in patients with chronic peripheral facial nerve palsy with synkinesis – A case-control-study. *PLoS One* 17(11), e0276152. https://doi.org/10.1371/journal.pone.0276152

Tesorieri, M. (2016) Differential diagnosis of pathologically induced upper and lower limb asymmetry in a burial from late medieval Ireland (CE 1439-1511). *International Journal of Paleopathology* 14, 46–54. https://doi.org/10.1016/j.ijpp.2016.04.003

Tomás, I., Marinho, J.S., Limeres, J., Santos, M.J., Araújo, L. and Diz, P. (2008) Changes in salivary composition in patients with renal failure. *Archives of Oral Biology* 53(6), 528–32. https://doi.org/10.1016/j.archoralbio.2008.01.006

Warinner, C., Herbig, A., Mann, A., Fellows Yates, J.A., Weiß, C.L., Burbano, H.A. and Krause, J. (2017) A robust framework for microbial archaeology. *Annual Review of Genomics and Human Genetics* 18(1), 321–56. https://doi.org/10.1146/annurev-genom-091416-035526

Watt, B. (2023) Loose smut of wheat or barley (*Ustilago tritici*). *Bugwood.org*, University of Maine. https://www.forestryimages.org/browse/detail.cfm?imgnum=5503666

Wunderle, J., Leclerque, J., Schaffrath, U., Slusarenko, A. and Koch, E. (2012) Assessment of the loose smut fungi (*Ustilago nuda* and *U. tritici*) in tissues of barley and wheat by fluorescence microscopy and real-time PCR. *European Journal of Plant Pathology* 133, 865–75.

Yamada, K., Tsuruta, A., Hanada, K. and Hayashi, T. (2004) Morphology of the articular eminence in temporomandibular joints and condylar bone change. *Journal of Oral Rehabilitation* 31(5), 438–44. https://doi.org/10.1111/j.1365-2842.2004.01255.x

Yang, C. and Perry, L. (2013) Identification of ancient starch grains from the tribe Triticeae in the North China Plain. *Journal of Archaeological Science* 40, 3170–7.

Zhang, W., Xu, L., Luo, T., Wu, F., Zhao, B. and Li, X. (2020) The etiology of Bell's palsy: a review. *Journal of Neurology* 267, 1896–905. https://doi.org/10.1007/s00415-019-09282-4

2

Disabled individuals in a Belgian medieval Cistercian monastic community: a palaeopathological perspective

Mathilde Daumas, Caroline Polet and Stéphane Louryan

Throughout human history, disabilities and impairments have been an inherent aspect of societies, deeply influenced by cultural contexts and by the diverse forms they manifest. Some physical and/or mental impairments leave skeletal traces detectable by palaeopathologists. This study presents notable cases of serious disorders, such as disproportionate dwarfism, bilateral calcaneus varus (clubfoot) with evidence of crutch use, and cerebral palsy. All these pathological individuals were buried at the Cistercian Abbey of the Dunes in Koksijde (Belgium), which was most active between the 12th and 15th centuries. Their grave locations are diverse, ranging from the cloister to the lay brothers' cemetery. There was no difference detected in their funerary treatment, except one 'adapted' burial caused by a physical deformity resulting from cerebral palsy.

Most of the pathologies identified are profoundly disabling, with many being congenital and detectable either at birth or during early childhood. Yet all the cases presented in this study reached adulthood, with some even surviving into advanced age (over 60 years), suggesting that they received care and support from a young age. Possible health challenges encountered, as well as the nature of the assistance received, will be discussed according to the nature of the pathology.

The prevalence of disabilities within this monastic community raises questions about the social consequences of visible congenital anomalies. Were these individuals placed in a religious institution due to their disabilities, or did they choose the monastic life for other reasons, as it might have offered opportunities to develop talents that did not rely on physical abilities? Interestingly, historical texts provide no specific mention of disabled individuals within the context of this abbey, making it challenging to determine whether they were fully integrated into the monastic community or were laypersons and/or benefactors.

Keywords: Palaeopathology, disability, Middle Ages, dwarfism, monastery.

Introduction

Disability has been present across all periods of history and among all societies. Though it appears intrinsic to the human condition, its manifestation and meaning vary widely depending on the social, cultural, economic and political context (Hubert 2000; Roberts 2000; Tilley 2015). Impairment itself is strictly anatomical and biological, representing a physical limitation. Disability, however, encompasses the social construction layered onto these biological realities – the way society perceives this condition and its consequences for the individual (Metzler 2011). For instance, infertility may be seen as a social disadvantage, but for someone like a monk who chooses a childless life, this 'impairment' would be neither problematic nor limiting (Middlemass and Tyers 2014). Careful consideration is therefore required when examining past disabilities: perceptions of health, illness and disability vary according to social contexts, and individuals experience their conditions differently. What one person considers a disabling condition might not be perceived in the same way by another (Tilley 2015).

The Middle Ages, for instance, provide valuable insights into the social constructs of disability through various sources, including iconography and architecture, as well as literary, ecclesiastical, medical or legal texts (Metzler 2011). Christianity introduced a new perspective on illness and poverty, portraying the disabled body as an archetype of the 'good poor', a catalyst for Christian charity and a means of spiritual salvation. However, physical or mental impairments carried dual meanings: they could symbolise divine punishment for sin or, conversely, the acceptance of one's fate as divinely ordained (Dubourg 2022a). This created a spectrum of social responses, ranging from acceptance to marginalisation.

Biological anthropology occasionally reveals signs of congenital or acquired pathologies through the analysis of skeletal remains, as in the present study. However, limitations inherent in osteological analysis mean that only certain physical conditions, especially those affecting the locomotor system, can be identified. Many forms of 'invisible' disability, such as mental or sensory impairments, generally leave no trace on bones, complicating the study of these conditions in historical populations (Roberts 2000; Delattre 2018; 2019). Nevertheless, skeletal evidence can highlight individuals with congenital disorders, long-standing infections, healed trauma or other diseases.

The discovery of severe pathologies in medieval burials is far from unusual and spans diverse contexts, whether cloistral, urban or rural (Metzler 2011; Bohling *et al.* 2023). The increasing amount of archaeological evidence therefore suggests that disabilities, especially congenital ones, were not inherently problematic for medieval societies (Billoin *et al.* 2010; Dubourg 2022a; Bohling *et al.* 2023). These examples reflect a degree of social integration, as evidenced by at least one person caring enough to provide necessary assistance to those affected. Yet, this raises an essential question: does this care reflect an institutionalised system of support, or is it more likely a result of close personal or familial relationships?

While hospitable institutions already existed in Antiquity, it was the Church and its mission of charity that inspired the establishment of numerous new foundations, particularly during the later Middle Ages (Chaouky 2019; Dubourg 2022a). In medieval Western Europe, monasteries played a crucial role in providing aid to the sick, with some institutions developing explicitly medical functions (Dubourg 2022a). Consequently, funerary assemblages associated with monastic establishments often reveal individuals who suffered from pathologies or may have received medical treatment.

Within this religious context, the present study examines three individuals interred at the Dunes Abbey in Koksijde (Belgium), whose skeletal remains exhibit evidence of severe or incapacitating pathologies. These individuals were selected for analysis due to the clear skeletal evidence of physical impairment and the exceptional preservation of their remains, which facilitated detailed palaeopathological examination and differential diagnosis. By focusing on these cases, this paper aims to explore both the physical consequences of their conditions – such as mobility and functional capacity – and the evidence of long-term care and social support they may have received. Moreover, it investigates how these conditions may have influenced their social integration and degree of inclusion within the religious community, during their lifetime and beyond death.

Material and methods

The archaeological context

The Dunes Abbey of Koksijde was founded in the early 12th century on the Belgian West Flanders coast. Initially a small religious community, it adhered to the Rule of St Benedict and then became affiliated to the Cistercian order in 1138, integrating the strict observance and communal life characteristic of Cistercian monasticism. By the 13th century, the abbey had expanded significantly, reaching its peak with over 120 monks and around 250 lay brothers, making it one of the largest Cistercian communities in the Low Countries. This period of growth saw the construction of a new abbey complex, which allows us to define specific burial zones linked to the position of the deceased in relation to the religious community. Thereafter, the abbey faced a gradual decline in the 15th and 16th centuries. A series of religious conflicts, outbreaks of disease, and adverse climate events, together with the decrease in vocations, resulted in a diminished presence of individuals opting to take religious vows. By the late 16th century, the monastery had been abandoned (Lehouck 2010).

The skeletal remains discussed in this paper were excavated during various seasons of archaeological excavation conducted during the second half of the 20th century. Due to the absence of an osteologist during these excavations, detailed *in situ* information concerning the graves is unavailable, with only minimal documentation provided. Some of the excavated individuals have been transferred to the collections of the Royal Belgian Institute of Natural Sciences, where the three cases have undergone palaeopathological analysis.

The locations of the burials

Case 1 (H32) was uncovered north of the abbey church (Fig. 2.1). Archaeological data suggest that the burial predates the construction of the church. However, due to the lack of distinguishable burial zones within the oldest section of the funerary area, the social status of this individual remains undetermined. In contrast, Cases 2 (36) and 3 (C25) were buried during the more recent period, dating between the 13th and 16th centuries, which allows for the identification of specific burial zones. Individual 36 was found in the northern gallery of the cloister (Fig. 2.1). This

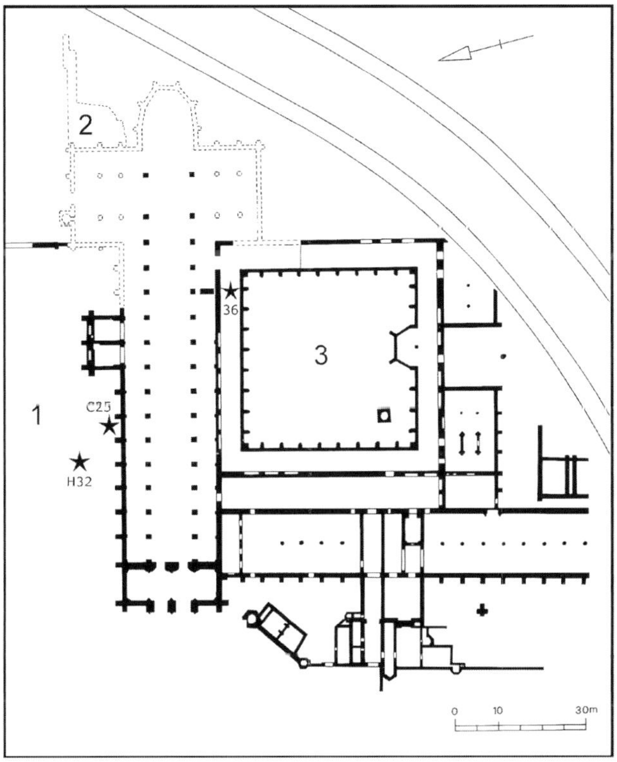

Figure 2.1. Excavation plan of the Abbey of the Dunes in Koksijde. Stars indicate the locations of the cases presented (case 1-36, case 2-C25 and case 3 H32). 1= lay brothers' cemetery; 2= excavated section of the monks' cemetery; 3= the main cloister. Adapted from Dewilde and Demeulemeester (1991).

area was dedicated to high-ranking religious dignitaries or lay benefactors, suggesting a high social status. Individual C25 was buried in the cemetery of the lay brothers, religious members of lower social status, near the church wall (Fig. 2.1).

Biological analysis of skeletal remains

Standard macroscopic methods were employed to analyse the skeletal remains. The primary method for sex determination was probabilistic sex diagnosis (DSP) (Brůžek *et al.* 2017). Where pathological conditions did not allow for DSP analysis, sex was estimated using cranial and pelvic dimorphism (Ferembach *et al.* 1980). Age at death was estimated using classical anthropological methods, including analysis of the auricular surface, pubic symphysis, the sternal end of the fourth rib, dental attrition or cranial sutures (İşcan *et al.* 1984; Brooks and Suchey 1990; Schmitt 2005). Additionally, cementochronology[1] was applied for Case 1 to further refine age estimation (Bertrand *et al.* 2016). The overall population is predominantly male, with only 4% of the individuals being female. Very few infants below 5 years old were recorded, but there is a notable presence of children, adolescents and young adults. Individuals who died before the age of 30 represent a significant proportion of the population (Daumas *et al.* 2016).

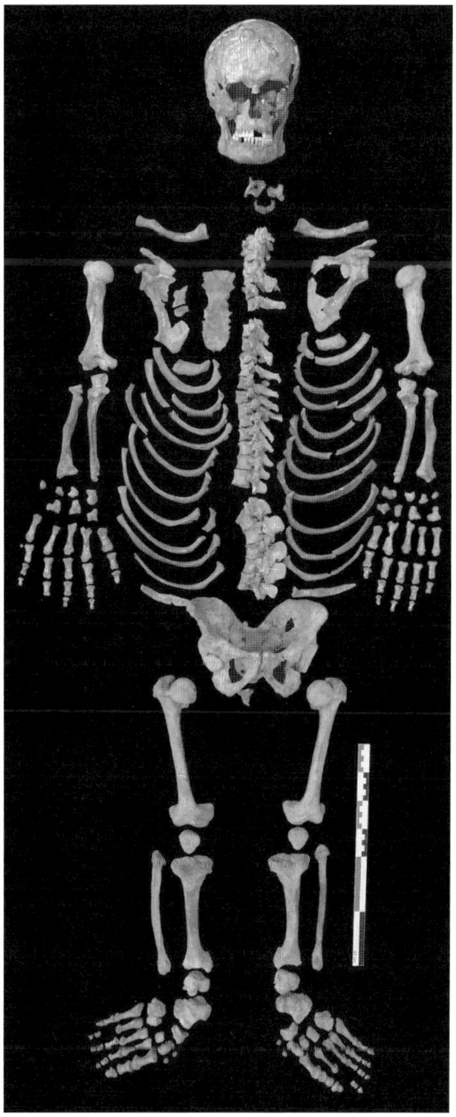

Figure 2.2. Case 1, individual H32 with achondroplasia (photo by C. Polet and E. Dewamme, Royal Belgian Institute of Natural Sciences).

Pathological analysis

Case 1 (H32)

Individual H32 is an adult male, estimated to have died around the age of 40. His skeleton, nearly complete and well-preserved (Fig. 2.2), has been extensively studied and published (Dambrain 1969; Susanne 1970; Polet and Orban 2011; Bertrand *et al.* 2016); therefore, we provide only a summary of his palaeopathological profile here. H32 exhibits a distinctly disproportionate body structure characteristic of achondroplasia, the most common form of dwarfism. His trunk is of normal size, yet his limbs are unusually short and robust, leading to an estimated stature of 132 cm – significantly below the average height for the Koksijde population, around 170 ± 6 cm (Bosmans

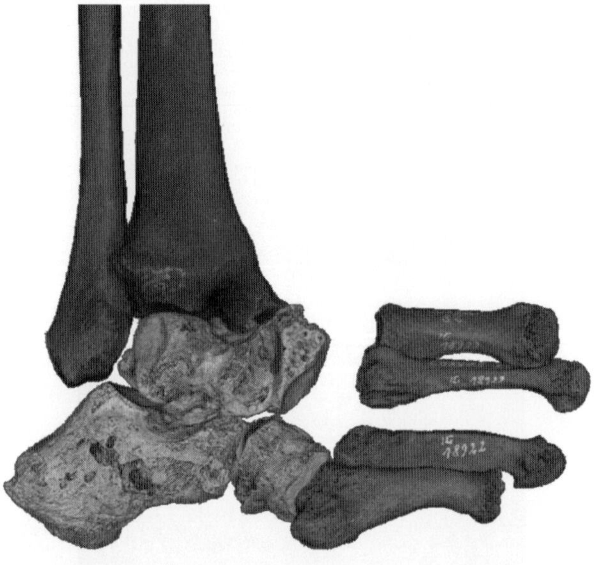

Figure 2.3. Case 2, partial photogrammetric model of the right foot of individual 36 in anterior view. The dorsal flexion and inward rotation of the feet designate a clubfoot in calcaneus varus position (photo by M. Daumas and E. Dewamme, Royal Belgian Institute of Natural Sciences).

et al. 2015). The individual's small stature, characterised by shortened limbs – especially in the proximal segments (humeri and femurs) – is typical of achondroplasia. Other notable features include a disproportionately large cranial vault relative to the face, prominent forehead, pronounced nasal concavity and enlarged orbits (Fig. 2.2).

Achondroplasia, present in approximately 1 in 10,000 to 1 in 30,000 live births, is caused by a genetic mutation on chromosome 4, resulting in a consistent phenotype among those affected (Horton *et al.* 2007). H32's skeletal features align closely with modern cases of achondroplasia, where average male stature is similarly around 132 cm. The condition has a long-standing presence in the archaeological record, with documented cases from the Neolithic period across various geographical regions (Africa, America, Europe).

Case 2 (36)

Individual 36 is an elderly adult male, probably over 50 or even 60 years old, depending on age estimation methods, with a nearly complete skeleton, missing only some phalanges from the feet and hands. His stature aligns with the average height of the population. However, his skeletal remains reveal significant, lifelong impairments related to severe foot deformities (Daumas *et al.* 2022). Notably, he displays extensive changes in the ankle and foot bones, including the tarsals and metatarsals, indicating a pronounced dorsal flexion and inward rotation of the feet (Fig. 2.3), a condition known as calcaneus varus (Fassier 2016). This congenital condition, commonly referred to as clubfoot, resulted in the feet flexing toward the legs and turning inward, forcing Individual 36 to walk solely on the outer edges of his toes. The bilateral manifestation of these deformities strongly suggests a congenital origin (Fassier 2016). Moreover, the precise positioning of his foot bones (Fig. 2.3), which could not be attributed to a post-polio paralysis, confirms the congenital nature of this deformity (Daumas *et al.* 2022).

Despite his significant physical limitations, Individual 36's skeleton reveals compelling evidence of maintained mobility throughout his life. His lower body displays extensive musculoskeletal changes, with deeply remodelled muscle attachment sites. His pelvic girdle shows advanced coxarthrosis with eburnation, signalling sustained, albeit impaired, movement in the hip joint. Additionally, the pronounced asymmetry of his upper limbs – where the right side is much more robust than the left – suggests intensive, unilateral use, probably due to reliance on a crutch for support. The extreme development of entheses and osteoarthritis in the upper body supports the theory that he adjusted his gait early on to accommodate his clubfoot, using his arms to aid in movement. Bone adaptation underscores the early onset of compensatory physical modifications, as these changes are likely to have begun in childhood.

Clubfoot is a common congenital condition nowadays, affecting 1 to 2 per 1,000 births and more prevalent among males (Fassier 2016). Although it is easily treated at birth in developed countries, clubfoot is considered to be the leading congenital cause of physical disability in developing countries (Penny 2005). The lack of corrective treatment indeed led to considerable mobility challenges and physical adaptation. In medieval societies, such a condition would also have imposed significant physical challenges.

Case 3 (C25)

Individual C25 is a mature adult male. The excavation notes indicate an atypical burial position, with his 'tibias folded under his femurs'. His skeletal remains reveal several distinct characteristics (Fig. 2.4), particularly in the lower limbs, which are slender despite having prominent pelvic muscle attachments. His knee demonstrates a pronounced valgus deformity (Fig. 2.4), while his right patella displays an unusually small, unique articular surface. The bones of his legs and feet are notably gracile, with more pronounced deformities on the left side. Additionally, his spine and rib cage show severe double-curve scoliosis, while his upper limbs present normal to robust morphology (Fig. 2.4).

The skeletal modifications observed indicate a neuromuscular impairment, probably paralytic, primarily impacting his lower limbs and ossification processes. C25 displays numerous traits associated with cerebral palsy as described in palaeopathological and clinical literature: lower limb atrophy (paraplegia), patellar deformities, knee flexion

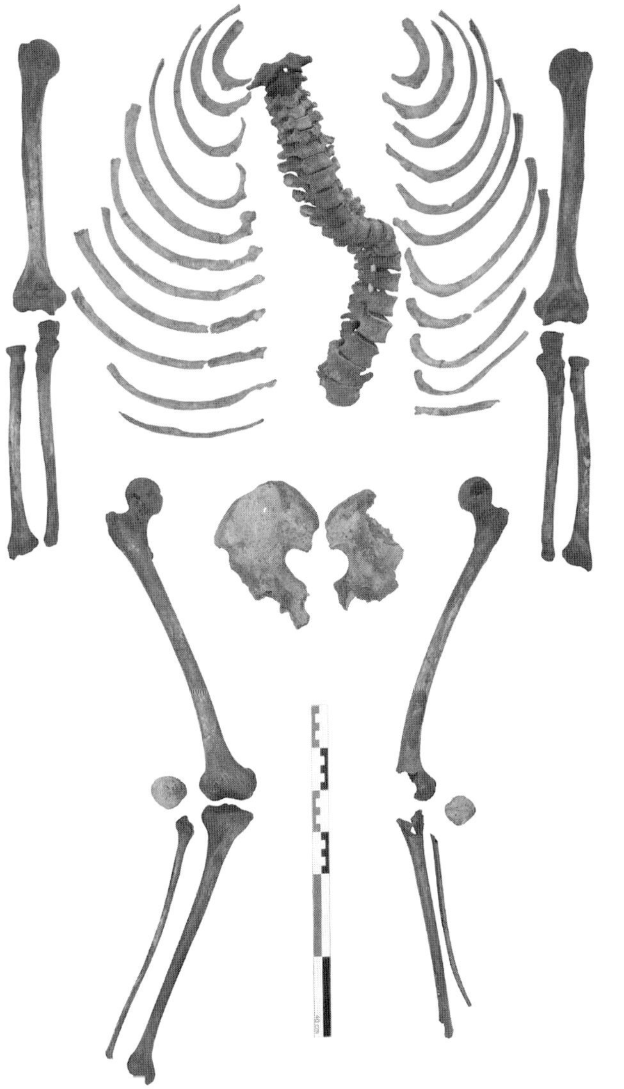

Figure 2.4. Case 3, main long bones of C25, an individual with probable cerebral palsy (photo by M. Daumas and E. Dewamme, Royal Belgian Institute of Natural Sciences).

contractures, scoliosis with vertebral asymmetry, widened metaphyses relative to slender diaphyses, probable foot malformations, multiple fractures and periodontal disease (Cheney and Palmer 1997; Koman *et al.* 2004; Koop 2009; Tesorieri 2016; Otjen *et al.* 2019).

The diagnosis of cerebral palsy suggests that C25 experienced motor development issues from an early age. Muscle spasticity would have resulted in valgus deformity, knee flexion, and flexed hip positions, as evidenced by deformities in his pelvic bones and femurs (Miller 2019; Otjen *et al.* 2019). Nevertheless, he appears to have retained a partial degree of mobility, adopting a crouch gait or similar modified walking style (Koman *et al.* 2004; Church *et al.* 2018; Otjen *et al.* 2019). Remodelling of his upper limbs, particularly the scapulae and humeri, along with signs of possible chronic trauma, suggests the use of a mobility aid, potentially a crutch with axillary support on his right side.

Cerebral palsy results from neurological damage occurring during the development of the central nervous system, leading to varied degrees of muscle weakness, spasticity, contractures and sensory impairments (Koman *et al.* 2004). Today, cerebral palsy affects approximately 1 to 5 per 1,000 live births, often linked to prematurity, low birth weight, or other prenatal or perinatal complications, including trauma, hypoxia, complicated delivery, infection, or genetic factors (Cheney and Palmer 1997; Koman *et al.* 2004). While similar risk factors existed historically, high infant mortality and lack of medical interventions in pre modern societies most likely limited the survival rates of individuals with this condition into adulthood. Therefore, it is challenging to assess the prevalence of cerebral palsy in past societies.

Discussion

The Dunes Abbey of Koksijde, as a Cistercian monastery, followed the medieval Christian principles of charity and compassion, yet it was never formally designated as a caregiving institution. Although medical support may have been accessible to its members, as evidenced by the presence of an infirmary (Vanclooster 2005), there is no indication or evidence that this structure was specifically dedicated to the care and treatment of the sick. Despite this, the presence of several individuals with significant physical impairments interred in the abbey's cemetery suggests a strong degree of social integration and care, implying that the monastery served as a place of support and acceptance. Notably, the three cases presented are not isolated; other burials at the site reveal signs of physical trauma, infectious or congenital diseases, and even therapeutic interventions like trepanations (Amato 2015; Bosmans *et al.* 2015; Simons 2018). The overall health profile of Koksijde's population reflects frequent respiratory infections, as well as a high prevalence of *cribra orbitalia* and enamel hypoplasia, suggesting several stresses (nutritional deficiencies or illness) in early life (Bosmans *et al.* 2015; Vanassche 2020). However, the relatively high average adult stature (170 cm ± 6 cm) suggests that adolescence was likely to have been spent in favourable living conditions, probably due to integration and life within the religious community.

The three cases presented here are distinguished by congenital pathologies that would have been apparent from birth or early childhood. Foot deformities, for instance, are identifiable from birth (Fassier 2016). Similarly, achondroplasia results in distinct physical characteristics, such as dwarfism and limb micromelia,[2] that are evident immediately after birth (Susanne 1970; Polet and Orban 2011). Cerebral palsy presents additional developmental challenges, including abnormal motor functions, potential speech, vision and hearing impairments, and, in some cases, partial epilepsy, all observable within the first few months of life (Koman *et al.* 2004; Tesorieri 2016). These visible differences are likely to have marked these children as 'unique' or 'different' within their community from an early age.

Understanding the level of assistance provided in early life is complex, as basic and advanced care are vital for any child's survival. For the individuals in this study, it appears that their evident differences did not interrupt the natural care given to all children. They received essential nutrition, supervision and protection during early childhood, much like their peers. Yet, the congenital conditions visible on skeletal remains of older individuals indicate that a continued level of care must have been implemented (Roberts 2000). Their survival into later life, despite severe disabilities, is particularly significant when considering the general health profile and mortality rates of the community, marking them as exceptions. This extended survival paradoxically underscores both their 'good health' within the scope of their condition and a supportive environment that fostered their well-being through sustained assistance. This longevity reflects not only continuous support but also a relatively favourable living environment that fostered their survival against challenging odds.

It is therefore necessary to consider the nature of the care and support that must have been provided to these individuals. Given their varying levels of disability, it is likely that the assistance they received was individualised. For example, adult achondroplasts, aside from a reduced stature, are generally autonomous. Support for daily needs such as access to food and water would have been minimal. Similarly, the two other individuals, with visibly functional upper limbs, probably retained the ability to feed themselves. The stature of individual 36, comparable to the population's average, even suggests the absence of nutritional deficiency during growth. Although the literature reports that people with paralysis can have swallowing and feeding problems (Megyesi *et al.* 2009; Tesorieri 2016), there is no evidence to suggest that individual C25 suffered from poor nutrition.

The mobility of individuals 36 and C25 was very likely limited but not entirely absent. Individual 36, despite the functional challenges posed by bilateral clubfoot, was still able to walk, probably thanks to the use of a walking device. Similarly, C25, who experienced partial paralysis in the lower limbs and scoliosis, may have adopted a crouched gait throughout life. Clearly, regardless of their significant physical limitations, these individuals retained a certain degree of autonomy, although their ability to travel longer distances would have been considerably restricted. This limited mobility would have constrained their participation in most social and economic activities, necessitating external support for essential needs such as the acquisition of food and water, most likely provided by the monastic community.

The overall health condition of individual C25, in particular, is likely to have required a more intensive level of care. Literature indicates that contemporary patients with long-term mobility impairments are often vulnerable to soft tissue complications, such as those affecting respiratory and circulatory functions (Megyesi *et al.* 2009; Tesorieri 2016), which are largely undetectable in osteological remains. However, it is possible that caregivers provided assistance through physical mobilisation or massage to mitigate such issues, as seen in other palaeopathological examples (Tilley and Oxenham 2011).

While it is clear that these individuals received support in life, it is also worth examining the nature of care extended to them at the time of their death. Despite limited archaeothanatological observations, burial practices appear consistent and do not show evidence of non-normative treatment, apart from the atypical positioning of individual C25 (Daumas and Polet *in press*). Funerary practices often reflect the physical condition of the deceased, especially in body positioning within the grave (Delattre and Sallem 2014; Ardagna *et al.* 2021b). Many cases reveal that Christian burial prescriptions, which require a normative body orientation, were adapted to accommodate physical deformities (Delattre and Sallem 2014). Occasionally, the grave structure itself was adjusted for the physical constraints posed by disability, such as by deepening or reshaping (Ardagna *et al.* 2021b). These atypical adjustments reflect physical necessity rather than intentional differentiation in funerary practices. While exceptions do exist, religious norms often had more influence over funeral practices than physical conditions (Bohling *et al.* 2022; 2023).

Beyond body positioning, the burial location itself can also reflect a certain status or role within the abbey community. At Koksijde, disabled individuals were found in both privileged and less privileged burial sectors, suggesting diverse social and economic status origins. Individual 36, for instance, was buried in the cloister, leading to two possible interpretations. He may either have gained access to this privileged burial zone as a benefactor or have been fully integrated into the monastic community perhaps as a high-ranking member of the order. Similarly, although C25's paralysis would have excluded him from agricultural labour or most crafts, he was buried in the lay brothers' sector, very close to the abbey church wall – a positioning that could be seen as privileged. This suggests that, despite physical disabilities, these individuals may have held respected roles within the monastic hierarchy. Therefore, disabled individuals were not marginalised in the burial space of the Dunes Abbey, suggesting a degree of integration and acceptance within the community (Daumas and Polet *in press*). However, this integration in death does not necessarily reveal details about where they lived. While some disabled individuals might have been cared for by the monastic community throughout their lifetimes, this may not have been the case for all. For instance, socially advantaged individuals like 36 may have lived and received care outside the abbey, only to be buried in the cloister when they died.

The recurring instances of disabled individuals buried within monastic structures raise questions. Some hypotheses centre on the medieval perception of disability as a potential spiritual trial or calling (Dubourg 2022a). Indeed, while performing the sacraments and delivering the Eucharist typically required clergy to embody physical 'perfection' and set an example for the faithful, ordination was not a requirement for becoming a monk (Dubourg 2022b; 2023). Thus, individuals with physical disabilities were not barred from monastic life (Dubourg 2022b). Other hypotheses offer a more socially grounded interpretation, suggesting that, for some individuals with disabling conditions, a religious life might have represented the only viable path for personal fulfilment or survival. This path allowed these individuals to find a place and an active role within society – a chance to contribute within the protective structure of monastic communities (Billoin 2010; Delattre 2010).

The presence of individuals with disabilities is particularly well documented in female monastic communities, with palaeopathological analyses frequently revealing motor impairments or congenital anomalies, such as congenital hip dislocations or early-life fractures. The proportion of women with such conditions in convents often exceeds that in parish cemeteries (Billoin *et al.* 2010; Sarkic 2017; Ardagna *et al.* 2021a; Gallien and Darton 2021). Historical records occasionally refer to nuns with cognitive impairments (Dubourg 2023), a disability that cannot be accessed through palaeopathology. These congenital or early-life physical conditions may have limited these women's marital prospects. A medieval song suggests that, sometimes, the monastery was the sole option for certain women: 'I was not good enough for man, so I am given to God' (Sarkic 2017). Such women, deemed unfit for marriage and procreation due to physical disorders, may have found a social function within religious orders. Yet, this religious vocation was not necessarily undervalued; in fact, their physical differences were often interpreted as a divine mark, signifying a path to spiritual perfection (Billoin 2010, Gallien and Darton 2021). Their ascetic rigour and reputed fervour were believed to yield particularly powerful prayers. Consequently, female monasteries, especially Cistercian ones in the 13th century, sometimes enjoyed a more revered status than their male counterparts (Coomans 2005).

These questions therefore remain open: were disabled individuals integrated as active members of the community, or was their presence more aligned with receiving physical and spiritual support from their monastic community – or, more likely, a combination of both? Given the interpretive limitations when relying solely on skeletal remains, caution is necessary, as palaeopathological analyses alone cannot fully answer this. For instance, the multiple fractures observed on individual C25 are challenging to interpret. These traumatic events might result from complications due to mobility issues, reveal potential mistreatment, or merely pathological fractures related to physiological effects of the ageing process.

Further research is needed to unearth the untold narratives of the disabled throughout history and more particularly inside monastic context. While this study highlights the reality of the care extended to individuals with disabilities, it cannot capture the socio-affective impacts they experienced within the community. History reminds us that disability is not a fixed or uniformly negative experience. Notably, medieval figures like Aregund (wife of Chlothar I) or Bertrada (mother of Charlemagne) reached prestigious positions despite their disabilities. This challenges the preconceived notion that disability in the medieval period condemned one to a life of poverty and begging, perhaps worse than death (Dubourg 2022a). Instead, it suggests a more nuanced view, one that acknowledges both the difficulties and the resilience of individuals who may have found acceptance and purpose within monastic life.

Acknowledgements

We are grateful to the Royal Belgian Institute of Natural Sciences for allowing access to the collection.

Notes

1 A histological method for assessing age at death, through evaluation of the dental cementum annual deposit.

2 Shortening of both proximal and distal segments of limb.

References

Amato, L. (2015) Letsels, verwondingen en medische behandelingen bij uiteenlopende sociale groepen in de middeleeuwse abdij Ten Duinen in Koksijde. *Novi Monasterii: jaarboek Abdijmuseum Ten Duinen* 14, 43–75.

Ardagna, Y., L'Hostis, A. and Ollivier, D. (2021a) Dysplasie de la hanche et moniales cisterciennes : l'exemple de l'Abbaye de Saint Pierre de l'Almanarre (XIIIe-XIVe, Hyères, Var, France). *Bulletins et mémoires de la société d'anthropologie de Paris*, 33(S). https://doi.org/10.4000/bmsap.6921

Ardagna, Y., Sperandio, E., Rigeade, C., Darton, Y. and Kacki, S. (2021b) La paléopathologie comme outil archéothanatologique: contribution à l'interprétation des positions d'inhumation atypiques. In S. Kacki, H. Réveillas and C. Knüsel (eds) *Rencontre autour du corps malade: prise en charge et traitement funéraire des individus souffrants à travers les siècles: actes de la 10e Rencontre du GAAF, Athénée municipal de Bordeaux, 23–25 mai 2018*, 201–13. Publication du GAAF, no. 10. Reugny, Groupe d'anthropologie et d'archéologie funéraire.

Bertrand, B., Robbins Schug, G., Polet, C., Naji, S. and Colard, T. (2016) Age-at-death estimation of pathological individuals: a complementary approach using teeth cementum annulations.

International Journal of Paleopathology 15, 120–27. https://doi.org/10.1016/j.ijpp.2014.04.001

Billoin, D. (2010) Les dominicaines de Sélestat (fin XIIIème-XVIIIème siècle): des filles handicapées au couvent. In V. Delattre and R. Sallem (eds) *Décrypter la différence: lecture archéologique et historique de la place des personnes handicapées dans les communautés du passé*, 185–86. Paris, CQFD (Les défis de civilisation).

Billoin, D., Gallien, V. and Périn, P. (2010) Réflexion sur la place des personnes handicapées à l'aube du Moyen-Âge. In V. Delattre and R. Sallem (eds) *Décrypter la différence: lecture archéologique et historique de la place des personnes handicapées dans les communautés du* passé, 79–84. Paris, CQFD (Les défis de civilisation).

Bohling, S., Croucher, K. and Buckberry, J. (2022) The bioarchaeology of disability: a population-scale approach to investigating disability, physical impairment, and care in archaeological communities. *International Journal of Paleopathology* 38, 76–94. https://doi.org/10.1016/j.ijpp.2022.05.006

Bohling, S., Croucher, K. and Buckberry, J. (2023) Understanding disability and physical impairment in early medieval England: an integration of osteoarchaeological and funerary evidence. *Medieval Archaeology* 67(1), 73–114. https://doi.org/10.1080/00766097.2023.2204666

Bosmans, S., Vercauteren, M. and Polet, C. (2015) De gezondheidstoestand in de middeleeuwse abdij Ten Duinen in Koksijde. *Novi Monasterii: jaarboek Abdijmuseum Ten Duinen* 14, 21–42.

Brooks, S. and Suchey, J.M. (1990) Skeletal age determination based on the os pubis: A comparison of the Acsádi-Nemeskéri and Suchey-Brooks methods. *Human Evolution* 5, 227–38. https://doi.org/10.1007/BF02437238

Brůžek, J., Santos, F., Dutailly, B., Murail, P. and Cunha, E. (2017) Validation and reliability of the sex estimation of the human os coxae using freely available DSP2 software for bioarchaeology and forensic anthropology. *American Journal of Physical Anthropology* 164(2), 440–9. https://doi.org/10.1002/ajpa.23282

Chaouky, H. (2019) Les hôpitaux et les lieux de soin dans l'Antiquité et au Moyen Âge. In A. Froment and H. Guy (eds) *Archéologie de la santé, anthropologie du* soin, 159–69. Paris, La Découverte (Recherches).

Cheney, P.D. and Palmer, F.B. (1997) Overview: cerebral palsy. *Mental Retardation and Developmental Disabilities Research Reviews* 3(2), 109–11. https://doi.org/10.1002/(SICI)1098-2779(1997)3:2<109::AID-MRDD1>3.0.CO;2-M

Church, C., Ge, J., Hager, S., Haumont, T., Lennon, N., Niiler, T., Hulbert, R. and Miller, F. (2018) Flexed-knee gait in children with cerebral palsy: a long-term follow-up study. *The Bone and Joint Journal* 100-B(4), 549–56. https://doi.org/10.1302/0301-620X.100B4.BJJ-2017-0732.R1

Coomans, T. (2005) Moniales cisterciennes et mémoire dynastique: églises funéraires princières et abbayes cisterciennes dans les anciens Pays-Bas médiévaux. In J. Hall and C. Kratzke (eds) *Sepulturae cistercienses: sépulture, mémoire et patronage dans les monastères cisterciens au Moyen Age = burial, memorial and patronage in medieval Cistercian monasteries = Grablegen, Memoria und Patronatswesen in mittelalterlichen Zisterzienserklöstern*, 87–145. Forges-Chimay, Cîteaux (Cîteaux, commentarii cistercienses. *Revue d'histoire cistercienne* 56; fasc. 1-4).

Dambrain, R. (1969) L'achondroplasie. Étude cranio-faciale. *Bulletin de la Société royale belge d'Anthropologie et de Préhistoire* 80, 69–86.

Daumas, M., Lefèvre, P., Polet, C., Orban, R., Beauthier, J.-P., Martrille, L., Rooze, M. and Louryan, S. (2016) De bevolkingsstructuur op de begraafplaats van de abdij Ten Duinen in Koksijde (12de-16de eeuw). *Novi Monasterii: jaarboek Abdijmuseum Ten Duinen* 16, 59–76.

Daumas, M., Polet, C., Mahieu, C. and Louryan, S. (2022) Sur les pas du premier cas archéologique d'un pied calcanéus varus (anciennement pied talus varus). *Morphologie* 106(354, Supplement), S31–S32. https://doi.org/10.1016/j.morpho.2022.06.038

Daumas, M. and Polet, C. (*in press*) Le nain et l'estropié de naissance, deux individus «hors normes» inhumés dans le cimetière médiéval de l'abbaye des Dunes de Coxyde. In Actes de la 46e Journée du RMBLF «Marges, marginaux et marginalités au Moyen Âge. Transgressions et expériences de la norme (Ve-XVe siècle)». *Revue Belge de Philologie et d'Histoire (RBPH).*

Delattre, V. (2010) Handicap et vie religieuse au Moyen-Âge: un chanoine pris en charge par sa communauté à l'abbaye Saint-Severin de Chateau-Landon (Seine-et-Marne). In V. Delattre and R. Sallem (eds) *Décrypter la différence: lecture archéologique et historique de la place des personnes handicapées dans les communautés du passé,* 177–79. Paris, CQFD (Les défis de civilisation).

Delattre, V. (2018) *Handicap: quand l'archéologie nous éclaire.* Paris, Cité des sciences et de l'industrie Éditions le Pommier Universcience (Le collège, 29).

Delattre, V. (2019) Handicaps et sociétés du passé. In A. Froment and H. Guy (eds) *Archéologie de la santé, anthropologie du soin*, 251–61. Paris, La Découverte (Recherches).

Delattre, V. and Sallem, R. (2014) Lecture archéo-anthropologique du statut social du «corps différent» dans les communautés passées. In A.-C. Gillis (ed.) *Corps, travail et statut social: L'apport de la paléoanthropologie funéraire aux sciences historiques*, 117–33. Quebec, Presses universitaires du Septentrion. https://doi.org/10.4000/books.septentrion.60891

Dewilde, M. and Demeulemeester, J. (1991) De opgravingscampagnes van 1987 en 1988 in de Onze-Lieve-Vrouw-ten Duinenabdij te Koksijde. *Archeologie in Vlaanderen* 1, 213–30. https://doi.org/10.55465/GYHI9532

Dubourg, N. (2022a) European medieval disability history: an overview. In M.H. Rioux, J. Viera, A. Buettgen and E. Zubrow (eds) *Handbook of Disability*, 1–18. Singapore, Springer Nature..https://doi.org/10.1007/978-981-16-1278-7_2-1

Dubourg, N. (2022b) L'incapacité dans les lettres de dispenses pontificales: Aller à l'encontre de la réglementation ecclésiastique médiévale pour franchir la clôture. In A. Burkardt and A. Roger (eds) *L'exception et la Règle: Les pratiques d'entrée et de sortie des couvents, de la fin du Moyen Âge au XIXe siècle*, 41–54. Rennes, Presses universitaires de Rennes. https://doi.org/10.4000/books.pur.161052

Dubourg, N. (2023) *Disabled Clerics in the Late Middle Ages: Un/suitable for Divine Service?* Amsterdam, Amsterdam University Press.

Fassier, A. (2016) Pied bot varus équin et autres malformations congénitales des pieds. *Journal de Pédiatrie et de Puériculture* 29(6), 310–23. https://doi.org/10.1016/j.jpp.2016.09.006

Ferembach, D., Schwidetzky, I. and Stoukal, M. (1980) Recommendations for age and sex diagnoses of skeletons. *Journal of Human Evolution* 9(7), 517–49. https://doi.org/10.1016/0047-2484(80)90061-5

Gallien, V. and Darton, Y. (2021) Le corps à l'épreuve des règles monastiques: gestion des maux et longévité (abbaye Sainte-Croix, Poitiers, XVIe-XVIIIe siècles). In S. Kacki, H. Réveilla, and C. Knüsel (eds) *Rencontre autour du corps malade: prise en charge et traitement funéraire des individus souffrants à travers les siècles: actes de la 10e Rencontre du GAAF, Athénée municipal de Bordeaux, 23-25 mai 2018*, 67–74. Publication du GAAF, no. 10. Reugny, Groupe d'anthropologie et d'archéologie funéraire.

Horton, W.A., Hall, J.G. and Hecht, J.T. (2007) Achondroplasia. *The Lancet* 370(9582), 162 72. https://doi.org/10.1016/S0140-6736(07)61090-3

Hubert, J. (ed.) (2000) *Madness, Disability, and Social Exclusion: The Archaeology and Anthropology of 'Difference'*. One World Archaeology 40. London and New York, Routledge,

İşcan, M.Y., Loth, S.R. and Wright, R.K. (1984) Age estimation from the rib by phase analysis: white males. *Journal of Forensic Sciences* 29(4), 1094–104.

Koman, L.A., Smith, B.P. and Shilt, J.S. (2004) Cerebral palsy. *The Lancet* 363(9421), 1619–31. https://doi.org/10.1016/S0140-6736(04)16207-7

Koop, S.E. (2009) Scoliosis in cerebral palsy, *Developmental Medicine and Child Neurology* 51(s4), 92–8. https://doi.org/10.1111/j.1469-8749.2009.03461.x

Lehouck, A. (2010) Een mysterie opgelost! Twaalfde-eeuwse sporen in de Cisterciënzerabdij O.L.V.-Ten Duinen in Koksijde (1107–1578/1627). In M. Dewilde, A. Ervynck and F. Becuwe (eds) *Cenulae recens factae: een huldeboek voor John De Meulemeester*, 255–84. Gent, Academia Press, Jaarboek Abdijmuseum Ten Duinen. Novi Monasterii. http://hdl.handle.net/1854/LU-909622

Megyesi, M.S., Tubbs, R.M. and Sauer, N.J. (2009) An analysis of human skeletal remains with cerebral palsy: associated skeletal age delay and dental pathologies. *Journal of Forensic Sciences* 54(2), 270–4. https://doi.org/10.1111/j.1556-4029.2008.00948.x

Metzler, I. (2011) Disability in the Middle Ages: impairment at the intersection of historical inquiry and disability studies. *History Compass* 9(1), 45–60. https://doi.org/10.1111/j.1478-0542.2010.00746.x

Middlemass, R. and Tyers, T. (2014) Disabling masculinity: manhood and infertility in the High Middle Ages. In S. Crawford and C. Lee (eds) *Social Dimensions of Medieval Disease and Disability*, 47–58. BAR International Series 2668. Oxford, Archaeopress.

Miller, F. (2019) Knee flexion deformity in cerebral palsy. In F. Miller, S. Bachrach, N. Lennon and M. O'Neil (eds) *Cerebral Palsy*, 1–22. Cham, Springer International Publishing. https://doi.org/10.1007/978-3-319-50592-3_136-1

Otjen, J.P., Sousa, T.C., Bauer, J.M. and Thapa, M. (2019) Cerebral palsy – beyond hip deformities. *Pediatric Radiology* 49(12), 1587–594. https://doi.org/10.1007/s00247-019-04519-w

Penny, J.N. (2005) The neglected clubfoot. *Techniques in Orthopaedics* 20(2), 153–66. https://doi.org/10.1097/01.bto.0000162987.08300.5e

Polet, C. and Orban, R. (2011) L'achondroplase de l'abbaye des Dunes de Coxyde (Belgique, XIIe–XVe siècle). In P. Charlier (ed.) *3ème colloque international de pathographie. Bourges, 3-5 avril 2009*, 125–62. Paris, De Boccard.

Roberts, C.A. (2000) Did they take sugar? The use of skeletal evidence in the study of disability in past populations. In J. Hubert (ed.) *Madness, Disability, and Social Exclusion: The Archaeology and Anthropology of 'Difference'*, 46–59. One World Archaeology 40. London and New York, Routledge.

Sarkic, N. (2017) *A Study of the Lifestyle and Health Parameters of Nuns from Convents on the Iberian Peninsula in Modern Times*. http://purl.org/dc/dcmitype/Text. Universidad Autónoma de Madrid. https://produccioncientifica.ucm.es/documentos/638560b5cdfe7d707fec1103?lang=fr

Schmitt, A. (2005) Une nouvelle méthode pour estimer l'âge au décès des adultes à partir de la surface sacro-pelvienne iliaque. *Bulletins et Mémoires de la Société d'Anthropologie de Paris* 17(1–2), 89–101.

Simons, J. (2018) *Étude de trépanations néolithiques et médiévales en Belgique: diagnostic positif et différentiel*. Mémoire de Master, Université de Bordeaux.

Susanne, C. (1970) L'achondroplase de la population d'âge franc de Coxyde (Belgique). *Bulletin de l'Institut Royal des Sciences Naturelles de Belgique – Bulletin van het Koninklijk Belgisch Instituut voor Natuurwetenschappen* 46(10), 1–78.

Tesorieri, M. (2016) Differential diagnosis of pathologically induced upper and lower limb asymmetry in a burial from late medieval Ireland (CE 1439–1511). *International Journal of Paleopathology* 14, 46–54. https://doi.org/10.1016/j.ijpp.2016.04.003

Tilley, L. (2015) *Theory and Practice in the Bioarchaeology of Care*. New York, Springer Science+Business Media, LLC.

Tilley, L. and Oxenham, M.F. (2011) Survival against the odds: modeling the social implications of care provision to seriously disabled individuals. *International Journal of Paleopathology* 1(1), 35–42. https://doi.org/10.1016/j.ijpp.2011.02.003

Vanassche, C. (2020) *La santé respiratoire de la population de l'Abbaye des Dunes (Coxyde, Belgique): étude des sinusites maxillaires chroniques*. Mémoire de Master, Université Libre de Bruxelles.

Vanclooster, D. (ed.) (2005) *De Duinenabdij te Koksijde: cisterciënzers in de Lage Landen*. Tielt, Lannoo.

3

A high-ranking cleric with arterial calcification: vascular disease in medieval England

David Bennett-Jones and Louise Loe

Examination of mummified human remains has shown calcific vascular disease to be an ancient disease with global distribution. Despite this, calcified blood vessels are hardly ever recognised during excavation and analysis of dry skeletal remains. Four segments of calcified artery from an elderly male with Diffuse Idiopathic Skeletal Hyperostosis (DISH), who was buried in a stone coffin in a 13th-century cathedral chapel, were imaged in microscopic detail, using focus-stacking photography. The images show ossification of the endothelial and medial layers of arteries.

The skeletal remains are most likely to be those of William of Mancetter, the long-serving and influential Dean of Lichfield Cathedral (1222–1254) and benefactor of the chapel. Contemporary ecclesiastical statutes confirm the exalted status of the Dean at Lichfield and that he received a generous stipendiary allowance, twice that of the canons. This is the first published archaeological example of calcified arteries found in association with the skeletonised remains of a named individual from medieval England. Any suggestion of a causal relationship between the Dean's supposed diet and the vascular disease from which he suffered is speculative.

Keywords: Vascular disease, arterial calcification, medieval, diet, episcopal statutes, focus-stacking photography.

Introduction

Arterial disease has been detected in mummified remains from ancient Egypt (Ruffer 1911; Sandison 1962; Thompson *et al.* 2013), Chalcolithic (Copper-Age) Italian Alps (Murphy *et al.* 2003), pre-Columbian South America (Marrazzini *et al.* 2015), pre-Columbian American Southwest (Thompson *et al.* 2013), Renaissance Naples (Gaeta *et al.* 2013), 17th-century South Korea (Kim *et al.* 2015) and 19th-century hunter-gatherers in the Aleutian Islands (Clarke *et al.* 2014). The disease is common in modern western society, especially in older people (Powell and Davies 2020), with a 'hallmark' complication of arterial calcification (Shioi and Ikari 2018, 294). Despite the prevalence of vascular disease in ancient and modern society, calcified blood vessels have rarely been recognised during archaeological excavation and analysis of dry skeletal remains (Binder and Roberts 2023, 214). Consequently, post-mortem morphology of calcified arteries following skeletonisation has been studied only in a contemporary context (Subirana-Domènech *et al.* 2012; Biehler-Gomez *et al.* 2018; Biehler-Gomez *et al.* 2019).

Except in unusual circumstances, such as the survival of soft tissues in anaerobic conditions (Van Beek *et al.* 2023), human remains from northern Europe consist only of skeletal elements. Although entombment may slow post-depositional decay, it does not prevent it and, as in soil (Henderson 1987, 51), skeletonisation is affected by temperature, humidity, acidity and the level of oxygen (Magni 2019). However, entombment, in contrast to inhumation, is likely to facilitate the archaeological retrieval of an entire skeleton, including smaller bones (Boddington 1987, 30) and calcified soft tissues (Lynnerup *et al.* 2015). Although Roberts (2023, 12) recently described archaeological retrieval of calcified atheroma as 'virtually absent', fragments of calcified blood vessels have been reported from 2nd-millennium BC Amara, Northern Sudan (Binder and Roberts 2014) and medieval

Britain (Rodwell 1983a; Waldron 2007; Loe *et al.* 2020, 309–13). The Amara specimens were examined photographically and with plain radiography, and selected elements were further examined with scanning electron microscopy (SEM) and energy dispersive X-ray spectroscopy for the presence of calcium.

Without the most assiduous excavation technique, which should include sieving (Biehler-Gomez *et al.* 2023, 232), such friable remains are easily overlooked (Loe *et al.* 2020, 311–13, 509–10). During excavation of a medieval cemetery at Stoke Quay, Loe and her colleagues routinely sampled soils in the proximity of the chest and pelvis. Calcified and ossified elements were retrieved and examined with a hand lens (×10 magnification). Selected elements were further examined with plain film radiography. As well as calcified cartilage, renal calculi, heart valves and pleural plaques, the authors identified calcified blood vessels from five individuals. Periodontitis, which is known from clinical studies to be associated with vascular disease (Herrera *et al.* 2020; Zardawi *et al.* 2021; Lu *et al.* 2024), was noted to be present in all five cases.

Soft tissue calcification can be divided into two categories (Dalinka and Melchior 1980): 1. metastatic calcification occurs when circulating blood calcium and/or phosphate levels are elevated (Stewart *et al.* 1983), causing calcium salts to be deposited in normal tissues; 2. dystrophic calcification, on the other hand, occurs when disease or necrosis prompts a regulated process of ossification of soft tissue by osteoblasts. Two layers of the arterial wall ossify during dystrophic calcification of diseased arteries (Deneke *et al.* 2001; Hayden *et al.* 2005; Fuery *et al.* 2018): the intimal layer, which consists of the internal endothelial lining of the artery, and the medial layer, which consists of muscle cells and elastic tissue (Mescher 2016, 220; Channon and Vallance 2020). Ossification of the intima is preceded by the pathological deposition of a fatty substance, atheroma, whilst medial ossification is an idiopathic process when vascular smooth muscle cells transform into osteoblast-like cells (Dovan *et al.* 2008).

The fatty substance, atheroma, which is deposited on the intimal surface of the blood vessel, protrudes into the lumen, causing reduced and turbulent blood flow (Makin *et al.* 2002). Such atheromatous deposits may ulcerate and become 'unstable' leading to thrombotic complications (Shah 2003). The reduced blood flow and activation of thrombotic factors (Lusis 2000; Spronk *et al.* 2004) causes ischaemia of the tissues beyond and is associated with stroke, heart attack (Falk 2006) and peripheral gangrene (Bell 1957; Conn *et al.* 1973). Although calcification of the media does not impinge on the arterial lumen, it alters the pulsatility of blood flow, causing heart damage (Lehto *et al.* 1996) and peripheral ischaemia (Fok and Lanzer 2018; Lanzer *et al.* 2021). Atheroma typically affects larger arteries whereas medial sclerosis, although it can affect any size of artery, mainly affects smaller arteries (St. Hilaire 2022). Calcification in either layer of the arterial wall is associated with older age, diabetes and renal failure (Kannel 1961; Kannel and McGee 1979). Regardless of underlying disease, it has been noted that the size of calcific deposits can vary (Shioi and Ikari 2018; Montanaro *et al.* 2021). Microcalcification, which is unlikely to be identified among excavated remains, is associated with vascular complications, whereas macrocalcification, with heavier calcium deposits in the arterial wall, is associated with plaque stabilisation and clinical stability.

Some individuals, referred to as bone-formers, have a propensity to ossify ligaments and cartilage (Rogers *et al.* 1997; Mays 2016). This can result in a clinical condition called Diffuse Idiopathic Skeletal Hyperostosis (DISH), in which excessive bone formation results in characteristic changes to the skeleton (Littlejohn 2022). Although this disease with ossification of ligaments alongside the vertebral column (Resnick and Niwayama 1976) is often asymptomatic, it is associated with obesity and diabetes (Kiss *et al.* 2002) and may be a risk factor for cardiovascular disease (Glick *et al.* 2020). DISH has been found in individuals from a medieval monastic context (Rogers and Waldron 2001), giving rise to the suggestion of dietary excess among members of the clerical community (Verlaan *et al.* 2007), although this association has been questioned (Mays 2009).

This study examined four arterial segments, found with a named individual from the medieval period, whose lifestyle was assessed.

Materials

Thirty-two fragments of calcified artery were found during systematic archaeological excavation of a discrete, partially disarticulated skeleton (LC82/47), from a stone tomb in St Chad's Chapel, Lichfield Cathedral in 1981–1982 (Rodwell 1983a). The skeleton, which was complete apart from the skull, had been disturbed in ancient times, possibly for the purpose of collecting relics. The calcified arteries lay at the east end of the tomb, with a 'scatter of toe and foot bones, together with some fragments of leather' (Rodwell 1983a, 6). Four of these arterial fragments form the material for this study. Based on the context of the burial and a pewter funerary chalice beside the body, the identity of this individual is likely to be William of Mancetter, Dean of Lichfield Cathedral from 1222 to 1254 (Rodwell 1983b).

Methods

The four fragments were examined with a X10 hand lens and then, in microscopic detail, using focus-stacking photography. A mirrorless, digital single-lens reflex (DSLR) camera (Canon EOS M6 MkII®) was moved towards the object and multiple photographic images obtained at successively closer range (Fig. 3.1). The optical components consist of

Table 3.1. Characteristics of microscope objective lenses and distance of stepwise camera travel on motorised focus-stacking rail

Objective lens	*Field of View (mm)*	*Quoted depth of focus (μm)*	*Distance between stacked photos (μm)*
×2	11.15×7.4	200	100
×5	4.46×2.96	40	20
×10	2.23×1.48	10	5

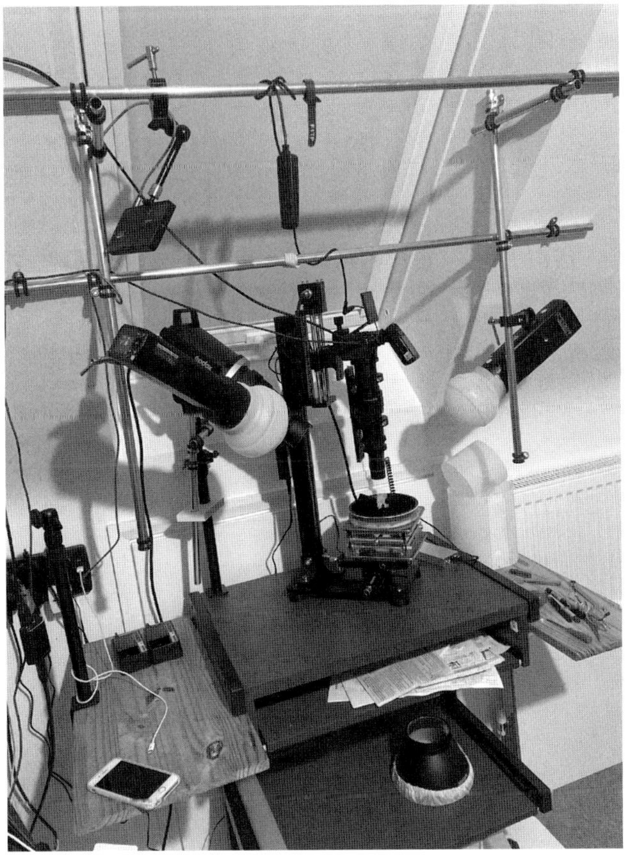

Figure 3.1. Focus-stacking photographic equipment mounted vertically.

a microscope objective lens (M-Plan ×2–×10), relay lens (Raynox DCR150®) and adjustable extension tube. The camera was secured to a motorised, focus-stacking rail (WeMacro®), which advances the camera stepwise towards the object being photographed. The distance between each photograph is determined by the depth of focus of the objective lens and is controlled with a cell phone app (WeMacro®). The number of images required is determined by the depth of the object to be photographed. For example, an object with a 4 mm depth, photographed with the ×2 objective, taken at 100-micron intervals, would require 41 successive images. The size of the camera sensor was 22.3 × 14.9 mm, so the field of view using the ×2 objective was approximately 11.1 × 7.4 mm. More powerful objectives have a proportionately smaller field of view (see Table 3.1). The portion of each image which is in focus was rendered using proprietary software (HeliconFocus®), to provide a single fully focused image. When a larger area was required for examination, adjacent, overlapping focus-stacked photographs were stitched together using proprietary software (Panorama Stitcher®).

Information about the status of the Dean of Lichfield Cathedral and his stipend in the 13th century was obtained from the statutes of Pattishall and Meulen, as documented in *Monasticon Anglicanum* (Dugdale 1849).

Results

The skeletal remains

The partially disarticulated remains of a single, discrete skeleton (LC82/47) were found in a stone coffin which was opened during building work in St Chad's Head Chapel, Lichfield Cathedral in 1981 and examined archaeologically in 1982 (Rodwell 1983a). The coffin lid formed the sill of the south window and was an integral part of the chapel's construction, although there had been disturbance in historical times. This occurred first when the window was reconstructed in the perpendicular style in the late 15th century, and later in the 18th or 19th century, when some builders' rubble, including dateable glass fragments, was inserted through a hole at the foot end of the tomb.

The skeleton, which was examined by Juliet Rogers, palaeopathologist (Loe and Rogers 2012), was confirmed to be that of a single, male individual and to be complete, apart from absence of the skull, although the mandible and two maxillary teeth remained. The first cervical vertebra and four other vertebrae were in the correct anatomical position, A block of six cervical and upper thoracic vertebrae were displaced but remained in correct juxtaposition (Rodwell 1983a). The position of the arm and leg bones and the pelvis suggested that they had been removed and subsequently replaced in the tomb, wrapped in a red-brown fabric (Fig. 3.2). This selection of bones, including the missing skull, were those commonly used as relics for display and veneration, prior to the English Reformation. It is assumed that the long bones were replaced in the tomb sufficiently soon after removal that memory of their original location persisted.

Smaller bones, such as those from the hands and feet, were comingled at the foot end of the sarcophagus, together

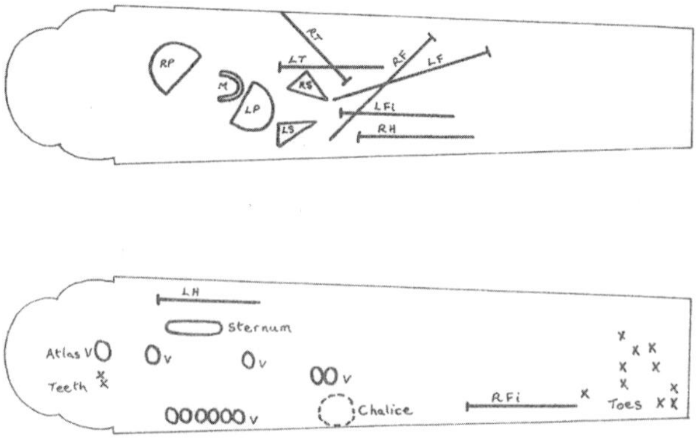

Figure 3.2. Distribution of skeletal remains and chalice as found in the tomb. The upper schematic plan shows the position of the ex-situ *bones, which had been moved. The upper ends of long bones are indicated with a bar. The lower schematic plan shows the position of the principal* in situ *bones, and the chalice. (Rodwell 1983a), with permission.*

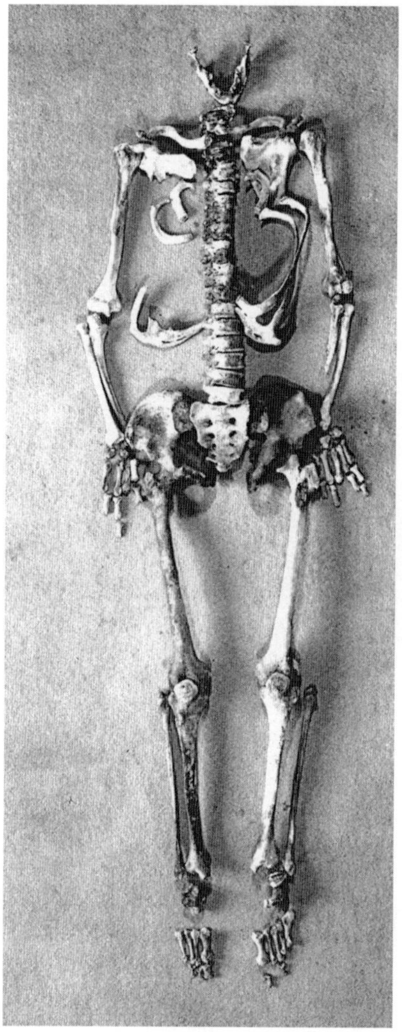

Figure 3.3. Skeletal remains laid out for examination, with evidence of DISH (Rodwell 1983a, Plate 2), with permission.

with some calcified tubular fragments, the calcified arteries, which were contaminated by a small amount of builders' debris. The remains of a pewter mortuary chalice were found to the right side of the tomb, indicating the ecclesiastical status of the deceased person. The skeletal remains were identified by Rodwell (1983b, 11) as most likely to be those of William of Mancetter, Dean of Lichfield Cathedral from 1222 to 1254.

The skeletal remains, which were reported to be in good condition for the most part, were those of an elderly individual (50–65 years), approximately 1.69 m (5 ft 6½ in) in height (Rodwell 1983a). From the mandible and two loose maxillary teeth it was noted that there was no dental caries, although two lower molar teeth had been lost before death, with evidence of periodontal disease. There was widespread, exuberant bone formation affecting the vertebrae, ligamentous insertions and costal cartilages, typical of DISH (Fig. 3.3). There was evidence of osteoarthritis, affecting several joints. Thirty-two fragments of calcified blood vessel were found with the skeletal remains, varying from 7 to 42 mm in length and from 1 to 8.5 mm in diameter, of which four are the focus of this article (Fig. 3.4) (Rodwell 1983a)

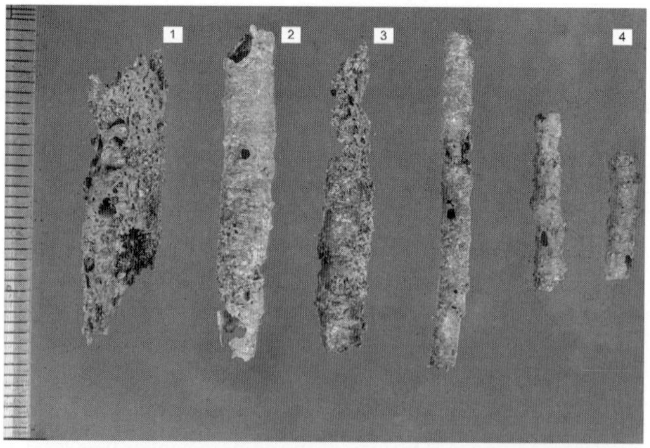

Figure 3.4. The fragments labelled 1–4 correspond to the fragments in this report (Photo from Juliet Rogers' teaching collection, University of Bristol. Slide labelled LC82/47).

The assemblage of calcified arteries

The four fragments were imaged using focus-stacking technology, with the following findings (Figs 3.5, 3.6).

Fragment 1 (28.0 × 7.2 mm) is a medium-sized artery showing a disorganised pattern of calcification, with extensive porosity similar to that of remodelling bone. There is irregular layering of the calcification, which is of variable thickness. The luminal surface has an irregular pattern of calcification, suggestive of atheroma deposition. There is a yellow colouration to the luminal and abluminal surfaces, with a paler colour where there is post-mortem breakage. There is no surviving evidence of medial calcification.

Fragment 2 (25.8 × 5.1 mm) is a small- to medium-sized artery with a relatively smooth abluminal surface, and circumferential fine striation along the entire length. Calcified material within the wall of the artery wall shows porosity like that in Fragment 1, although the thickness of the arterial wall is relatively consistent, at approximately 1 mm.

Fragment 3 (28.0 × 4.8 mm) is a small- to medium-sized artery with circumferential striation of the abluminal surface. There are numerous calcified bud-like excrescences from the external surface. The calcified material within the arterial wall shows only minor porosity. The luminal surface is mostly smooth, with a small area of irregular roughness. There are areas of yellow plaque-like discolouration of the luminal surface and a whitish-yellow, waxy component to the arterial wall.

Fragment 4 (11.6 × 2.8 mm) is a small artery with a smooth, finely striated abluminal surface, with minor budding excrescences. There is no porosity within the arterial wall and the luminal surface is smooth.

The statutes of Pattishall and Meulen

The exalted ecclesiastical status of the Dean of Lichfield Cathedral was specified in the statutes of Bishop Pattishall (Dugdale 1849, 1257), probably drawn up by Mancetter himself (Savage 1921, 20):

> *Item quotienscunque chorus incensatur, ex quacunque parte fuerit, semper erit decanus primus incensatus, et primus textum deosculibatur, et primus pacem ad missam recipient.*[1]

However, the pre-eminence of a dean's status within the cathedral hierarchy extended to more worldly considerations, as outlined in statutes published during the later episcopacy of Bishop Meulen (Dugdale 1849, 1260):

> *Decanus autem, qui continuam facit in ecclesiâ residentiam, duplicem canonici recipiat communam, in diurnis distributionibus.*[2]

Discussion

Although the antiquity of calcific vascular disease has been confirmed from radiographs and autopsy of Egyptian mummified remains (Ruffer 1911), calcified arteries are 'essentially absent' from archaeological excavation of dry skeletal remains (Roberts 2023, 12). This may reflect a failure of excavation technique and post-excavation analysis of these fragile items, which are liable to break down easily, as much as being due to taphonomic diagenesis. For example, despite their significance being highlighted in preparatory study sessions, the presence of calcified arteries placed in association with skeletal remains was overlooked by all the archaeologists involved, including experienced excavators (Biehler-Gomez *et al.* 2023, 250). On the other hand, 66 calcified non-skeletal elements were found in 10/669 (1.5%) skeletons during excavation at Stoke Quay following routine collection and processing of targeted soil samples which were subsequently processed and scanned in the laboratory. These consisted of calculi, probably renal, pleural plaques, heart valves and blood vessels (Loe *et al.* 2020, 509).

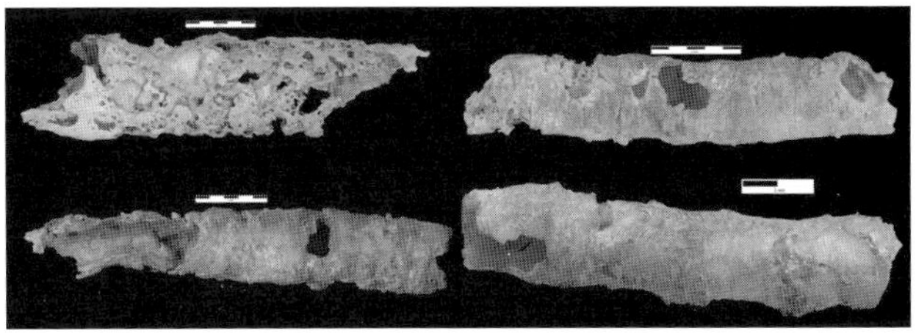

Figure 3.5. Top left: Fragment 1 (see Figure 3.4), showing calcified atheroma. Top right: Fragment 2, showing circumferential striation, suggesting medial sclerosis. Bottom left: Fragment 3, showing thickening of arterial wall, irregularity of the endothelial surface and superficial calcified excrescences on the external wall. Bottom right: Fragment 4, a small artery, showing evidence of medial sclerosis and calcified superficial excrescences on the external wall (Photo by David Bennett-Jones).

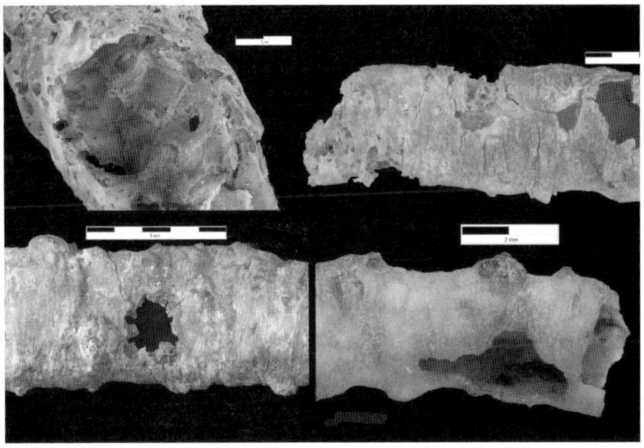

Figure 3.6. Arterial fragments 1–4, positioned as in Figure 3.5, to show detail of luminal and abluminal surfaces (Photo by David Bennett-Jones).

When calcified arteries have been retrieved in the past, they may be mentioned in the excavation report only in passing, as little more than a curiosity (*e.g.* Rodwell 1983a; Waldron 2007). This is partly explained by limited clinical understanding of calcific vascular disease at that time, the pathophysiology of atheroma and thrombotic complications having been elucidated more recently (*e.g.* Makin *et al.* 2002). Another factor which may explain the limited discussion of such material at that time was the lack of analytical methods between inspection with a hand lens and expensive imaging techniques such as SEM, micro-CT or destructive histology. Focus-stacking provides an accessible, affordable intermediate level of visualisation for detailed appreciation of the morphology.

Because such finds are so scarce, it is not possible to determine the prevalence of vascular disease in historic or archaeological populations. This problem is compounded by selective representation, since mummification in Egypt was available only for wealthier individuals (Antoine *et al.* 2023, 121). In this context, it is notable that two examples of preservation and retrieval in the British literature (Rodwell 1983a; Waldron 2007) were entombed individuals, presumably of high status. Correlation with dietary intake is also confounded by a lack of reliable data. Although the type of diet, meat-/plant-based and terrestrial/marine, can be objectively determined by measurement of stable carbon and nitrogen isotopes in bone (Schoeninger *et al.* 1983), this gives no information about the calorie intake. Contemporary literary accounts such as Geoffrey Chaucer's *Canterbury Tales* and historical analysis (Bishop 1983, 189; Harvey 1993, 70) indicate a rich, not to say excessive, diet among the clergy and monastics. However, adherence to the Benedictine Rule will have varied over the centuries, and literary accounts may be biased.

Turning to bioarchaeological evidence, a higher prevalence of DISH has been identified among individuals buried in high-status locations in monastic cemeteries (Waldron 1985; Rogers and Waldron 2001; Verlaan *et al.* 2007). Whilst the presence of DISH is rarely of clinical importance *per se*, it may point to a higher prevalence of obesity, diabetes and vascular disease (Kiss *et al.* 2002; Glick *et al.* 2020). However, Mays (2009) points out that assumptions concerning burial location may be misleading, and he comments that different criteria for diagnosing DISH makes comparison between studies difficult. To address the problem, he compared the findings from each study separately with those from an impoverished lay community, Wharram Percy. He concluded that, while further study is needed, the prevalence of DISH in monastic and secular locations respectively does not support the argument for obesity in the monastic community.

The presence of DISH in this individual is based on conventional diagnostic criteria (van der Merwe *et al.* 2012), with the associated finding of calcified enthesophytes and costal cartilage. However, it is not possible to know whether these findings represent acquired disease due to diet, obesity and, possibly, type 2 diabetes, or represent the inherent tendency of a 'bone-former' to ossify non-skeletal tissues (Mays 2016). Unlike DISH, diagnostic criteria for the two different forms of arterial calcification in the archaeological record are non-existent, given the absence of samples. However, the morphology of the four fragments of calcified artery in this study do permit some speculative comments to be made about both intimal and medial ossification.

Fragment 1, the largest of the four arterial segments, has a disorganised, circumferential morphology with layers of material deposited, one upon another. The extensive porosity is similar to the appearance of Haversian canals in bone remodelling (Ascenzi and Roe 2012). The layering and the colouration of the material correspond closely to the appearance of contemporary calcified atheromatous deposits from autopsy material between 2014 and 2017, and from cemeterial contexts buried for more than ten years (Biehler-Gomez *et al.* 2019).

The external, abluminal surface of fragments 2 and 3 is similar in each case, being smooth with a fine striation, compatible with circumferentially disposed calcified muscle or elastin fibres of the tunica media (Mescher 2016, 220). This appearance is compatible with medial sclerosis. On the external surface of both fragments, particularly Fragment 3, protuberances of calcified material are of uncertain significance, but may represent sprouting angiogenesis, due to ischaemia (Dor and Keshet 1997). On the luminal surface of the vessel wall, particularly Fragment 2, calcified material, with porosity like that in Fragment 1, can be seen. The luminal surfaces of these two fragments are variably smooth, suggesting areas of normal endothelium, and rough, compatible with patchy deposition of atheroma.

Fragment 4, the smallest artery, has the smooth external surface suggestive of medial sclerosis, with none of the disorganised calcification with porosity seen in the larger fragments. The luminal surface, where it can be seen, is smooth. Budding can be seen over much of the abluminal surface. The explanation for this is speculative, but the possibility of ischaemia-induced angiogenesis (Dor and Keshet 1997) should be considered.

It is rare in palaeopathology to be able to make a direct link between the pathological changes found in skeletal remains and a specific diagnosis or symptom burden, and this case is no exception. Vascular calcification, both intimal and medial, is currently of interest, because of the high associated morbidity and mortality. Atheroma is now a very common finding in older people (Powell and Davies 2020), although it occurs at a younger age in those with diabetes and chronic renal failure. Medial sclerosis, while it may occur due to old age alone, is most typically seen in type 2 diabetes and chronic renal failure (Manzoor *et al.* 2018), when it may involve the entire vascular tree. In these four fragments, there is a distinction between the layered atherosclerotic calcification, as seen in Fragment 1, and smooth, arteriosclerotic calcification of the media, seen most clearly in Fragment 4. Fragments 2 and 3 show an intermediate pattern, with features of both pathologies.

The degree of calcification in this case must be regarded as exceptional, not least for it to have survived diagenetic processes for over 750 years. However, recent work suggests that the size of calcific deposits has an impact on prognosis (Shioi and Ikari 2018; Montanaro *et al.* 2021). Microcalcification, with calcific deposits <0.5 mm which cannot, in the early stages, be identified even with modern radiological techniques such as CT, is associated with greater risk of thrombosis. Macrocalcification, which is readily seen using CT, can stabilise atheromatous plaque. From the appearance of these arterial segments, it seems likely that William of Mancetter had macrocalcification which may have contributed to his relative longevity for the time, despite unquestionable widespread vascular disease.

Despite the potential severity of known complications, individuals with either or both types of vascular pathology may be asymptomatic. For this reason, even with modern knowledge, there may be little reason to alter a lifestyle that is associated with progressive or worsening vascular disease. Such may have been the situation for William of Mancetter, who exerted political influence during a prolonged period of significant challenge (Britton 1836), at a time of great rivalry between the episcopal sees of Lichfield and Coventry (Savage 1921) – likely to be a source of considerable stress. Dean Mancetter is believed to have been instrumental in drawing up the statutes of Pattishall (Savage 1921, 20–1), which emphasised the exalted status of the dean, second only to that of the bishop. Another set of statutes was published under Bishop Meulen, after the death of Dean Mancetter, but quite probably reflecting the stipendiary arrangements during his tenure. These stated that the dean would receive a double allowance of commons (food) each day, in recognition of his continuous presence at the cathedral throughout the year (Dugdale 1849). It seems likely that this reflected a practice that had developed during Dean Mancetter's tenure, rather than representing an enhanced remuneration package in a new job description.

This single case of intimal and medial calcification cannot be attributed to diet, lifestyle, personality or, indeed, any of the risk factors known to increase susceptibility to vascular disease. It is unlikely that a sufficient sample of affected individuals from a non-selective environment could ever be collected to calculate the prevalence of severe vascular disease in a medieval context. However, there are features, such as the yellowish colouration and the evidence of possible vascular budding, which could be investigated further with laboratory analysis. The yellowish colouration of these blood vessels may, theoretically, represent cholesterol and other fatty substances incarcerated within the calcified artery wall. If the surface budding excrescences of the smaller blood vessels represents ischaemic sprouting angiogenesis, as they might, then proteomic analysis for chemical mediators such as vascular endothelial growth factor (VEGF) (Ho and Shanahan 2016) might be rewarding. For these reasons, greater efforts should be made during archaeological excavation to identify and retrieve calcified vascular elements, which belong in the biomolecular laboratory, rather than the cabinet of curiosities.

Acknowledgements

We are grateful to Professor Warwick Rodwell for his helpful information about the provenance of the arteries and for advice, and for permission to publish Figures 3.2 and 3.3. We also thank Clare Townsend, Library Manager at Lichfield Cathedral for copies of Professor Rodwell's two reports. Professor Rajesh Thakker provided helpful advice about the analytical approach to these calcified remains. We are very grateful to Ruth Edy for help with the translation of the medieval Latin quotes from the statutes.

Notes

1 Whenever the choir is incensed, from whatever part he is, the dean will always be incensed first and he will kiss the (holy) text first and he will be the first to receive the Peace at Mass.

2 And the dean, who makes a continuous residence in the church, shall receive double the allowance of the canons, in the daily distributions.

References

Antoine, D., Vandenbeusch, M., Whiting, R. and Moreno, B. (2023) Cardiovascular disease in Nile Valley mummies. In M. Binder, C. A. Roberts and D. Antoine (eds) *The Bioarchaeology of Cardiovascular Disease*, 98–129. Cambridge, Cambridge University Press.

Ascenzi, M-G. and Roe, A. (2012) The osteon: the micromechanical unit of compact bone. *Frontiers in Bioscience* 17(1), 1551.

Bell, E.T. (1957) Atherosclerotic gangrene of the lower extremities in diabetic and nondiabetic persons. *American Journal of Clinical Pathology* 28(1), 27–36.

Biehler-Gomez, L., Cappella, A., Castoldi, E., Martrille, L. and Cattaneo, C. (2018) Survival of atherosclerotic calcifications in skeletonized material: forensic and pathological implications. *Journal of Forensic Sciences* 63(2), 386–91.

Biehler-Gomez, L., Maderna, E., Brescia, G., Caruso, V., Rizzi, A. and Cattaneo, C. (2019) Distinguishing atherosclerotic calcifications in dry bone: implications for forensic identification. *Journal of Forensic Sciences* 64(3), 839–44.

Biehler-Gomez, L., Maderna, E. and Cattaneo, C. (2023) The challenging diagnosis of cardiovascular disease in skeletal remains: identifying atherosclerotic calcifications from modern documented individuals. In M. Binder, C. A. Roberts and D. Antoine (eds) *The Bioarchaeology of Cardiovascular Disease*, 229–45. Cambridge, Cambridge University Press.

Binder, M. and Roberts, C.A. (2014) Calcified structures associated with human skeletal remains: possible atherosclerosis affecting the population buried at Amara West, Sudan (1300–800BC). *International Journal of Paleopathology* 6(1), 20–29.

Binder, M. and Roberts, C.A. (2023) Absence of evidence is not evidence of absence. In M. Binder, C. A. Roberts and D. Antoine (eds) *The Bioarchaeology of Cardiovascular Disease*, 214–26. Cambridge, Cambridge University Press.

Bishop, M. (1983) *The Pelican Book of the Middle Ages*. Harmondsworth, Penguin. http://archive.org/details/pelicanbookofmid0000bish

Boddington, A. (1987) Chaos, disturbance and decay in an Anglo-Saxon cemetery. In A. Boddington, A.N. Garland and R.C. Janaway (eds) *Death, Decay and Reconstruction. Approaches to Archaeology and Forensic Science*, 27–42. Manchester, Manchester University Press.

Britton, J. (1836) *The History and Antiquities of the See and Cathedral Church of Lichfield.* London, M. A. Nattali.

Channon, K. and Vallance, P. (2020) Blood vessels and the endothelium. In J. Dwight, J. Firth, C. Conlon and T. Cox (eds) *Oxford Textbook of Medicine*, 0. Oxford, Oxford University Press. https://doi.org/10.1093/med/9780198746690.003.0338

Clarke, E.M., Thompson, R.C., Allam, A.H., Wann, L.S., Lombardi, G.P., Sutherland, M.L., Sutherland, J.D., Cox, S.L., Soliman, M.A.-T., Abd el-Maksoud, G., Badr, I., Miyamoto, M.I., Frohlich, B., Nur el-din, A.-H., Stewart, A.F.R., Narula, J., Zink, A.R., Finch, C.E., Michalik, D.E. and Thomas, G.S. (2014) Is atherosclerosis fundamental to human aging? Lessons from ancient mummies. *Journal of Cardiology* 63(5), 329–34.

Conn, J., Krumlovsky, F.A., Greco, F.D. and Simon, N.M. (1973) Calciphylaxis: etiology of progressive vascular calcification and gangrene? *Annals of Surgery* 177(2), 206–10.

Dalinka, M.K. and Melchior, E.L. (1980) Soft tissue calcifications in systemic disease. *Bulletin of the New York Academy of Medicine* 56(6) 539–63.

Deneke, T., Langner, K., Grewe, P.H., Herer, E. and Müller, K.-M. (2001) Ossification in atherosclerotic carotid arteries. *Zeitschrift für Kardiologie* 90(15), III106–III115.

Dor, Y. and Keshet, E. (1997) Ischemia-driven angiogenesis. *Trends in Cardiovascular Medicine* 7(8), 289–94.

Dovan, D., Florentiu, A. and Lichiardopol, R. (2008) Medial arterial calcification (Monckeberg's sclerosis) in diabetic patients and non-diabetic patients. *Romanian Journal of Diabetes Nutrition and Metabolic Diseases* 15(4), 1–11.

Dugdale, W. (1849) *Monasticon Anglicanum.* J. Caley, H. Ellis and B. Bandinel (eds). London, T. G. March. http://archive.org/details/monasticonanglic0008vari

Falk, E. (2006) Pathogenesis of atherosclerosis. *Journal of the American College of Cardiology* 47(8), C7–C12.

Fok, P-W. and Lanzer, P. (2018) Media sclerosis drives and localizes atherosclerosis in peripheral arteries. M.C. Vinci (ed.) *PLOS ONE* 13(10), p.e0205599.

Fuery, M.A., Liang, L., Kaplan, F.S. and Mohler, E.R. (2018) Vascular ossification: pathology, mechanisms, and clinical implications. *Bone* 109, 28–34.

Gaeta, R., Giuffra, V. and Fornaciari, G. (2013) Atherosclerosis in the Renaissance elite: Ferdinand I King of Naples (1431–1494). *Virchows Archiv* 462(5), 593–5.

Glick, K., Novofastovski, I., Schwartz, N. and Mader, R. (2020) Cardiovascular disease in diffuse idiopathic skeletal hyperostosis (DISH): from theory to reality—a 10-year follow-up study. *Arthritis Research and Therapy* 22(1), 190.

Harvey, B.F. (1993) *Living and Dying in England, 1100–1540: The Monastic Experience*. Oxford, Clarendeon Press. http://archive.org/details/livingdyinginengl0000harv

Hayden, M.R., Tyagi, S.C., Kolb, L., Sowers, J.R. and Khanna, R. (2005) Vascular ossification – calcification in metabolic syndrome, type 2 diabetes mellitus, chronic kidney disease, and calciphylaxis – calcific uremic arteriolopathy: the emerging role of sodium thiosulfate. *Cardiovascular Diabetology* 4(4), 1–22.

Henderson, J. (1987) Factors determining the state of preservation of human remains. In A. Boddington, A.N. Garland and R.C. Janaway (eds) *Death, Decay and Reconstruction. Approaches to Archaeology and Forensic Science*, 43–54. Manchester, Manchester University Press.

Herrera, D., Molina, A., Buhlin, K. and Klinge, B. (2020) Periodontal diseases and association with atherosclerotic disease. *Periodontology 2000* 83(1), 66–89.

Ho, C.Y. and Shanahan, C.M. (2016) Medial arterial calcification. *Arteriosclerosis, Thrombosis, and Vascular Biology* 36, 1475–82.

Kannel, W.B. (1961) Factors of risk in the development of coronary heart disease—six-year follow-up experience: the Framingham study. *Annals of Internal Medicine* 55(1), 33.

Kannel, W.B. and McGee, D.L. (1979) Diabetes and cardiovascular risk factors: the Framingham study. *Circulation* 59(1), 8–13.

Kim, M.J., Kim, Y-S., Oh, C.S., Go, J-H., Lee, I.S., Park, W-K., Cho, S-M., Kim, S-K. and Shin, D.H. (2015) Anatomical confirmation of computed tomography-based diagnosis of the atherosclerosis discovered in 17th century Korean mummy. M. Torriani (ed.) *PLOS ONE* 10(3), p.e0119474.

Kiss, C., Szilágyi, M., Paksy, A. and Poór, G. (2002) Risk factors for diffuse idiopathic skeletal hyperostosis: a case-control study. *Rheumatology* 41(1), 27–30.

Lanzer, P., Hannan, F.M., Lanzer, J.D., Janzen, J., Raggi, P., Furniss, D., Schuchardt, M., Thakker, R., Fok, P.-W., Saez-Rodriguez, J., Millan, A., Sato, Y., Ferraresi, R., Virmani, R. and St. Hilaire, C. (2021) Medial arterial calcification. *Journal of the American College of Cardiology* 78(11), 1145–65.

Lehto, S., Niskanen, L., Suhonen, M., Rönnemaa, T. and Laakso, M. (1996) Medial artery calcification. A neglected harbinger of cardiovascular complications in non-insulin-dependent diabetes mellitus. *Arteriosclerosis, Thrombosis, and Vascular Biology* 16(8), 978–83.

Littlejohn, G.O. (2022) Historical bioarcheological aspects of diffuse idiopathic skeletal hyperostosis in humans. *Seminars in Arthritis and Rheumatism* 52, 1–7.

Loe, L., Dean, B., Farber, E., Gibson, M., Rose, A. and Webb, H. (2020) The human skeletal remains: an overview of the osteology and palaeopathology. In R. Brown, S. Teague, L. Loe, B. Sudds and E. Popescu (eds) *Excavations at Stoke Quay, Ipswich: Southern Gipeswic and the Parish of St Augustine*, 273–339. East Anglian Archaeology 172. Norwich, East Anglian Archaeology.

Loe, L. and Rogers, L. (2012) Juliet Margaret Rogers (1940–2001). In J. Buikstra and C. Roberts *The Global History of Palaeopathology. Pioneers and Prospects*, 179–85. Oxford, Oxford University Press.

Lu, L., Zhao, D., Li, C., Sun, Y., Geng, F., Zhang, S., Li, W., Wang, S. and Pan, Y. (2024) The role of periodontitis in the development of atherosclerotic cardiovascular disease in participants with the components of metabolic syndrome: a systematic review and meta-analysis. *Clinical Oral Investigations* 28(6), 339.

Lusis, A.J. (2000) Atherosclerosis. *Nature* 407(6801), 233–41.

Lynnerup, N., Roberts, C. and Binder, M. (2015) Survival of calcified atheromata in the archaeological record – the effect of taphonomy, excavation and curation strategies on preservation and analysis. *American Journal of Physical Anthropology* 208–9.

Magni, P.A. (2019) A taphonomic examination of inhumed and entombed remains in Parma Cemeteries, Italy. *Global Journal of Forensic Science and Medicine* 1(4). https://irispublishers.com/gjfsm/fulltext/a-taphonomic-examination-of-inhumed-and-entombed-remains-in-parma-cemeteries-italy.ID.000518.php

Makin, A., Silverman, S.H. and Lip, G.Y. (2002) Peripheral vascular disease and Virchow's triad for thrombogenesis. *QJM* 95(4), 199–210.

Manzoor, S., Ahmed, S., Ali, A., Han, K.H., Sechopoulos, I., O'Neill, A., Fei, B. and O'Neill, W.C. (2018) Progression of medial arterial calcification in CKD. *Kidney International Reports* 3(6), 1328–35.

Marrazzini, A., Gaeta, R. and Fornaciari, G. (2015) Paleopathology of pre-Columbian mummies at the Museum of Anthropology and Ethnology in Florence. *Journal of History of Medicine* 27(2), 481–502.

Mays, S. (2009) The osteology of monasticism in medieval England. In C. Knusel (ed.) *The Social Archaeology of Funerary Remains*, 179–89. Oxford, Oxbow Books.

Mays, S. (2016) Bone-formers and bone-losers in an archaeological population. *American Journal of Physical Anthropology* 159(4), 577–84.

Mescher, A.L. (2016) *Junqueira's Basic Histology: Text and Atlas*. New York, Mcgraw-Hill Education. http://archive.org/details/junqueirasbasich0000mesc

Montanaro, M., Scimeca, M., Anemona, L., Servadei, F., Giacobbi, E., Bonfiglio, R., Bonanno, E., Urbano, N., Ippoliti, A., Santeusanio, G., Schillaci, O. and Mauriello, A. (2021) The paradox effect of calcification in carotid atherosclerosis: microcalcification is correlated with plaque instability. *International Journal of Molecular Sciences* 22(1), 395.

Murphy, W.A., Nedden, D.Z., Gostner, P., Knapp, R., Recheis, W. and Seidler, H. (2003) The Iceman: discovery and imaging. *Radiology* 226(3), 614–29.

Powell, J. and Davies, A. (2020) Peripheral arterial disease. In J. Dwight, J. Firth, C. Conlon and T. Cox (eds) *Oxford Textbook of Medicine*, 6 edn (Oxford, 2020; online edn, Oxford Academic). https://doi.org/10.1093/med/9780198746690.003.0371

Resnick, D. and Niwayama, G. (1976) Radiographic and pathologic features of spinal involvement in Diffuse Idiopathic Skeletal Hyperostosis (DISH). *Radiology* 119(3), 559–68.

Roberts, C.A. (2023) Exploring the sources of indirect evidence for cardiovascular disease in bioarchaeology. In M. Binder, C. A. Roberts and D. Antoine (eds) *The Bioarchaeology of Cardiovascular Disease*, 7–38. Cambridge, Cambridge University Press.

Rodwell, W. (1983a) *Lichfield Cathedral: St Chad's Head Chapel*. Lichfield, Lichfield Cathedral.

Rodwell, W. (1983b) *The Skeleton in the Wall*. Lichfield, Lichfield Cathedral.

Rogers, J., Shepstone, L. and Dieppe, P. (1997) Bone formers: osteophyte and enthesophyte formation are positively associated. *Annals of the Rheumatic Diseases* 56(2), 85–90.

Rogers, J. and Waldron, T. (2001) DISH and the monastic way of life. *International Journal of Osteoarchaeology* 11(5), 357–65.

Ruffer, M.A. (1911) On arterial lesions found in Egyptian mummies (1580 B.C.–525 A.D.). *The Journal of Pathology and Bacteriology* 15(4), 453–62.

Sandison, A. (1962) Degenerative vascular disease in the Egyptian mummy. *Medical History* 6(1), 77–81.

Savage, H.E. (1921) *Lichfield Cathedral: William de Mancetter, Dean: An Address*. Lichfield, F.H. Bull & E. Wiseman.

Schoeninger, M., DeNiro, M. and Tauber, H. (1983) Stable nitrogen isotope ratios of bone collagen reflect marine and terrestrial components of prehistoric human diet. *Science* 220(4604), 1381–3.

Shah, P.K. (2003) Mechanisms of plaque vulnerability and rupture. *Journal of the American College of Cardiology* 41(4), S15–S22.

Shioi, A., and Ikari, Y. (2018) Plaque calcification during atherosclerosis progression and regression. *Journal of Atherosclerosis and Thrombosis* 25(4), 294–303.

Spronk, H.M., van der Voort, D. and ten Cate, H. (2004) Blood coagulation and the risk of atherothrombosis: a complex relationship. *Thrombosis Journal* 2(1), 12.

St. Hilaire, C. (2022) Medial arterial calcification: a significant and independent contributor of peripheral artery disease. *Arteriosclerosis, Thrombosis, and Vascular Biology* 42(3), 253–60.

Stewart, V.L., Herling, P. and Dalinka, K. (1983) Calcification in soft tissues. *Journal of the American Medical Association* 250(1), 78–81.

Subirana-Domènech, M., Borondo-Alcáza, J.C., Armentano-Oller, N., Valsecchi, G.F., Vicente, I.G. and Castellà-García, J. (2012) Arteriosclerosis in paleopathology. Are macroscopic findings well known? *International Journal of Paleopathology* 2, 246–8.

Thompson, R.C., Allam, A.H., Lombardi, G.P., Wann, L.S., Sutherland, M.L., Sutherland, J.D., Soliman, M.A.-T., Frohlich, B., Mininberg, D.T., Monge, J.M., Vallodolid, C.M., Cox, S.L., Abd el-Maksoud, G., Badr, I., Miyamoto, M.I., el-Halim Nur el-din, A., Narula, J., Finch, C.E. and Thomas, G.S. (2013) Atherosclerosis across 4000 years of human history: the Horus study of four ancient populations. *The Lancet* 381(9873), 1211–22.

Van Beek, R., Quik, C., Bergerbrant, S., Huisman, F. and Kama, P. (2023) Bogs, bones and bodies: the deposition of human remains in northern European mires (9000 BC–AD 1900). *Antiquity* 97(391), 120–40.

van der Merwe, A.E., Maat, G.J.R. and Watt, I. (2012) Diffuse idiopathic skeletal hyperostosis: diagnosis in a palaeopathological context. *HOMO* 63(3), 202–15.

Verlaan, J.J., Oner, F.C. and Maat, G.J.R. (2007) Diffuse idiopathic skeletal hyperostosis in ancient clergymen. *European Spine Journal* 16(8), 1129–35.

Waldron, T. (1985) DISH at Merton Priory: evidence for a 'new' occupational disease? *British Medical Journal* 291(6511), 1762–3.

Waldron, T. (2007) *St Peter's, Barton-upon-Humber, Lincolnshire: a Parish Church and its community Vol. 2 - The Human Remains*. Oxford, Oxbow Books.

Zardawi, F., Gul, S., Abdulkareem, A., Sha, A. and Yates, J. (2021) Association between periodontal disease and atherosclerotic cardiovascular diseases: revisited. *Frontiers in Cardiovascular Medicine* 7, 1–17.

Part 2

Life pathways in monastic contexts

4

Unveiling the past: bioanthropological insights into life and death at the friary of Santa Caterina (Barcelona) in the 13th to 15th centuries

Andrea Sanz Boscà, Andreu Falcó, Lluís Lloveras and Carme Rissech

Santa Caterina was one of the most important friaries in Barcelona (Spain) during the medieval and post-medieval periods (1223–1836). Founded in 1223, Santa Caterina belonged to the Order of Preachers (Dominicans). The main goal of this study is to provide information by means of bioanthropological analysis on the life and health conditions of a sample of individuals buried in this friary. The material analysed comes from a total of 24 graves located in the cloister gallery (13th–15th centuries) and in the area outside the church behind the apse (13th–14th centuries). The results indicate 34 individuals, of whom 18 came from the cloister gallery, corresponding to 13 adults (5 females, 6 males and 2 indeterminate) and 5 immatures, while 16 came from outside the church, corresponding to 13 adults (5 females, 5 males and 3 indeterminate) and 3 immatures. The oldest females (41 to 65 years) and males (20 to 64 years) were located in the cloister gallery. In contrast, only young male and female adults were found outside the church. In addition, in the cloister gallery, immatures were always more than 7 years old, while outside the church, immatures included younger individuals, including newborns. The musculoskeletal stress markers and pathologies observed point to a higher physical activity in the individuals from the cloister gallery, particularly in females. Our results suggest that the group interred in the cloister gallery is diverse, corresponding to monastic and lay people closely related to the friary. In contrast, individuals outside the church represented a lay cohort able to afford burial in this location. According to previous studies and to the available archaeological data, the friars were buried in both the cloister and the cloister galleries. Lay members of the highest social class were buried inside the church, and less well-off people were interred in the area outside the church, in this case behind the apse.

Keywords: Santa Caterina, monasteries, preachers, bioanthropology, palaeopathology.

Introduction

Archaeologically excavated cemeteries have been and will continue to be among the most valuable sources of information in order to better understand the lives of our ancestors and the societies in which they lived. From osteological studies of human remains, a wide variety of data can be obtained, ranging from individualised personal information, such as height, age or sex, to other aspects like health status, diseases and even the cause of death. Additionally, these studies allow us to make inferences about the social and economic status of individuals through occupational markers or the study of their diet, and by comparing with other burial grounds, we can appreciate the differences between different population groups separated by time and geography.

Thus, the aim of this study is to provide information about the lifestyle, health and socio-economic status of the individuals buried in the friary of Santa Caterina of Barcelona, particularly those buried in the cloister gallery (13th–15th centuries) and in the area outside the church behind the apse (13th–14th centuries), by means of bioanthropological methods.

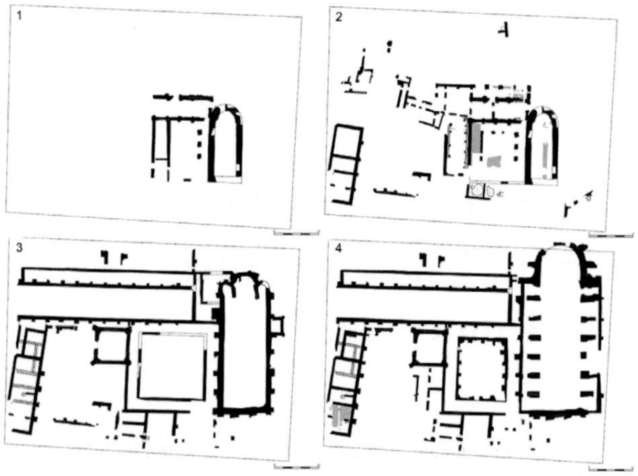

Figure 4.1. Plans of the different phases of occupation. 1. Phase Va; 2. Phase Vc; 3. Phase VIb; 4. Phase VIc (Image modified from Huertas and Aguelo 2006).

The friary of Santa Caterina

The former friary of Santa Caterina in Barcelona was occupied by the Dominican Order from its foundation in 1223. The Order of Preachers (Dominicans) is believed to have arrived in Barcelona in 1219, invited by Bishop Berenguer de Palou, three years after the founding of the order by Domingo de Guzmán (Aguelo *et al.* 2005). The Dominicans quickly gained popularity in the city and among the urban elites, which led to the accumulation of wealth and influence (Moliner 1974; Batlle and Casas 1980; Zaldívar 2012).

Soon afterwards, in 1223, due to the exponential growth of the order and the need for a larger space to carry out their activities, the community received the donation of a plot of land located outside the late Roman city walls, in an area that is still known as the Santa Caterina neighbourhood. They resided there until the closure of the friary, following the disentailment of Mendizábal in 1835 and its subsequent demolition two years later. The present-day Santa Caterina market, formerly the Isabel II Market, was later built on the same site (Aguelo *et al.* 2005).

Thanks to the renovation of the market and the consequent excavation of the site carried out by SAPIC-CODEX Arqueologia i Patrimoni S.C.C.L. (1999–2003), the evolution of the friary of Santa Caterina was uncovered (Huertas and Aguelo, 2006). The excavations revealed an architectural complex in constant evolution, divided into various phases. However, in this introduction, only the first phase of the friary's occupation will be described.

Once established on the new site, the Dominicans began an architectural construction and remodelling project (Fig. 4.1). At the time of their arrival, the plot appears to have been occupied by a pre-existing church (Phase Va), but during the first 20 years, the Preachers carried out building works to adapt the space, maintaining the church and expanding and renovating the rest of the premises, such as the cloister (Fig. 4.1), while also constructing new buildings (Phase Vc) (Aguelo *et al.* 2005).

In 1243, the Order decided to undertake a more ambitious remodelling and construction programme that affected all the existing structures, which is generally considered the beginning of what we know as the friary of Santa Caterina. At this time, a new church was built on the site of the previous one, measuring 13.5 m wide and about 20 m long, with the construction of a tripartite east end (Fig. 4.1). The cloister was renovated, and a rectangular one was built, measuring 29 m by 30 m with 3-metre-wide galleries around the central space, as well as dormitories, a chapter house and other rooms of unknown function, which could have been kitchens, the refectory, the infirmary or the library, known from historical sources (Phase VIa) (Aguelo *et al.* 2005).

This phase of occupation lasted until the first half of the 14th century, at which point the church was again modified in monumental style, following Mediterranean Gothic models, replacing the three previous semicircular apses with one polygonal apse. This expansion led to changes in the cloister, displacing the southern gallery as well as other parts of the monastic complex (Phase VIc) (Fig. 4.1) (Aguelo *et al.* 2005).

The funerary world at the friary of Santa Caterina

Burials near and around churches have been recorded since before the 9th century, becoming more common with the appearance of the 'sagrera', a protected area that, starting in the 11th century, was organised around churches. By the 12th century, it was common practice to be buried in church cemeteries, although, subsequently, the tendency was to prefer burial inside the church itself. A bull of Pope Gregory IX in 1227 gave the faithful the right to be buried in friaries (Aguelo *et al.* 2005).

In the case of the friary of Santa Caterina, burial was not allowed unless one belonged to the community, a restriction that did not change until the mid-13th century, when the friary received a donation of land to be used as a cemetery (Aguelo *et al.* 2005).

Archaeological excavations have identified several graveyards. During the early years of Dominican occupation, the burial area was located outside the first church (Phase Vc). When this phase ended, and the three-apsed church began to be built, the cemetery was moved to behind the apse, where a high density of simple grave pits has been recorded. Tombs made of masonry have also been found attached to the exterior of the church and in the cloister courtyard (Trujillo *et al.* 2023) (Phase VIb).

Again, with the adaptation of the architectural complex to Gothic models, the necropolis located around the apses was buried under the new polygonal apse and moved to

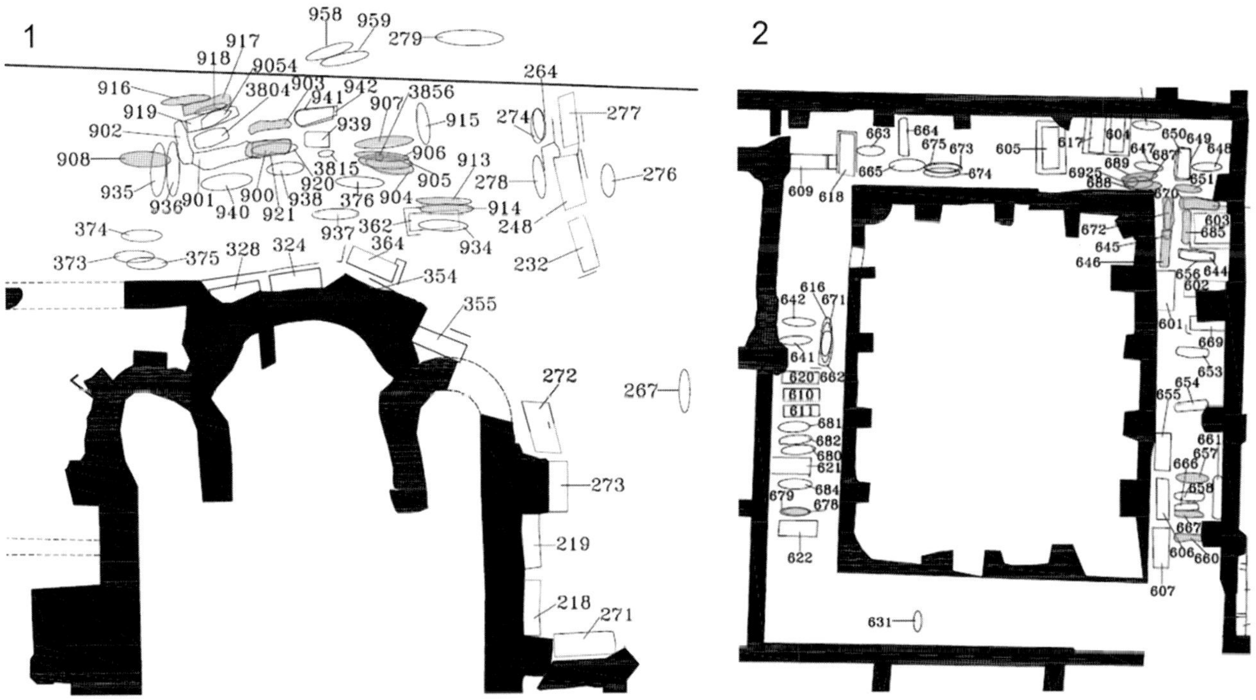

Figure 4.2. Plans and UF excavated behind the tripartite apse (1) and in the Gothic cloister (2) (Image modified from Huertas and Aguelo 2006).

outside the opposite end of the church. For this phase (VIc), an extensive set of tombs has been recorded in the cloister gallery, while around the church, between the buttresses and attached to the exterior wall, masonry tombs also continued to be constructed (Aguelo *et al.* 2005), such as funerary units (UF–'unitats funeràries' in the original excavation reports), UF 221 (Cevallos *et al.* 2023; 2024) or UF 228 in Chapter 6 of this volume.

Materials and methods

The material analysed in this study comes from 24 funerary units (UFs) from two different occupation phases: UF 11 located behind the tripartite apse (Phase VIb) and UF 13 from the galleries of the Gothic cloister (Fig. 4.2), mainly in the southern gallery (Phase VIc). These burials are dated from the 13th to 14th century and from the 14th to 15th century, respectively (Huertas and Aguelo, 2006). Currently, they form part of the collections of the Museu d'Història de Barcelona (MUHBA).

Methods

The bioanthropological analysis of this material was conducted in the laboratories of the Human Bone Laboratory of the Barcelona City History Museum (Museu d'Història de Barcelona) (MUHBA) in the Zona Franca (Barcelona). The material arrived in boxes containing individualised skeletons. However, isolated skeletal elements that could not be attributed to a specific burial were grouped (based on their area of origin) according to anatomical groups, matching them by size, degree of bone maturation, sex, colour, shape and anomalies (Brothwell 1987). Joint congruence, pathological grouping, and the symmetry of paired bones were also considered (Brothwell 1987). The skeletal material was cleaned and, when possible, reconstructed using adhesive tape, which was removed after study. For each individual, this bioanthropological study includes the following: skeletal description, preservation assessment, age and sex estimation, and a metrical analysis of the skeletal elements based on standard measurements. The different procedures are described below.

The degree of preservation was calculated using the three indexes of preservation (IP) (IP1, IP2, IP3) proposed by Walker *et al.* (1988) and modified by Alesan *et al.* (1990). Age estimation was based on nine possible age categories: perinatal (36–40 weeks gestation), infants (1–12 months), early childhood I (1–6 years), early childhood II (7–12 years), juvenile (13–20 years), young adult (21–30 years), adult (31–40 years), mature adult (41–59 years), and elderly (> 60 years). Age was estimated using maturation rates for immature individuals (Weaver 1979; Krogman and Işcan 1986; Brothwell 1987) and bone growth rates (Alduc-Le

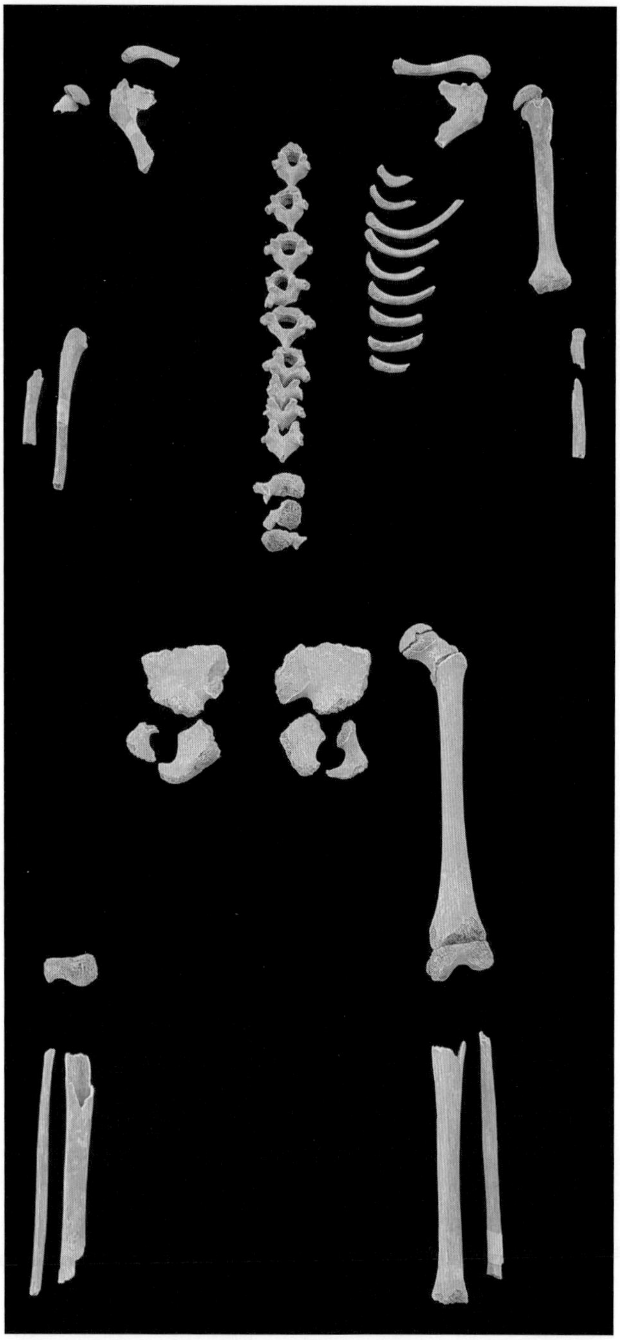

Figure 4.3. Skeleton from the best-preserved UF in the sample, corresponding to individual 917-1 (90.09%).

Bagouse 1988; Scheuer and Black 2000; Schaefer *et al.* 2009; Rissech *et al.* 2013; Rissech 2016). For adults, classic methods were prioritised, such as morphological changes in the pubic symphysis (Todd 1920; McKern and Stewart 1957; Gilbert and McKern 1973), auricular surface changes of the ilium (Lovejoy *et al.* 1985), acetabular changes (Rissech *et al.* 2006, 2007, 2019), morphological changes in the sternal end of the fourth rib (Işcan and Loth 1986; Krogman and Işcan 1986), tooth wear (Brothwell 1987), and the degree of cranial suture closure (Masset 1982).

In relation to sex estimation, the existence of a potential lack of correlation between trait expression and sex in immature individuals, particularly in those under 11 years of age is known (Cardoso and Saunders 2008); for this reason, the sex of immature individuals was estimated by Schutkowski's method (1993) exclusively in individuals over 11 years of age. For adults, the primary criteria used were the morphological characteristics of the skull, mandible, and especially the os coxae (Ferembach *et al.* 1980; Rissech and Malgosa 1991, 1997). When the remains were highly fragmented, discriminant functions based on Mediterranean populations were applied, as defined by Silva (1995), Alemán *et al.* (1997), Safont *et al.* (2000), Barrio *et al.* (2006), Mastrangelo *et al.* (2011), Peckmann *et al.* (2016), and Saldías *et al.* (2016), using long bones, and for the innominate following Rissech and Malgosa (1997).

For anthropometric analysis (cranial and postcranial), the measurements from Martin and Saller (1957, 529–97) and Olivier (1960) were taken. Skeletal indices were calculated from these to describe the shape and size of the bones. Stature estimation was carried out using Pearson's method (1899), which is based on Mediterranean populations. However, due to the absence of femurs and long bones in general in the sample analysed, the methodology of Pablos *et al.* (2013) was also applied in some cases. Nevertheless, these results are considered approximate since Pablos' method is based on a white North American reference sample, and the present case corresponds to a Mediterranean population.

The palaeopathological study followed the guidelines of Ortner (2003) and considered non-metric traits and anatomical variations (Mann *et al.* 2016). When necessary, radiographs were taken using the Intech For View CR system 2 at the Sagrada Família Veterinary Clinic. Regarding oral pathologies, Cobo's recommendations (2022) were followed, focusing primarily on the analysis of cavities, dental calculus, alveolar recession, fistulas and enamel hypoplasia. Alterations that could be the result of oral pathology, such as antemortem and postmortem tooth loss, were also considered.

For the presence of musculoskeletal stress markers, the protocols indicated by Capasso *et al.* (1999) and Fernández (2008) were followed. We registered their presence or absence. Finally, statistical analysis was carried out using IBM SPSS 27.0 (IBM Corp. 2020). Potential sex and lateralisation differences were analysed using the Mann-Whitney U test.

Results

Table 4.1 shows the data obtained for each of the individuals analysed. It should be noted that material fragmentation was one of the factors that influenced this study, with the state of preservation being extremely poor, displaying an average total IP3 of 25.53% (Table 4.1). Only one individual exceeded 75% preservation (90.09%, UF 917-1) (Table 4.1; Fig. 4.3).

The results indicate a total of 34 individuals, 26 adults, and 8 immature individuals (Table 4.1). Of these, 18 came from the cloister galleries (GC) and 16 from the apse (A). Of the 18 from the GC, 13 were adults (5 female, 6 male and 2 indeterminate), and 5 were immature individuals aged over 9 years. Of the 16 from the apse (A), 13 were adults (5 female, 5 male and 3 indeterminate), and 3 were immature individuals older than 2 years (Table 4.1).

Table 4.1. Individuals (ID) from Santa Caterina friary, indicating the preservation index (IP3), sex (Sx: M, male; F, female; I, indeterminate), age, height (H), pathologies, activity markers, epigenetic markers, chronology (Chr) and location (Loc: GC, cloister galleries; A, apse)

ID	*IP3*	*Sx*	*Age*	*H*	*Pathologies*	*Physical activity marker*	*Epigenetic characteristics*	*Chr*	*Loc*
660-1	54.54	M	20–23	173	Exostosis on ribs	R fibula interosseous mb		14th–15th	GC
685	54.54	M	20–24	166	Possible inflammation of sacroiliac joint	Brachialis m insertion R radius		13th	GC
652	22.72	M	54–64	–	Exostosis on vertebrae & ribs	L calcaneus tendon insertion Interosseous crest on R radius	Tubercle of fibula to L calcaneus	14th–15th	GC
670-2	4.54	M	Adult	171	Osteoarthritis MTT1L		Double facet on MTT1L. Fusion of subtalar facets of R calcaneus Tubercle of R calcaneus	14th–15th	GC
667	4.54	M	Adult	168		L calcaneus tendon insertion	Fusion of subtalar facets of L calcaneus	14th–15th	GC
645-1	18.18	M	Adult	168	Hand phalanx eburnation	Insertion of long plantar ligament of L calcaneus Insertion of calcaneal tendon (bilateral) Vaginae fibrosae digitorum manus enthesopathy	Os trigonum of L talus Fusion of subtalar facets of L calcaneus Bipartite transverse foramen of a cervical vertebra	13th	GC
670-1	50	F	41–60	158	Osteophytes & exostoses in ribs & vertebrae. Periostitis in R tibia	Deltoid m in L clavicle. Costoclavicular ligament in both clavicles Coracoclavicular ligament in L clavicle Brachialis & supinator ms Left ulna Calcaneus tendon insertion (bilateral) Interosseous membrane of tibia (bilateral)	Bipartite transverse foramen of a cervical vertebra Double facet on 1st metatarsal (bilateral)	14th–15th	GC
678	63.63	F	43–58	155	Possible osteoma in MTT2R Osteoarthritis in R radius Osteophyte in coronoid process of R ulna Rib exostoses Osteoarthritis in MTC1R and MTC1L Exostoses in the vertebrae	Deltoid m in R clavicle. Biceps brachii m in R radius. Vaginae fibrosae digitorum manus enthesopathy		14th–15th	GC

(Continued)

Table 4.1. Individuals (ID) from Santa Caterina friary, indicating the preservation index (IP3), sex (Sx: M, male; F, female; I, indeterminate), age, height (H), pathologies, activity markers, epigenetic markers, chronology (Chr) and location (Loc: GC, cloister galleries; A, apse). (Continued)

ID	*IP3*	*Sx*	*Age*	*H*	*Pathologies*	*Physical activity marker*	*Epigenetic characteristics*	*Chr*	*Loc*
646	36.36	F	>60	–	Oseophytes in olecranon fossa L humerus, R radius, coronoid process R ulna &MTC1R & MTC1L. Possible osteoma in MTT2R Rib & vertebra exostoses	.		13th	GC
689	22.72	F	Adult	–	Possible Schmörl's nodes in thoracic vertebra. Exostoses at top of sacrum. Osteoarthritic ridge in glenoid cavity of R scapula.	Brachioradialis m in R humerus	Circumflex groove of R scapula	14th–15th	GC
672-1	45.45	F	Adult	160	Rib exostoses. Possible osteoarthritic pathology in proximal phalanx of MTT1L (radiography)	Soleus m of L tibia Femoral head ligament in R femur Brachialis & supinator ms L ulna	Os trigonum of L talus.	13th	GC
657	9.09	I	Adult	–				14th–15th	GC
672-2	13.63	I	Adult	–		Brachialis & supinator ms L ulna		13th	GC
670-3	9.09	I	15+–	–				14th–15th	GC
688	50	F	12–14	–		Costoclavicular ligament of R clavicle		14th–15th	GC
651	68.18	I	9–10	–			Supracondylar process of L humerus	14th–15th	GC
660-2	18.18	I	9–10	–	*Cribra femoralis* on R femur			14th–15th	GC
645-2	4.54	I	9–10	–				14th–15th	GC
904	22.72	M	24–28	177	Periostitis on R & L fibula Arthritic ridge in distal phalanx.	Interosseous mb R fibula. Tendon of L calcaneus	Tubercle of fibula to calcaneus (bilateral)	13th–14th	A
917-2	0	M	Adult	–				13th–14th	A
916	13.63	M	Adult	–		Calcaneal tendon (bilateral) Right patellar ligament		13th–14th	A
908	0	M	Adult	178			Os trigonum L talus.	13th–14th	A
907	0	M	Adult	–				13th–14th	A

(Continued)

Table 4.1. Individuals (ID) from Santa Caterina friary, indicating the preservation index (IP3), sex (Sx: M, male; F, female; I, indeterminate), age, height (H), pathologies, activity markers, epigenetic markers, chronology (Chr) and location (Loc: GC, cloister galleries; A, apse). (Continued)

ID	*IP3*	*Sx*	*Age*	*H*	*Pathologies*	*Physical activity marker*	*Epigenetic characteristics*	*Chr*	*Loc*
917-1	90.9	F	20–24	158	Possible discs prolapse in thoracic vertebra Osteoarthritis on R & L ulna in sigmoid notch & in MTC1R	Linea aspera femur Brachioradialis m on R humerus Vaginae fibrosae digitorum manus enthesopathy. Brachialis m in L ulna Interosseous mb L radius.	Circumflex groove of R scapula	13th–14th	A
900-1	50	F	20–24	–	Osteochondritis dissecans in humerus (bilateral) Thoracic vertebra with exostoses Arthritic relief in trochlear notch of R ulna Osteoarthritis MTC1R and MTC1L R & L humerus with arthritic relief on upper edge of trochlea R & L ulna with arthritic crest on coronoid process	Vaginae fibrosae digitorum manus enthesopathy.		13th–14th	A
903	18.18	F	25–35	–	Possible pleuritis on ribs	Vaginae fibrosae digitorum manus enthesopathy.		13th–14th	A
914-1	31.81	F	25–35	167	Possible pleuritis on ribs		Bilateral lateral squatting facets on tibia.	13th–14th	A
905	18.18	F	Adult	162	Osteochondritis dissecans in III wedge of R foot Periostitis in L tibia & fibula, & R tibia	Fibula interosseous mb (bilateral)	Fibular tubercle of L calcaneus Os trigonum of L talus	13th–14th	A
906	13.63	I	Adult	–				13th–14th	A
913	13.63	I	Adult	–		Interosseous mb L fibula Tendon of R calcaneus		13th–14th	A
900-3	0	I	Adult	–				13th–14th	A
914-3	13.63	I	<3–4	–				13th–14th	A
900-2	0	I	<4–5	–				13th–14th	A
914-2	4.54	I	9–10	–				13th–14th	A

Abbreviations: L-left, R-right, MTT- metatarsal, MTC-metacarpal, mb-membrane, m-muscle

Demography

Regarding the ages of the adult individuals (Table 4.1), it can be observed that both men (aged 20 to 64) and women (aged 41 to 65) from the cloister galleries were in a more mature stage compared to those from the apse, where the men were aged 24 to 28 and the women were aged 20 to 35, all within childbearing age. As for the immature individuals, five were located in the cloister galleries and three in the apse area. The distribution shows that in the cloister galleries, there were individuals from the infantile II (7–12 years) and juvenile (13–20 years) categories. In the apse, we found Infantile I (0–6 years) and Infantile II (7–12 years) individuals.

Table 4.2. Descriptive statistics of postcranial indices from the friary of Santa Caterina (GC and A) considering sex and laterality. Mann-Whitney U Test.

Bone	*Postcranial indices*	*Laterality*	*Male*			*Female*			*U*	*p*
			n	*Mean*	*SD*	*n*	*Mean*	*SD*		
Clavicle	Clavicular robusticity index	L	–	–	–	1	22.55	–	–	–
		R	–	–	–	2	21.12	–	–	–
	Clavicular Terry index	L	–	–	–	3	12.57	–	–	–
		R	–	–	–	4	13.58	–	–	–
Scapula	Glenoidal index	L	–	–	–	1	68.81	–	–	–
Ulna	Platolenia index (Verneau).	L	–	–	–	5	63.86	13.38	–	–
		R	–	–	–	2	75.76	11.05	–	–
	Ulnar robusticity index	L	–	–	–	2	19.9	0.53	–	–
		R	–	–	–	1	20	–	–	–
Radius	Radial diaphyseal index in the middle	L	1	70.44	–	–	–	–	–	–
	Radial robusticity index	L	1	19.9	–	–	–	–	–	–
Femur	Platimetria index	L	–	–	–	1	88.8	–	–	–
		R	–	–	–	1	97.17	–	–	–
Tibia	Tibial cnemic index	L	–	–	–	5	66.88	8.33	–	–
		R	–	–	–	3	68.31	4.03	–	–
	Tibial mid-diaphyseal index	L	–	–	–	1	74.55	–	–	–
Talus	Index height	L	1	49.78	–	1	54.27	–	–	–
		R	–	–	–	1	54.89	–	–	–
	Index width	L	1	68.38	–	1	70.63	–	–	–
		R	–	–	–	1	75.63	–	–	–
Calcaneus	Index height	L	4	46.49	3.07	2	46.54	1.56	4	1
		R	2	45.19	5.42	–	–	–	–	–

Postcranial anthropometry

Due to the sample sizes of the two burial groups, GC (cloister galleries) and A (apse), and in order to increase statistical significance, the individuals from both areas were combined into a single sample. The results show that, in general, the male anthropometric values are higher than the female ones, but no significant sexual differences were found, probably due to the small sample size in this study. Regarding the anthropological indices (Table 4.2), the data reveal robust and platolenic ulnae, eurymeric femurs, and mesocnemic tibiae, with more pronounced values on the right side (Table 4.2). These findings suggest a certain degree of flattening in the ulna, probably due to muscular activity in the arm, while the femurs and tibiae are rounded, indicating less activity in the lower limbs (Olivier 1960).

The robustness index of the radius in males was identified (L = 19.9), as was the robustness index of the clavicle (L = 22.55/R = 21.12) and the ulna (L = 19.9/R = 20) in females, indicating slight robustness in both the upper and lower limbs, suggesting some physical activity, mainly in the arms.

The height estimation for adult individuals showed a marked differentiation by sex, with an average male height of 171.57 cm and an average female height of 160 cm (Table 4.1). Based on the burial location, there was a clear difference in the average height between individuals from the apse (177.5 cm♂ and 162.33 cm♀) and those from the cloister galleries (169.2 cm♂ and 157.66 cm♀) (Table 4.1). However, these values are not comparable to the heights of their contemporaries in Santa Caterina because the heights in this study were estimated using Pablos' method, which is based on a North American population and the values obtained are higher than those calculated by Pearson's method for this population.

Musculoskeletal stress markers (MSM)

Regarding the physical activity markers in adults, these have been registered in 11 adult individuals from the cloister galleries, including five men (UFs 652, 685, 667, 660-1, 645-1), five women (670-1, 678, 689, 646, 672-1) and one indeterminate individual (UFs 672-2), as well as in 7 adult individuals from the apse, among whom there are four

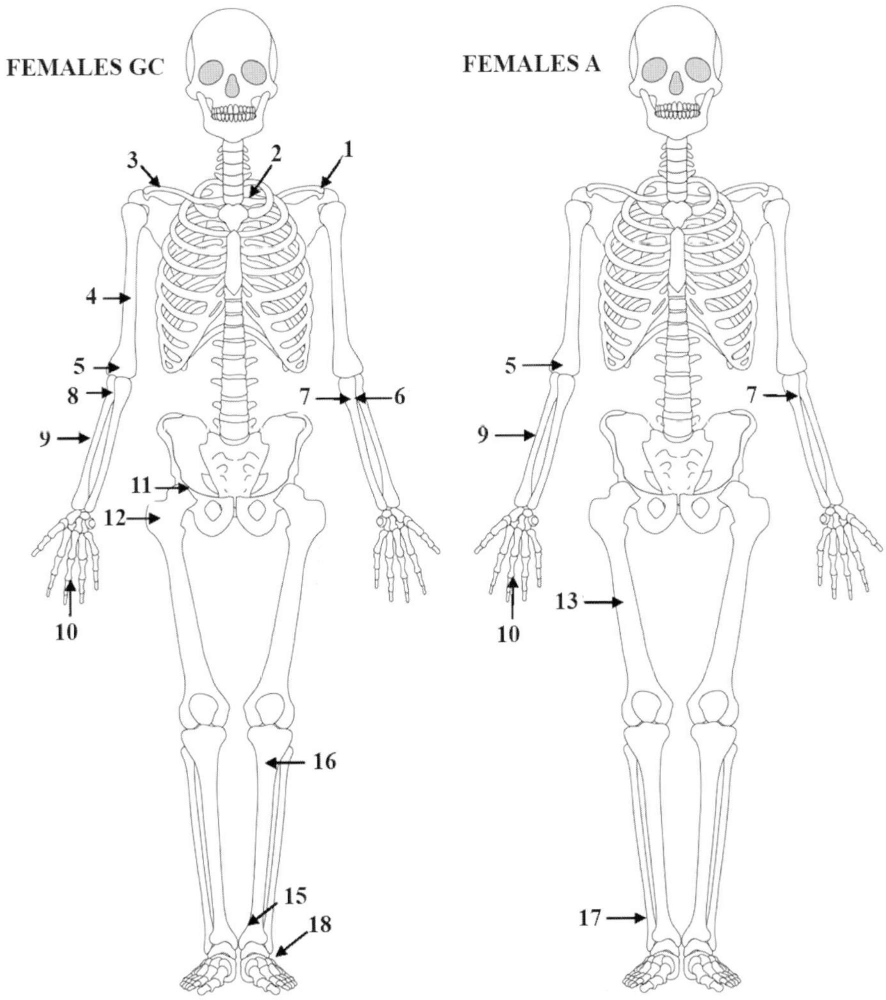

Clavicle
1. Deltoid muscle attachment
2. Costoclavicular ligament
3. Conoid ligament

Humerus
4. Deltoid muscle attachment
5. Brachioradial muscle attachment

Ulna
6. Supinator muscle attachment
7. Brachial muscle attachment

Radius
8. Biceps brachii muscle attachment
9. Interosseous crest

Hands
10. Phalanges of the hand

Femur
11. Round ligament
12. Obturator muscles
13. Muscles of the linea aspera

Patella (kneecap)
14. Patellar ligament

Tibia
15. Interosseous ligament
16. Soleus muscle

Fibula
17. Interosseous ligament

Calcaneus
18. Achilles tendon
19. Plantar ligament

Figure 4.4. Comparison of the MSM observed in the women buried in the cloister galleries (GC) and the apse (A) at the friary of Santa Caterina, distributed by skeletal elements.

women (UFs 917-1, 905, 900-1, 903), two male individuals (UFs 904 and 916), and one indeterminate individual (UF 913). In these individuals MSM have been identified in both the upper limb (clavicle, humerus, ulna, radius and hands) and the lower limb (femur, patella, tibia, fibula and calcaneus), being more numerous in the individuals from the cloister galleries than in those from the apse (Figs 4.4 and 4.5). In the women from the cloister galleries, MSM have been identified in both the upper limb (clavicle, humerus, radius, ulna and phalanges of the hands) and the lower limb (femur, tibia, calcaneus) (see Fig. 4.4). The women from the apse displayed MSM in the upper limb (humerus, ulna, radius and phalanges of the hands) and the lower limb (femur, fibula). In the men, MSM have been identified in both the upper limb (radius and phalanges of the hands) and the lower limb (fibula and calcaneus) in those from the cloister galleries. In the men from the apse, MSM have only been observed in the lower limb (patella, fibula and calcaneus) (see Fig. 4.5).

The areas most affected by MSM are the forearm and hands, with a total of 18 registrations. These are followed by the calcaneus (8) and tibia and fibula (6). The most frequently recorded markers include the capsular ligament complex in the phalanges (6), the interosseous ligament between the tibia and fibula (5), the tendo calcaneus (7) and the brachial muscle attachment (4). When comparing the physical activity markers between men and women without considering the funerary area of origin (GC and A), women exhibit a higher number of physical activity markers (28) than men (12) (see Fig. 4.6). Thus, women exhibit much more physical activity in comparison with men, both in the upper and lower limb. This activity in women in the upper limb is mainly related to rotational and abduction movements of the arm, elbow flexion-extension, forearm pronosupination, and the ability to manipulate and hold objects. In the lower limb it is associated with repetitive muscle abduction and adduction, regularly walking long distances, and carrying heavy loads (Fernández 2008).

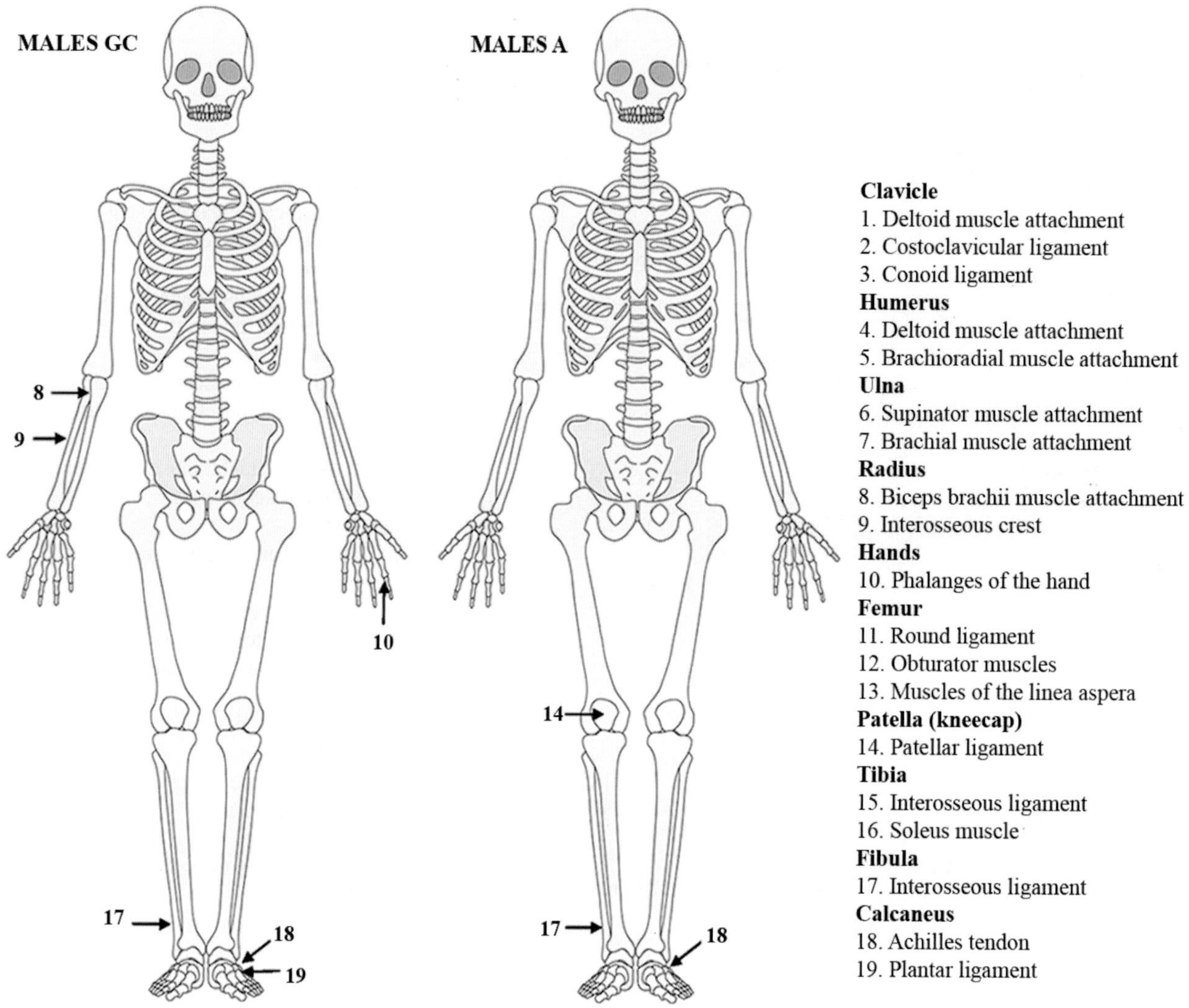

Figure 4.5. Comparison of the MSM observed in the men buried in the cloister galleries (GC) and the apse (A) at the friary of Santa Caterina, distributed by skeletal elements.

It is worth noting an MSM identified in the costoclavicular ligament of an immature female individual aged between 12 and 14 years (UF 688) located in the cloister galleries.

Pathology

Taking into account the two burial areas (GC and A), osteo-articular pathologies, periostitis, possible bone inflammations, Schmörl's nodes, cribra and an osteoma have been observed in both adult and immature individuals.

Regarding arthropathies, they are more frequent in the cloister area than in the apse, particularly in the upper limb and the spine, and more pronounced in female individuals (Table 4.1).

It is noteworthy that several cases of periostitis have been identified, all located in the lower limb (UF 670-1, 904, 905), and two possible inflammations: one on the inner surface of a rib (UF 914-1), compatible with pleuritis, and another in the sacroiliac joint area (UF 685). Concerning vertebral lesions, a Schmörl's node was found in a lumbar vertebra (UF 689), and a prolapse in a thoracic vertebra (UF 917-1). Also noteworthy is the identification of a *cribra femoralis* in an immature individual (FU 660-2). Traditionally, cribra have been considered a sign of biological stress, related to a deficient nutritional state in iron (iron deficiency), intestinal parasites, a lack of vitamin B12, or thalassemia, leading to problems of anaemia (Campillo 1993, 149; Ortner 2003, 22). An osteoma was also observed (UF 678) (Table 4.1).

Non-metric traits

Non-metric traits, or epigenetic traits, are anatomical variations within normality that have a certain genetic component influenced by environmental factors (Vives 1990; Pujol-Bayona *et al.* 2011). Due to the genetic component, they are important for assessing familial relationships within the population group. In this regard, a palatine torus (UF 917-1), bipartite transverse foramina in cervical vertebrae (UFs 670-1, 645-1), the fusion of the subtalar facets of the calcaneus (UFs 670-2, 645-1, 667), the tubercle of the

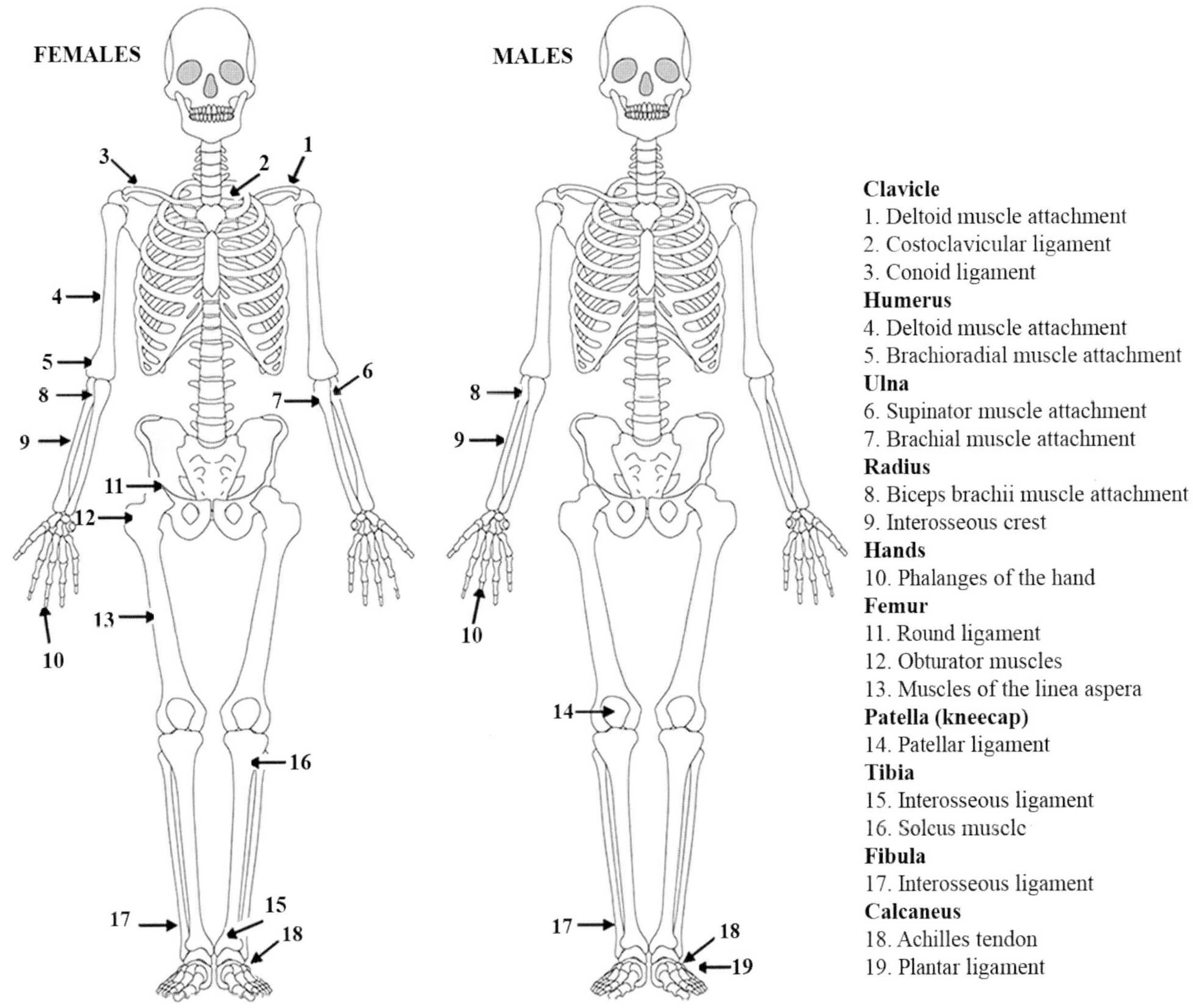

Figure 4.6. Comparison of the MSM observed in men and women buried in the cloister galleries (GC) and the apse (A) at the friary of Santa Caterina, distributed by skeletal elements.

fibula on the calcaneus (UFs 652, 670-2, 904), the accessory navicular (UFs 672-1, 645-1, 908, 905), squatting facets at the distal epiphysis of the tibia (UF 914-1), supernumerary facets on the metatarsals (UFs 670 1, 670-2), a circumflex groove on the lower edge of the scapula (UFs 689, 678, 917-1) and a supracondylar process (UF 651) have been identified.

Oral Pathology

All the dental pieces analysed correspond to adult individuals. A total of 50 dental pieces have been studied from the two burial spaces, of which 5 belong to the cloister galleries and 45 to the apse.

Regarding caries, a frequency of 40% is observed in the cloister galleries and 37.77% in the apse area. These are usually associated with poor dental hygiene and high carbohydrate intake (Ortner 2003). On the other hand, individuals from the cloister galleries show a higher frequency of dental calculus (60.00%) compared to the apse area (46.66%). The presence of calculus is associated with poor oral hygiene and a diet based on animal protein (Cobo 2022).

As for alveolar recession, which is closely linked to the presence of subgingival calculus, both burial areas studied present relatively high frequencies, with 60.00% in the cloister galleries and 24.44% in the apse area. Regarding hypoplasia, a condition considered a sign of nutritional or health stress that affects the growth of the individual and occurs due to nutritional deficits or other conditions such as rickets (lack of vitamin D), avitaminosis, parathyroid conditions, local infections, trauma, or even poisoning (Campillo 2001), it is only observed in the series interred in the apse (24.44%).

Discussion

This work presents the bioanthropological study conducted on 24 burials (13th to 14th centuries) from the cloister galleries and the exterior of the apse of the friary of Santa

Caterina in Barcelona. The results indicate a total of 34 individuals: 26 adults and 8 immature individuals (Table 4.1).

Regarding the cloister galleries, 13 adult individuals are presented, corresponding to 6 men (2 identified in the age range of 20–24 years, 1 between 54 and 64 years old and 3 of indeterminate age), 5 women (2 of them between 41 and 60 years old, 1 over 60 years and 2 of indeterminate age), 2 indeterminate individuals, and 5 immature (3 located between 9 and 10 years of age, 1 between 12 and 14 years and 1 around 15 years old). On the exterior of the tripartite apse, 13 adult individuals were identified, of whom 5 are men (1 identified between 24 and 28 years of age and 4 of indeterminate age), 5 women (2 between 20 and 24 years, 2 between 25 and 35 and 1 of indeterminate age), 3 indeterminate individuals, and 3 immatures (1 situated in the age range of 9 to 10 years and 2 between 0 and 5 years of age).

The presence of both men and women in both burial spaces, with relatively equal frequencies, is in accordance with the archaeological data, which indicate that, the cloister galleries were mainly a burial space reserved for members of the monastic community. However, over time, burials related to lay individuals began to be introduced (Huertas and Aguelo 2006).

Considering the morphological characteristics of the two samples analysed in this study, the individuals are observed to be not very robust, similar to those from the courtyard and cloister galleries and the apse area analysed by Trujillo (2023), contrasting with those from UF 221 from the medieval period studied by Cevallos (2023), which are very robust. It is significant that the men and women from the cloister galleries in the present study show a higher number of physical activity markers and more indicators of arthritis than those from the apse, particularly the women.

Although the tasks of the friars mainly focused on preaching, teaching, catechesis, confession and spreading devotion, following the concept of *Laudare, benedicere, praedicare*, in the 13th century the mendicant orders opted for a model of subsistence based on the renunciation of the possession of goods and income, in which everyday economic support had to be obtained through alms or the practice of humble trades (Reglero 2021). This fact could explain the MSM observed in the upper and lower limbs of the men from the cloister galleries. In this regard, archaeological data (Huertas and Aguelo 2006) also indicate the presence of a large garden within the friary, where the brothers of the monastic community could work and cultivate their food, implying the presence of physical activity markers. From this data, the hypothesis that the men from the cloister galleries may be linked to the members of the monastic community gains strength.

The women from the cloister galleries exhibit a higher frequency and number of activity markers than the men from the same area. Furthermore, they also show a higher frequency and number of activity markers than the men and women from the apse. Thus, the women from the cloister galleries could be related by family ties or with the paid workers who worked within the friary, as the presence of lay individuals or servants responsible for agricultural operations, cooking, laundry, etc. has been documented in monastic contexts (Vila 2016). The men and women from the apse area, on the other hand, seem to have had a physically less active lifestyle compared to those from the cloister galleries. However, those from the apse appear to have experienced developmental difficulties, as they present a 24.44% prevalence of enamel hypoplasia, one of the highest values found so far in the friary of Santa Caterina. This fact indicates that the individuals from the apse were likely to have been laypeople from families of a certain social status, able to afford the fees to be buried in the friary of Santa Caterina friary but not wealthy enough to be buried inside the church itself or in the chapels of the outer wall of the friary. The presence of a high frequency of hypoplasia would further reinforce this observation.

Aside from poor oral hygiene, the frequencies of caries and calculus indicate a diet rich in carbohydrates and animal protein for both female and male individuals in the two burial areas, especially for individuals from the cloister galleries. Although the Dominican friars followed the Rule of Saint Augustine, characterised by a diet low in meat, fasting and abstinence, meat, while not consumed all year round, did play an important role in many monastic congregations (Vila 2016; Lloveras *et al.* 2022). Moreover, the Dominicans are a mendicant order and, therefore, rely on donations, which means they must accept the food that is offered to them.

Conclusions

Although we have a small sample size, which requires caution to be exercised before making any assertions, the demographic analysis shows that in both groups, the number of male and female individuals is balanced, but with older individuals in the cloister galleries, indicating they likely had a better (longer) life. Based on the demographic, morphological and palaeopathological data, the individuals in the cloister galleries probably correspond to the monastic community and laypeople associated with the friary, who could be family members or working as servants within this community. The immature individuals over 7 years old are probably what are known as oblates. In the Middle Ages, an oblate was a child offered by the parents to a monastery, friary, or nunnery to be consecrated to God or educated there.

The individuals from the cloister galleries do not appear to have had growth problems (no hypoplasia) and had a diet based on animal protein and some carbohydrates. The apse seems to correspond more to a group of relatively well-off laypeople, able to pay the burial fee in this area of the friary, but not wealthy enough to be buried inside the friary church or in the chapels on its outer wall. Although it seems that these individuals are likely to have enjoyed relatively good nutrition, with an abundance of carbohydrates, the study also shows that they faced growth problems, as they have the highest frequencies of hypoplasia found at Santa Caterina.

Acknowledgements

Museum of History of Barcelona (MUHBA) for granting us access to this material, especially to Emili Revilla and Laura Suau. Philip Banks for his interesting and helpful comments.

SGR Social, Cultural, and Biological Evolution in the Pleistocene (StEP), Ref: 2021 SGR 01239.

MonBones research project: PID2020-118194RJ-100 AEI/MINECO.

References

Aguelo, J., Huertas, J. and Puig, F. (2005) Santa Caterina de Barcelona: assaig d'ocupació i evolución. *Quarhis: Quaderns d'Arqueologia i Història de la Ciutat de Barcelona* Època II 1, 13–43.

Alduc-Le Bagouse, A. (1988) Estimation de l'âge des non-adultes: maturation dentaire et croissance osseuse. Donnés comparatives pour deux nécropoles médiévales bas-normandes. In *Actes des 3èmes Journées Anthropologiques. Notes et Monographies Techniques* 24, 81–103. Paris, Centre National de la Recherche Scientifique, CNRS.

Alemán, I., Botella M.C. and Ruiz, L. (1997) Determinación del sexo en el esqueleto postcraneal. Estudio de una población mediterrània actual. *Archivo Español de Morfología* 2, 7–17.

Alesan, A. (1990) *Estudi d'una població subadulta de l'Edat del Ferro: demografia, antropometria i creixement*. Master's thesis, Universitat Autònoma de Barcelona.

Barrio, P., Trancho, G. and Sánchez, J. (2006) Metacarpal Sexual Determination in a Spanish Population. *Journal of Forensic Sciences* 51, 990–95.

Batlle, C. and Casas, M. (1980) La caritat privada i les institucions benèfiques de Barcelona (segle XIII). In *La pobreza y la asistencia a los pobres en la Cataluña medieval: volumen misceláneo de estudios y documentos*, 117–90. Barcelona, Consejo Superior de Investigaciones Científicas, Departamento de Estudios Medievales.

Brothwell, D. (1987) *Desterrando huesos*. Mexico, Fondo de cultura económica.

Campillo, D. (2001) *Introducción a la paleopatología*. Barcelona, Edicions Bellaterra.

Capasso, L., Kennedy, K. and Wilczak, C. (1999) *Atlas of Occupational Markers on Human Remains*. Teramo (Italy), Edigrafital S.p.A..

Cardoso, H.F.V. and Saunders, S.R. (2008) Two arch criteria of the ilium for sex determination of immature skeletal remains: a test of their accuracy and an assessment of intra- and inter-observer error. *Forensic Science International* 78, 24–29

Cevallos, A., Rissech, C., Huertas, J. and Lloveras, L. (2023) Estudio bioantropológico de los restos humanos procedentes de la unidad funeraria (UF) 221 del antiguo convento de Santa Caterina (1243–1836), Barcelona. *Rodis. Journal of Medieval and Post-Medieval Archaeology* 6, 171–216.

Cevallos, A., Tomàs, X., Lloveras, L. and Rissech, C. (2024) Reconstructing a lay individual's elbow fracture at Santa Caterina Friary, Barcelona (15th–16th Century): the contribution of paleopathology to the valorization of bioarcheological heritage. *Heritage* 7, 4182–92.

Cobo, S. (2022) *Estudio Dental de los Individuos del Hipogeo Neolítico Final-Calcolítico del Carrer París de Cerdanyola del Vallès (Barcelona)*. Master's thesis, Universidad de Barcelona.

Ferembach, D., Schwidetzky, I. and Stloukal, M. (1980) Recommendations for age and sex diagnoses of skeletons. *Journal of Human Evolution* 9, 517–49.

Fernández, M. (2008) *Marcadores óseos de actividad ocupacional en las poblaciones históricas de Palat de Rey (León) y de Gormaz (Soria)*. PhD thesis, Universidad de León.

Gilbert, B.M. and McKern, T.W. (1973) A method for aging the female os pubis. *American Journal of Physical Anthropology* 38, 31–8.

Huertas, J. and Aguelo, J. (2006) *Memòria de la intervenció arqueològica al solar del mercat de Santa Caterina, Ciutat Vella (Barcelona, Barcelonès)* Barcelona, Servei d'Arqueologia de la Ciutat de Barcelona.

Işcan, M.Y. and Loth, S.R. (1986) Determination of age from the sternal rib in white females. A test of the phase method. *Journal of Forensic Sciences* 31, 990–9.

Krogman, W.M. and Işcan, M.Y. (1986) *The Human Skeleton in Forensic Medicine*. Springfield, Illinois, Charles C. Thomas Ltd.

Lloveras, L., Cruells, J., Castellano-Tresserra, A., Riera, S. and Nadal, J. (2022) Monastic meals: the role of birds in the diet of the Poor Clares at Monastery of Santa Maria de Pedralbes (Barcelona, Spain) from medieval to modern periods (14th–19th centuries AD). *International Journal of Osteoarchaeology* 33(4), 1–16.

Lovejoy, C.O., Meindl, R.S., Pryzbeck, T.R. and Mensforth, R.P. (1985) Chronical metamorphosis of the auricular surface of the ilium. A new method for the determination of adult skeletal age at death. *American Journal of Physical Anthropology* 68, 15–28.

Mann, R., Hunt, D. and Lozanoff, S. (2016) *Photographic Regional Atlas of Non-Metric Traits and Anatomical Variants in the Human Skeleton*. Springfield, Illinois, Charles C. Thomas Ltd.

Martin, R. and Saller, K. (1957) *Lehrbuch der anthropologie*. Stuttgart, G. Fischer.

Masset, C. (1982) *Estimation de l'âge au décès par les sutures crâniennes*. Thèse de Sciences Naturelles Multigraphée. Université Paris VII.

Mastrangelo, P., De Luca, S., Alemán, I. and Botella, M. (2011) Sex assessment from the carpals bones: discriminant function analysis in a 20th century Spanish sample. *Forensic Science International* 206, 216.e1-216.e10.

McKern, T.W. and Stewart T.D. (1957) *Skeletal Age Changes in Young American Males*. Headquarters; Quartermaster Research and Development Command Technical Report, EP-45. Natick, Massachusetts.

Moliner, J.M. (1974) *Espiritualidad medieval: los mendicantes*. Burgos, Monte Carmelo.

Olivier, G. (1960) *Pratique Antropologique*. París, Ed. Vigot Frères.

Ortner, D. (2003) *Identification of Pathological Conditions in Human Skeletal Remains*, 2nd edition. San Diego, Academic Press.

Pablos, A., Olivencia, A., Pérez, A., Martínez, I., Lorenzo, C. and Arsuaga, J. (2013) From toe to head: use of robust regression methods in stature estimation based on foot remains. *Forensic Science International* 226, 299.e1–299.e7.

Pearson, K. (1899) Mathematical contributions to the theory of evolution. V. On the reconstruction of the stature of prehistoric races. *Philosophical Transactions of the Royal Society* 192, 169–244.

Peckmann, T., Meek, S., Dilkie, N. and Rozendaal, A. (2016) Determination of sex from the patella in a contemporary Spanish population. *Journal of Forensic and Legal Medicine* 44, 84–91.

Pujol-Bayona, A., Rissech, C., Prida, D. and Turbón, D. (2011) Estudio bioantropológico del conjunto funerario de época tardorromana Carrer Ample 1 de Barcelona. *Revista Española de Antropología Física* 32, 1–19.

Reglero, C. (2021) *Monasterios y monacato en la España medieval*. Madrid, Marcial Pons Historia.

Rissech, C. (2016) La estimación de la edad de individuos adultos y subadultos a partir de material esquelético. In C. Sanabria Medina (ed.) *Patología y antropología forense de la muerte: la investigación científico-judicial de la muerte y la tortura, desde las fosas clandestinas, hasta la audiencia pública*, 237–97. Bogotá, Forensic Publisher.

Rissech, C. and Malgosa, A. (1991) Importancia relativa de la longitud del pubis y la anchura del ilion en el estudio del dimorfismo sexual de los coxales. *Boletín de la Sociedad Española de Antropología Biológica* 12, 29–43.

Rissech, C. and Malgosa, A. (1997) Sex prediction by discriminant function with central portion measures of innominate bones. *Homo* 48, 22–32.

Rissech, C., Estabrook, G.F., Cunha, E. and Malgosa, A. (2006) Using the acetabulum to estimate age of death in adult males. *Journal of Forensic Sciences* 51, 213–29.

Rissech, C., Estabrook, G.F., Cunha, E. and Malgosa, A. (2007) Estimation of age at death for adult males using the acetabulum, applied to four Western European collections. *Journal of Forensic Sciences* 52, 774–8.

Rissech, C., Márquez-Grant, N. and Turbón, D. (2013) A collation of recently published western European formulae for age estimation of subadult skeletal remains: recommendations for forensic anthropology and osteoarchaeology. *Journal of Osteoarchaeology* 58, S163–S168.

Rissech, C., Winburn, A.P., San-Millán, M., Sastre, J. and Rocha, J. (2019) The acetabulum as an adult age marker and the new IDADE2 (the IDADE2 web page). *American Journal of Physical Anthropology* 169, 757–64.

Safont, S., Malgosa, A. and Subirà, E. (2000) Sex assessment on the basis of long bone circumference. *American Journal of Physical Anthropology* 113, 317–28.

Saldías, E., Malgosa, A., Jordana, X. and Isidre, A. (2016) Sex estimation from the navicular bone in Spanish contemporary skeletal collections. *Forensic Science International* 267, 229.e1–229.e6.

Schaefer, M., Black, S. and Scheuer, L. (2009) *Juvenile Osteology, A Laboratory and Field Manual*. Amsterdam, Elsevier–Academic Press.

Scheuer, L. and Black, S. (2000) *Developmental Juvenile Osteology*. London, Elsevier– Academic Press.

Schutkowski, H. (1993) Sex determination of infant and juvenile skeletons. I. Morphognostic features. *American Journal of Physical Anthropology* 90, 199–205.

Silva, A.M. (1995) Sex assessment using the calcaneus and the talus. *Antropologia Portuguesa* 13, 107–19.

Todd, T. (1920) Age changes in the pubic bone. I. The male white pubis. *American Journal of Physical Anthropology* 3, 285–334.

Trujillo, J., Rissech, C., Huertas, J. and Lloveras, L. (2023) El convento de Santa Caterina: estudio bioantropológico de una muestra de las inhumaciones del claustro gótico y la cabecera del ábside. *Rodis. Journal of Medieval and Post-Medieval Archaeology* 6, 217–42.

Vila, P. (2016) Aspectes de la cuina i l'alimentació a l'època del barroc, al convent de Sant Domènec de Girona (1600–1612). *Annals de l'Institut d'Estudis Gironins* 57, 187–234.

Vives, E. (1990) *La població catalana medieval. Origen i evolució.* Vic, Eumo Editorial.

Walker, P., Johnson, J.R. and Lambert, P.M. (1988) Age and sex biases in the preservation of human skeletal remains. *American Journal of Physical Anthropology* 76(2), 183–8.

Weaver, D.S. (1979). Application of the likelihood ratio test to age estimation using the infant and child temporal bone. *American Journal of Physical Anthropology* 50(2), 263–70.

Zaldívar, A.M. (2012) Patricians' embrace of the Dominican Convent of St. Catherine in thirteenth-century Barcelona. *Medieval Encounters* 18(2), 174–206.

5

The early medieval necropolis of the Former Municipal Courts of Barcelona: a unique discovery

Marta Merino and Carme Rissech

The archaeological intervention at the Antics Jutjats Municipals of Barcelona (Former Municipal Courts of Barcelona) led to a necropolis dating from the 9th and the 10th centuries being discovered. A total of 115 individuals were recovered from this cemetery. This study aims to elucidate the lifestyle and health status of these individuals buried at the Antics Jutjats Municipals site. Standard anthropological and palaeopathological methods were used to carry out the study. The results indicated 71 non-adult individuals (68 indeterminate, 2 males and 1 female) aged between 0 and 24.99 years and 44 adults (24 females, 18 males and 2 indeterminate) aged between 25 and 50 years. The deceased were buried in an extended supine position, with the head to the east and the feet to the west, without any grave goods, consistent with the Christian burial ritual of the period. The demographic information from this sample indicates a short life expectancy (27.45 years) and a high infant mortality rate (57.4%). Anthropometric data revealed an average height of 166.6 cm for males and 153.6 cm for females. Moreover, a range of skeletal indicators, including cribra orbitalia *and* cribra femoralis *and subperiosteal new bone formation were observed. This subperiosteal new bone formation was particularly evident in individuals under 5 years of age. All these indicators indicate some type of external or internal stressful event occurring to the individual, such as, for example, nutritional deficiencies or avitaminosis, possibly due to the absence of fresh products and lack of vitamin D. In addition, postcranial pathologies, antemortem and perimortem traumas, and occupational stress markers were also described, with a high rate of prevalence. These findings collectively point to a physically demanding lifestyle and lead us to consider the likelihood of physical violence having taken place. The layout of the funerary complex did not reveal any specific pattern or distribution according to sex, age, burial area or pathologies. This Christian cemetery in Barcelona is important, because, from the 6th century onwards, the Christian funeral world was usually structured around religious buildings; it seems that this necropolis could have broken with this tendency.*

Keywords: Medieval, Christian burials, avitaminosis, infant mortality.

Introduction

The necropolis of the Antics Jutjats Municipals of Barcelona in Catalonia, Spain, dates from the 9th to the 10th century, according to the ^{14}C dating carried out by Beta Analytic [individual UF2 (39–40 y old) 766–898 cal AD; individual UF3 (21–24 y old) 887–1013 cal AD**;** individual UF4 (5–6 y old), 862–994 cal AD; individual UF13 (40–60 y old) 768–900 cal AD; individual UF19 (27–29 y old) 768–900 cal AD; individual UF21 1 (25-29 y old) 771–903 cal AD; individual UF26 (2–3 y old), 892–1014 cal AD; individual UF38 (5–6 y old), 862–994 cal AD; individual UF54 (±30 y old) 862–994 cal AD and individual UF95 (1–2 y old) 676–779 cal AD] (Serra-Molinos 2023). During this period, Barcelona was an expanding city, which was beginning to grow beyond its Roman defences, although the medieval walls had not yet been constructed. At this time, the necropolis was located outside the Roman walls, in the *suburbium* (suburban area), close to the *territorium* (the surrounding

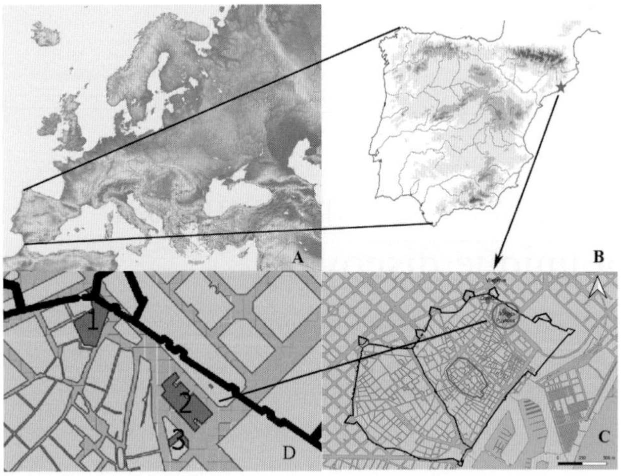

*Figure 5.1. Map showing the location of: Spain in the Iberian Peninsula and Europe (A); the city of Barcelona (B); the location of the Antics Jutjats Municipals in Barcelona (*Barcino*) (C) in relation to the Roman wall (light grey) and the medieval and modern wall (dark grey). Enlargement of the map of Barcelona where number 2 indicates the location of the necropolis (D).*

countryside). At the end of the 13th century, with the construction of the medieval defences of Barcelona, the necropolis was included within the walls and fell into disuse (Fig. 5.1). Today, it is located between Passeig Pujades, 3; Passeig Lluis Companys, 1–5; Passeig Picasso, 1; and Plaça de Vicenç Albert Ballester, in the Ciutat Vella district of Barcelona (Fig. 5.1) (Merino and Rissech 2022).

The archaeological excavation of the Antics Jutjats Municipals of Barcelona was carried out between 2018 and 2019 as part of a preventive excavation by the company ABANS Serveis Culturals. A total of 123 burials were located, of which 115 were totally excavated; the remaining 8 could not be excavated as they lay outside the limits of the excavation (Merino and Rissech 2022). It was estimated that there could be more than 460 burials in the entire plot, although the precise limits of the necropolis could not be recorded. The burials followed a strict spatial organisation, respecting each other, with an east–west orientation following the Christian ritual of this time, characterised by simplicity and the absence of grave goods. This necropolis represents a large and significant burial space in the area known as the eastern suburb (Beltrán de Heredia 2010), which does not seem to have been linked to any religious centre. This last characteristic differentiates this cemetery from other funerary spaces of this period (Ripoll and Molist 2012), because, from the 6th century onwards, the Christian funerary world was predominantly structured around urban religious buildings, basilicas or monasteries (Beltrán de Heredia 2010). This cemetery seems to have been a planned burial area, located at a considerable distance from the Late Roman city wall. To date, the Antics Jutjats site is unique in Barcelona. Apart from its organisation and the number of burials and type of funeral ritual, it represents a large funerary space of great importance in the area known as the eastern *suburbium* (Beltrán de Heredia 2010), which appears not to have been linked to any religious centre. In addition, there is very little information on the population of Catalonia during the early medieval period (Bonnassie 1975–1976; Banks 1987). Therefore, the aim of the study is to provide information on the lifestyle and health status of the population buried in the necropolis of the Antics Jutjats Municipals of Barcelona during the 9th and 10th centuries through bioanthropological and palaeopathological analyses.

Materials and methods

The material used for this study corresponds to the human skeletal remains recovered from the necropolis of the Antics Jutjats Municipals of Barcelona during the 2018–2019 excavation. The lead anthropologist for the excavation was the first author of this article (M.M.), working through the company ABANS Serveis Culturals. The anthropological study was initiated during fieldwork and continued in the anthropological laboratory of the MUHBA (Museu d'Història de Barcelona). The material was washed and, if fragmented, reconstructed. Reconstruction was carried out with painter's tape or with a type of transparent synthetic resin adhesive (Imedio glue) which, if necessary, can be removed with an acetone-based solvent. Subsequently, anatomical attribution and individualisation of the remains was carried out.

The taphonomic study and funerary ritual analysis were based on the information collected in the anthropological field records, together with the plans drawn and photographs taken during the excavation. Based on these data, the funerary ritual and the spatial relationship between the tombs were described.

The state of preservation of the sample was analysed according to the IP3 preservation index (Alesan 1990), which shows the percentage of preserved skeletal elements, the bone being considered as present, regardless of its preservation and integrity, in relation to a total of 22 bones (long bones, plus girdles, plus the skull, maxilla and mandible) of the human skeleton.

Biological sex estimation was performed using the usual criteria for the skull, mandible and hip bone (Ferembach *et al.* 1980; Rissech and Malgosa 1991, 1997) in adults and non-adults above 15 years of age. In adults, the robustness of the skeleton and discriminant functions based on long bones (Olivier 1960; Black 1978; Alemán *et al.* 1997; Safont *et al.* 2000) and hip bone (Rissech and Malgosa 1997) were also applied. This multiplicity of methods for adult sex estimation is especially useful when the skeleton is incomplete. In these cases, particularly for individuals whose remains consist mainly of limbs, preference was given to discriminant functions for skeletal elements and fragments.

The age at death of adult individuals was estimated from the morphology of the auricular surface, pubic symphysis, acetabulum and the sternal end of ribs (Brickley and McKinley 2004; Rissech *et al.* 2006). In addition, the skeletal maturation process of the iliac crest, sacrum and clavicle were also included (Brickley and McKinley 2004).

For non-adult individuals, criteria of dental development (Crétot 1978; Ubelaker 1989), growth rates and bone development (Stloukal and Hanáková 1978; Weaver 1979; Alduc-Le Bagouse 1988; Rissech *et al.* 2003; 2008; Rissech and Malgosa 2005; 2007; Rissech and Black 2007), and epiphyseal bone fusion (Ferembach *et al.* 1980; Krogman and Işcan 1986; Brothwell 1987; Rissech *et al.* 2001) were used.

For the anthropometric analysis, the standard measurements collected by Martin and Saller (1957) were used. Stature was calculated according to Pearson's (1899) formulae, using the femur because it is the skeletal element most closely related to stature and sensitive to the environment (Rissech *et al.* 2008). Pearson's formulae were applied because they are based on a Mediterranean population.

The palaeopathological analysis was based on macroscopic observation, following the reference texts of Ortner (2003) and Lewis (2018). In addition, non-pathological aspects, such as physical activity markers, were analysed to help understand the lifestyle of the individuals (Capasso 1998; Jordana 2007; Fernández, 2008).

For oral pathologies, the descriptions of Oms (2016) and Cobo (2022) were considered, along with environmental stress markers such as hypoplasia and dietary indicators such as the presence of dental calculus and caries. The record used for their registration was that of our laboratory (Cobo 2022) and that of Hillson (2001).

Results and discussion

Demographic profile

The results (Table 5.1) indicate a total of 115 individuals, comprising 44 adults and 71 non-adults. Among the 44 adults, 24 were identified as females, 18 as males, and the sex of 2 could not be determined due to insufficient data. Of the 71 non-adults, 2 were identified as males, 1 as a female and 68 could not be determined. The ages of these individuals are given in Table 5.1.

The preservation index values (IP3) of the sample analysed (Table 5.1) are high, indicating good preservation. The presence of a part of a bone is considered as the entire bone, and because of this, it does not inform us about the physical state of the bone, but it is useful to inform us about the number of bones preserved. Specifically, of the 71 non-adults, 32 exhibit an IP3 above 90%, and of these, 18 cases are 100%. Among the 44 adults, 39 have an IP3 greater than or equal to 90%, and of these, 18 have an IP3 of 100%. The individuals with low IP3 values are those affected by later structures. The good preservation of the remains is attributable to three main factors: (i) the type of sediment that filled the graves; (ii) the depth at which the individuals were buried in relation to the contemporary ground surface; and (iii) the absence of vegetation in the area.

Taphonomy

A wide variety of burial types were recorded, including cists, uncovered anthropomorphic grave-pits, anthropomorphic grave-pits covered with stone slabs, indeterminate-shaped grave-pits covered with stone slabs, uncovered indeterminate-shaped grave-pits, simple covered and uncovered grave-pits, and simple tile-covered grave-pits. All the burials were individual, oriented east–west and without grave goods (Fig. 5.2), following the Christian burial ritual.

The space was organised around the burials of the adults, which were placed in different rows with a separation of 1.5–2 m between them. The non-adults were placed between the adult graves, respecting the space of each grave. This suggests that the pits may have been marked in some way, although no such markers have been preserved. The individuals were recorded lying supine, with some variation in the position of the arms. Ninety-seven percent of the pits were completely filled, and 3% were partially filled, covering the remains of the individual to a certain extent. Most of the anatomical connections were tight, although some alterations were observed, such as in UF 71, where the entire thorax had been disturbed by small rodents.

The funerary assemblage obtained during the excavation does not appear to show spatial distribution or distinctions by biological sex or age at death. Given the considerable extent of the site of the Antics Jutjats Municipals and the proximity of similar burials in carrer Comerç, it is suggested that the cemetery covered a far larger area than could be excavated.

Demographics

As mentioned above, 115 individuals were recovered. Although the sample studied was not large enough, and therefore not representative, frequencies by age and sex were calculated, grouping individuals at 5-year intervals (Fig. 5.3, A and Table 5.1). The intention was to obtain an approximation of the possible mortality patterns and to calculate the life table of the group analysed in order to compare them, always approximately, to other population samples.

Fig. 5.3, A, shows the population grouped in age ranges of 5 years. High infant mortality was observed in the first five years of life (46.9% 54/115), especially at the age of 4 (Fig. 5.3, A and Table 5.1), possibly related to periods of weaning, which was usually between 3 and 3.9 years of age (Corruccini *et al.* 1985; Krenz-Niedbala 2001; Humphrey 2010). At this age, infants do not have a fully developed immune system and weaning results in the loss of passive immunity (O.N.U. 1984; Kent 2007). In pre-industrial populations, infant mortality usually ranges between 40 and 45% (Buchet 1983), but in the Antics Jutjats Municipals necropolis it is higher. Compared with other populations of

Table 5.1. Age, Sex and Preservation Index of each individual from the Antics Jutjats Municipals cemetery in Barcelona

Non–adults								*Adults*			
UF	*Age (years)*	*Sex*	*IP3*	*UF*	*Age (years)*	*Sex*	*IP3*	*UF*	*Age (years)*	*Sex*	*IP3*
4	5–6	ND	80%	74	3–4	ND	88%	1	+21	Male	59%
6	3–4	ND	80%	77	4–5	ND	92%	2	39–40	Male	100%
10	1.5–2	ND	96%	79	2–3.5	ND	88%	3	21–24	Female	95%
20	1.5–2	ND	96%	80	1.5–2.5	ND	76%	5	21–24	Male	95%
22	2.5–3.5	ND	100%	82	14–15	ND	100%	7	25–30	Male	100%
23	1–1.5	ND	69%	83	2–3	ND	80%	8	25–27	Female	95%
24	3–4	ND	100%	84	0.25–0.5	ND	65%	9	33–35	Female	100%
25	1–1.5	ND	73%	86	6–7	ND	100%	12	35–45	Female	100%
26	2–3	ND	92%	89	18–20	Female	100%	13	40–60	Female	95%
27	0.4–0.5	ND	42%	90	5–6	ND	100%	14	21–25	Female	95%
32	1–2	ND	46%	91	1–2	ND	100%	15	+21	ND	27%
34	2.5–3.5	ND	80%	92	0.4–0.5	ND	88%	16	33–45	Male	100%
35	1–1.5	ND	23%	93	0.5–0.75	ND	92%	17	25–27	Female	100%
36	0.75–1	ND	96%	94	38–40 W	ND	84%	19	27–29	Male	90%
37	5–6	ND	96%	95	1–2	ND	96%	21	25–29	Male	100%
38	5–6	ND	92%	96	18–20	Male	100%	28	45–49	Female	90%
39	1–1.5	ND	61%	98	3.5–4.5	ND	80%	29	33–40	Female	95%
43	3–4	ND	84%	99	10–11	ND	100%	30	35–45	Male	95%
44	1–1.5	ND	80%	101	1–2	ND	73%	31	25–35	Female	85%
45	ND	ND	11%	102	3–4.5	ND	69%	33	21–25	Male	95%
46	0–0.1	ND	80%	103	2–3	ND	80%	40	25–30	Female	90%
48	1–1.5	ND	84%	104	2–3	ND	42%	41	21–25	Female	90%
49	19–20	Male	100%	106	4–5	ND	92%	42	30–35	Female	95%
50	5–6	ND	100%	107	4–5	ND	96%	47	25–35	Male	100%
51	1–1.5	ND	73%	108	3–4	ND	100%	52	35–45	Female	95%
53	0.75–1	ND	73%	110	3–4	ND	100%	54	+/–30	Female	100%
56	3–4	ND	50%	112	38–40 W	ND	3%	55	37–43	Male	100%
58	1–2	ND	73%	113	1–2	ND	84%	57	18–25	Male	100%
59	1.5–2	ND	78%	114	0.75–1	ND	80%	60	35–45	Male	100%
61	0.25–0.5	ND	76%	115	8–9	ND	100%	62	25–30	Female	100%
66	0.9–1.2	ND	73%	116	0.75–1	ND	53%	63	23–25	ND	72%
67	3–4	ND	92%	117	1.5–2.5	ND	86%	64	35–40	Male	95%
68	1–1.5	ND	88%	118	6–7	ND	100%	65	25–30	Female	100%
69	3–4	ND	92%	119	2.5–3.5	ND	92%	70	45–55	Male	90%
71	11–13	ND	100%	123	0.75–1	ND	100%	73	35–45	Female	90%
72	12–14	ND	100%					75	25–33	Female	100%
								78	25–35	Male	95%
								87	25–35	Male	100%
								88	25–35	Female	95%
								100	25–35	Female	91%
								105	25–30	Female	90%
								111	+21	Female	63%
								120	35–50	Male	100%
								121	25–35	Male	100%

the same period, such as the necropolis of Ordoñana (Álava) with a rate of 31.8% (Fernandez del Prado 1978) and La Torrecilla in Granada with 22.7% (Souich 1979), the infant mortality rate of individuals up 12 years of age in this study is much higher (57.4% 66/115).

In the present study, juvenile mortality (13 to 20 years) drops drastically to 4.3% (5/115) (Fig. 5.3, A and Table 5.1). This stage is represented by 6 individuals, so it is underrepresented in the sample.

The highest mortality rate in adulthood is between the ages of 20 and 34.99, while the incidence decreases after the age of 35 due to the low number of individuals reaching adulthood (12%) or old age (0%). During the highest adult mortality phase, there is a significant sex difference, with higher mortality among women between 18 and 26 years of age, which is probably related to problems during childbirth.

However, these demographic inferences should be taken with caution, because of the small sample size and since complete population information is not available. However, considering the sample studied here as a relatively random selection of the cemetery's population, the results, although not accurate and with limitations, may be indicative of the demographic dynamics of this human group.

Life expectancy at birth is estimated at 27.45 years (Table 5.2). This value is similar to that of other populations in the Iberian Peninsula, such as S'Illot des Porros (Mallorca) with 28.3 years (Alesan *et al.* 1999) or the medieval population of Plaça Vella, Terrassa (Jordana and Malgosa 2002) at 26 years. These values are close to those of unindustrialised populations, ranging around 30 years for the Kung! (hunter-gatherers) and 20 years for Yanomama of the Amazon rainforest (Milner *et al.* 1989). Thus, these data could indicate hard living conditions for the population represented in the Antics Jutjats Municipals cemetery.

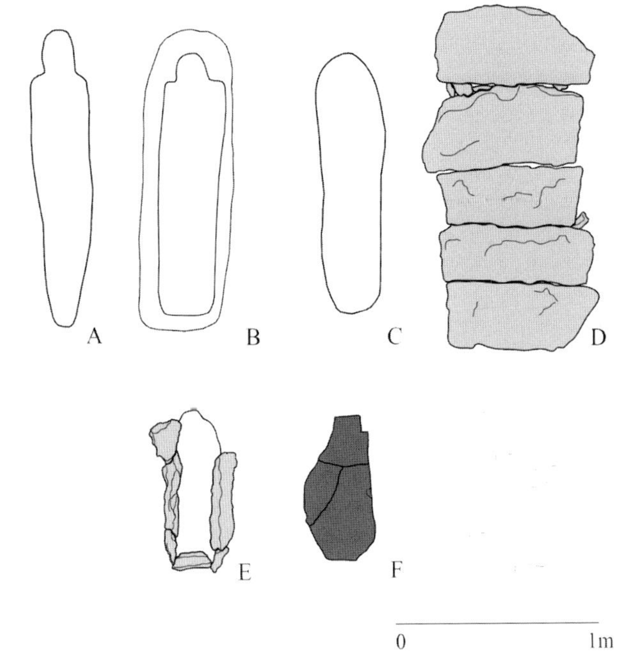

Figure 5.2. Tomb typologies: A, anthropomorphic grave-pit; B, anthropomorphic grave-pit with 2 levels of edging; C, simple grave-pit; D, covered with stone slabs; E, cist; F, tegula cover.

Anthropometry

The skeletal characteristics of individuals, specifically their morphology, are the result of the interaction between the genotype and the environment (Cavalli-Sforza and Bodmer 1971; Guglielmino-Matessi *et al.* 1979). In order to physically characterise the sample analysed and provide information on their lifestyle, standard cranial and postcranial anthropometric measurements were taken (Tables 5.3 and 5.4). Anthropometric analysis is divided into cranial and postcranial sections. Cranial morphology is used for the study of evolution, differentiation and classification between different populations, as it is less influenced by the environment. Postcranial morphology is mainly used to analyse the physical activities and lifestyle of the individuals.

Tables 5.3 and 5.4 show the indices resulting from the cranial (Table 5.3) and postcranial (Table 5.4) absolute values. Results indicate that the craniums have the typical

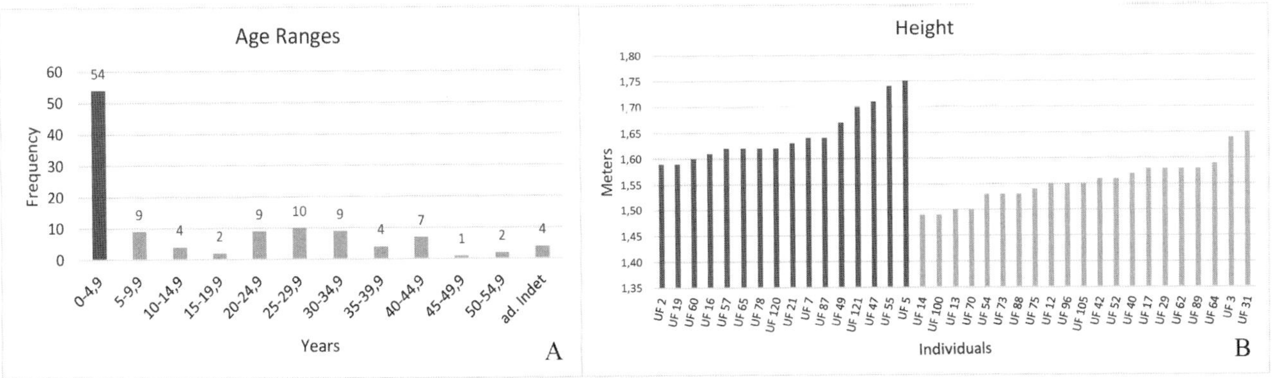

Figure 5.3. A. graph of the distribution of mortality in the population of the Antic Jutjats Municipals cemetery in Barcelona (Spain); B. graph of the height of adult individuals.

Table 5.2. Calculation of Life Expectancy at Birth for remains interred in the Antics Jutjats Municipals cemetery grouped by 5-year intervals

Age interval (x)	*Total no. years lived (Lx)*	*No. deaths (Dx)*	*Survivors (lx)*	*Survival probability (Ix)*	*Life expectancy(ex)*
0	1582.12	55	56	0.5045045	27.45810982
5	1431.50	63	48	0.43243243	22.5474
10	1360.00	67	44	0.3963964	17.9849
15	1310.00	69	42	0.37837838	14.098554
20	1229.00	78	33	0.2972973	11.074728
25	1090.50	88	23	0.20720721	8.498068
30	609.50	97	14	0.12612613	6.2381
35	542.50	101	10	0.09009009	4.16667
40	190.00	108	3	0.02702703	2.67
45	147.00	109	2	0.01801802	1
50	0.00	111	0	0	0

Table 5.3. Mean values and Standard Deviations (SD) of cranial indices. Statistical test, Student's t-Test

	Males			*Females*			
Cranial indices	*n*	*Mean*	*DS*	*n*	*Mean*	*DS*	*p*
Cranial I.	4	70.8	3.1	6	70.2	2.6	0.592
Vertical-longitudinal I	2	73	1.1	5	74.1	2.9	0.691
Vertical-transversal I	2	103	2.8	5	104.6	4.4	0.699
Auriculo-longitudinal I (R)	4	67.8	0.9	6	53.8	20.6	0.289
Auriculo-longitudinal I (L)	4	68.6	2.4	5	59.1	17.8	0.157
Fronto-parietal transversal I	4	83.5	9.2	5	81.1	3.6	0.480
Foramen magnum I	3	89.7	14.2	3	84.7	6.5	0.564
Cranial-facial transversal I	3	64.8	3.2	6	71.1	1.3	0.248
Fronto-sagittal I	7	90.2	5.2	7	95.4	27.5	0.564
Parietal sagittal I	7	91.8	4.5	10	89.5	2.7	0.197
Occipital sagittal I	7	82.2	32.7	10	94.9	10.5	0.248
Total sagittal I	7	63.9	9.9	7	62.1	3.2	0.767
Occipital squama sagittal I	9	91.5	10.9	9	92.1	10.8	0.564
Total facial I	3	98.0	5.2	5	91.3	10.4	0.180
Upper facial I	3	66.3	8.2	5	58.8	2.8	0.137
Jugomandibular I	3	61.1	13.0	5	72.8	17.0	0.770
Orbital I (R)	3	77.9	13.2	6	86.2	10.8	0.373
Orbital I (L)	1	86	–	5	86.1	13.2	0.373
Palatal I	1	352.8	–	5	128.9	114	0.157
Mand. ramus I (R)	4	68.1	43.5	8	43.8	16.9	0.157
Mand. ramus I (L)	5	66.1	37.8	11	49.8	11.2	0.157
Mand. I (Thompson)	1	131.3	–	5	88.9	6.3	0.157
Mand. robusticity I (at symphysis)	9	46.0	13.5	14	70.8	52.3	0.157
Mand. robusticity I (mental foramen R)	13	37.6	7.7	16	52.0	47.1	0.157
Mand. robusticity I (mental foramen L)	12	41.1	11.1	17	53.7	51.1	0.717

Abbreviations: Mand. = mandible; I: Index; R: right; L: left

Mediterranean morphology (Table 5.3). This means long craniums and long faces.

Regarding the postcranial measurements of the upper limbs (Table 5.4), in general, humeri exhibit euribrachy (round diaphyses); they have a mid-diaphyseal index greater than 76.5, while femurs display a subtle *linea aspera*, revealing a weak pilaster (pilastric index below 109) (Olivier 1960). The pilastric index refers to the degree of protrusion of the *linea aspera*, which indicates the use of the lower limb (insertions of the vastus internus, vastus externus and biceps femoris). On the other hand, these individuals show eurinecmic tibiae, indicating a null degree of flattening of the tibia (Table 5.4). These indices inform us about the specific structural characteristics related to the cross-sectional geometric properties of the humeral, femoral and tibial diaphyses respectively, which in turn are related to moments of inertia and forces applied on the bones (individual biomechanics) (Ruff 1987; Trinkaus *et al.* 1998; Chevalier 2013), providing information on their physical activity (Ruff 1987; Trinkaus *et al.* 1998; Chevalier 2013). In general, round diaphyses indicate low physical activity (Ruff 1987; Trinkaus *et al.* 1998; Chevalier 2013). Therefore, the results obtained in this sample indicate moderate physical activity in these individuals.

Comparing these data with other population series, the upper and lower limb indices in the sample from the Antics Jutjats Municipals cemetery are slightly lower than those observed at Carrer Ample 1 (Pujol-Bayona *et al.* 2011) and are more consistent with populations such as the Plaça de la Vil·la de Madrid (Jordana *et al.* 2003) both in Barcelona. This indicates a similar robustness and low physical activity in the Antics Jutjats Municipals and the Plaça de la Vil·la de Madrid series. The Plaça de la Vil·la de Madrid burials were of a higher economic status than those at carrer Ample 1.

Height

Individual height is primarily determined by an individual's genes, which provide the individual growth potential; however, the final individual height is the result of the interaction of genetic factors along with environmental, nutritional and cultural factors, which are specific to each population (Cavalli-Sforza and Bodmer 1971; Samaras and Elrick 2002).

Height is commonly used as a health status indicator, for example it is usually employed in research on historic populations to evaluate their welfare trends and to assess the long-term impacts of childhood environment (Cameron 2023). It is a useful pointer as an approximation to children's nutritional status and disease environment (Cameron 2023; Turbón and Rissech 2024).

The estimated average height for individuals buried in the Antics Jutjats Municipals cemetery is 153.60 cm for women and 166.60 cm for men, with significant sex differences (U=1.000, p=0.000). Figure 5.3, B illustrates the heights of the 39 individuals analysed, with 21 females depicted in light grey and 18 males in dark grey. These heights are similar to those observed in early medieval individuals analysed in other studies (Pujol-Bayona *et al.* 2011).

Growth

Analysis of the growth of ancient populations is limited to cross-sectional studies. In the present study, it was decided to analyse the development of the individuals buried at the Antics Jutjats Municipals site in relation to other populations for which the most exhaustive information possible was available.

The scatter plot (Fig. 5.4) shows femur diaphysis values in relation to age for non-adult individuals from the Antics Jutjats Municipals and the Portuguese reference collections from Lisbon, Coimbra, and the British St Bride's

Table 5.4. Mean values and Standard Deviations (SD) of postcranial indices. Statistical test, Student's t-Test

	Males			Females			
Postcranial index	*n*	*Mean*	*DS*	*n*	*Mean*	*DS*	*p*
Diaphyseal index at humerus midshaft (R)	16	81	7.3	22	83.5	21.4	0.663
Diaphy. index at humerus midshaft (L)	17	83.6	6.3	26	83.4	15.8	0.952
Humerus robustness I. (R)	12	20.6	1.7	11	19.8	2.4	0.366
Humerus robustness I. (L)	11	20.6	2.4	12	20.3	2.9	0.797
Radius robustness I. (R)	12	21.3	19.2	10	16.4	1.3	6.103
Radius robustness I. (L)	8	25.7	21.9	11	16.8	0.9	0.192
Diaphy. index at radius midshaft (R)	16	68.5	15.5	25	71.3	5.9	0.430
Diaphy. index at radius midshaft (L)	14	77.7	7.5	22	71.8	6.3	0.015*
Ulna robustness I. (R)	12	19.9	2.4	8	18	2.2	0.080
Ulna robustness I. (L)	12	19.3	1.9	10	17.9	1.2	0.084

(Continued)

Table 5.4. Mean values and Standard Deviations (SD) of postcranial indices. Statistical test, Student's t-Test (Continued)

	Males			Females			
Postcranial index	*n*	*Mean*	*DS*	*n*	*Mean*	*DS*	*p*
Platolenia ulna I. (R)	13	91.4	7.3	18	82.1	15.8	0.058
Platolenia ulna I. (L)	12	88.4	8.1	15	85.3	26.2	0.691
Glenoidal scapular I. (R)	6	75.5	11.8	9	71.9	4.3	0.415
Glenoidal scapular I. (L)	5	71.6	3.6	8	68.5	2.5	0.091
Clavicle robustness I. (R)	2	27.5	3.5	3	26.3	2.5	0.689
Clavicle robustness I. (L)	5	97.1	162.4	7	23.9	2.4	0.252
Terry clavicle I. (R)	2	19.5	2.1	3	16.3	0.6	0.077
Terry clavicle I. (L)	2	19.5	2.1	4	15.5	1.3	0.039*
Femoral robusticity Index (R)	13	19.7	0.6	14	19.1	1.5	0.221
Femoral robusticity index (L)	14	19.8	0.8	12	19.3	1.5	0.288
Femoral robusticity II index (R)	13	12.5	0.5	13	12.6	1.6	0.873
Femoral robusticity II index (L)	13	13.1	1.3	11	12.7	1.5	0.551
Pilastric Index (R)	17	105.1	12.4	26	108.9	15	0.389
Pilastric index (L)	17	106.3	10.2	25	106.1	10.5	0.942
Platymeric femoral Index (R)	17	89.9	15.5	20	86.4	16.3	0.498
Platymeric femoral index (L)	16	88.3	17.7	21	84.3	15	0.470
Diaphyseal index at tibia midshaft (R)	17	75.3	7.9	24	73.3	7.9	0.432
Diaphyseal index at tibia midshaft (L)	17	73.6	8.6	23	71.1	7.5	0.332
Cnemic index (R)	16	72.3	5.7	23	68.8	8	0.144
Cnemic index (L)	17	73.9	6.9	23	68.7	8.9	0.052
Length-width tibial index (R)	11	20.5	1.4	12	19.5	1.1	0.060
Length-width tibial index (L)	13	20.4	1.1	13	19.3	1.3	0.035*
Length-width fibular index (R)	6	9.8	0.9	10	10.3	1.1	0.341
Length-width fibular index (L)	5	9.4	0.6	8	9.9	1.2	0.395
Cotilociatic index(R)	8	131.1	37.6	6	78.8	17.8	0.009*
Cotilociatic index(L)	8	132.5	37	6	90.7	18.1	0.026*
Talus width index (R)	12	82.7	13.8	17	84.7	22.9	0.795
Talus width index (L)	14	77.6	5.6	20	84.1	19.9	0.249
Talus height index (R)	14	49.1	6.2	19	47.8	6.1	0.557
Talus height index (L)	16	48.5	5.1	20	47.3	4.8	0.502
Calcaneus height index (R)	12	45.2	9.5	16	45.5	8.4	0.914
Calcaneus heightindex (L)	11	45.3	8.8	17	45.3	8.4	0.978
Crural index (R)	9	82.7	2.2	11	82.6	2.5	0.957
Crural index (L)	11	82.6	1.5	9	82.8	2.1	0.886
Humerus-femoral index (R)	11	72.4	1.9	9	70.4	2.4	0.047*
Humerus-femoral index (L)	9	69.8	2.8	10	70.1	1.9	0.795
Intermembral index (R)	12	78	24	10	66.9	23.9	0.291
Intermembral index (L)	12	60.2	28.1	13	63.4	21.7	0.747
clavicle-humeral index (R)	2	27.7	23	3	43.7	1.2	0.281
clavicle-humeral index (L)	4	47	2.8	4	44.5	2.4	0.225
Brachial index (R)	10	74.8	1.9	9	76.9	1.5	0.017*
Brachial index (L)	5	77.4	3.4	8	75.9	3.1	0.440

R: right; L: left; * statistical significance

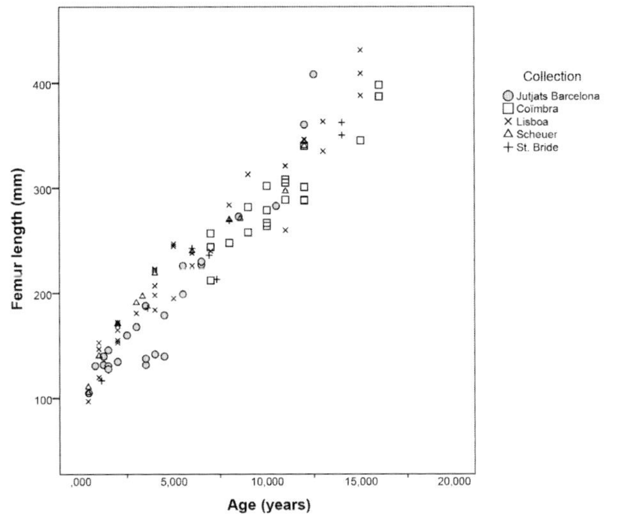

Figure 5.4. Growth graph of non-adult individuals, based on the length of the femoral shaft and comparing them with the collections from Coimbra, Lisbon, Scheuer and St Bride.

and Scheuer collections, all dating from the 18th to 20th century (Rissech *et al.* 2008). These reference collections were chosen as a good population reference for the non-adult individuals buried in the Antics Jutjats Municipals cemetery because they are well-documented and contextualised collections. There is both social and biological information, including sex, chronological age, biological origin, pathologies suffered, the work that the parents did, family relationships, and the social and economic context in which the individuals lived. These data can be found in the original works that contextualised them (Rocha 1995; Cardoso and Garcia 2006). Studies carried out on these documented collections indicate that the individuals belong to a lower-middle social class, whose conditions were not very favourable for child or youth growth (Cardoso and Garcia 2009).

The skeletal element chosen for comparative analysis was the femur, because it is one of the elements most sensitive to environmental and health conditions and does not present significant sexual differences until after the pubertal growth spurt, allowing a single sexual series up to the age of approximately 16 (Rissech *et al.* 2008).

Figure 5.4 shows that the values of the femur diaphysis of the individuals from the Antics Jutjats Municipals cemetery, in general, are well integrated within the values of the other documented series, indicating a similar growth between the individuals from the Antics Jutjats Municipals site and these populations. These similarities in growth between individuals from the Antics Jutjats Municipals and the documented collections associated with harsh growth conditions for subadult individuals from Antics Jutjats Municipals indicate a similarity in their life history.

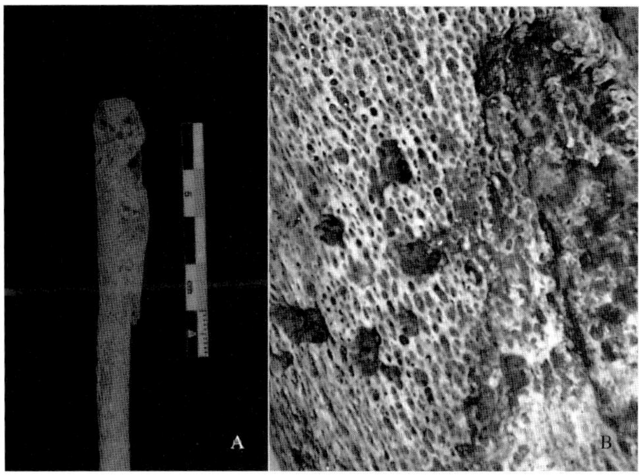

Figure 5.5. A. subperiosteal new bone formation on the right fibula, individual UF118. B. detail of the subperiosteal new bone formation.

Pathological conditions

The pathological alterations observed in these individuals were varied. Metabolic pathologies and nutritional deficiencies were observed, particularly *cribra orbitalia*, *cribra humeralis* and *cribra femoralis*. In 44 adult individuals only two cases of *cribra orbitalia* were found. In contrast, 20 non-adult individuals (28% non-adults 20/71) manifested certain types of *cribra*, such as *cribra orbitalia* (8 individuals), *cribra femoralis* (8 individuals) and *cribra humeralis* (1 individual) or a combination of *cribra orbitalia* and *femoralis* (2 individuals) and *cribra femoralis* and *humeralis* (1 individual). Nevertheless, in both adults and non-adult individuals from this site, subperiosteal new bone formation (Fig. 5.5) was also observed (in 10 adults and 16 non-adults). This new bone formation can be related to scurvy, rickets or osteitis. All these pathologies indicate biological stress, which, if the previous observations on the adults are taken into account, indicates that this biological stress could have occurred among this population relatively recently. The subperiosteal new bone formation is significant because the presence of active lesions is associated with lower survival rates for the individuals affected (DeWitte 2014).

In addition, four possible cases of intracranial tumours were recorded. (UF7, UF55, UF63 and UF121). These are manifested as macroporosities in the endocranium, affecting the spongy part of the bone without extending to the exocranium.

As for joint pathologies, in the sample analysed osteoarthritis is infrequent (Merino and Rissech 2022). Vertebral osteoarthritis is mainly incipient in 21 adult individuals and remarkable in 3 (UF 12, 16 and 70), showing a maximum degree of 3, following the classification of Luna *et al.* (2017). In addition, 5 individuals displayed Schmörl's nodes (UF89, 96, 19, 52 and 55). However, one case of carpometacarpal ankylosis affecting the second to fifth metacarpal bones (UF2) is highlighted, diagnosed as a probable case of Still's disease.

Traumatic pathologies were observed, particularly in the adult population, more precisely in 28 adult individuals (Fig. 5.6 A), although there were also three non-adults with traumatic injuries. In the 28 adult individuals, both consolidated and non-consolidated fractures were recorded, with rib fractures being the most common. The traumatisms exhibited in these adults were 18 skulls, 27 ribs, 1 patella, 5 radii, 5 ulnae, 1 humerus, 3 vertebrae, 1 clavicula, 1 first metacarpal, 1 third metacarpal, 1 fourth metacarpal, 3 fifth metacarpals, 1 femur and 1 calcaneus. It should be noted that, of the 31 individuals (28 adults plus 3 non-adults) with trauma, 18 of them have multiple traumas (all of them adults). As a result, 27% (31/115) of the total population have some type of trauma and 16% (18/115) have multiple traumatisms. It is relevant that of the 31 individuals with trauma, 19 had cranial trauma of penetrating stab wounds, including one of the three non-adult individuals. Of the three non-adult individuals (UF37, 5 to 6 years old, UF110, 3 to 4 years old and UF38, 5 to 6 years old), they show consolidated fractures on the eleventh right and second left ribs, and a perimortem penetrating stab wound in the occipital, respectively. The very young age of these non-adults is noteworthy. These types of traumatisms in children are rare, and they generally indicate violent interactions.

Congenital malformations were also present. These were found in four non-adult individuals (UF71, UF77, UF101 and UF107) and one adult (UF2). These malformations were observed in both the cranial and postcranial skeleton, corresponding to the non-fusion of the anterior arch of the atlas (UF71), fusion between the first two (first and second) ribs (UF77), an ossification defect in the pars basilaris (UF101), a defect in the posterior arch of the atlas (UF107), and a bifid distal phalanx in the third digit of the right hand (UF2).

Occupational stress markers were underrepresented in the population observed, indicating low levels of physical activity among these individuals (see Table 21 in Merino and Rissech 2022). These findings are consistent with the results of anthropometric analysis, which also suggested a low level of physical activity. Differences were observed in the development of enthesis between men and women, indicating general differences in muscle development, revealing men's greater development. However, the development of enthesis in these individuals is probably related to muscle development to compensate for reduced mobility caused by severe trauma.

As for oral pathologies (Fig. 5.6 B), the adult subjects (1008 teeth recovered) presented 69% dental calculus, 6% caries, 2.3% fistulae, 24.4% alveolar setback and 29.4% hypoplasia (all types). In 930 (deciduous) teeth of non-adult individuals recovered, 7.1% displayed calculus, 0.002% caries and 7% hypoplasia. Although these are not significant in the general population, some individuals exhibited a high prevalence of these conditions, such as the adult individual in UF2. Morphological variations in the oral region were also observed, including maxillary exostosis (UF29) and mandibular exostosis (UF12), diastema due to agenesis (UF21), dental malposition (UF12, UF13 and UF96), non-alimentary tooth wear (UF5, UF21, UF28, UF55, UF57, UF62 and UF105), and a bifid root first left incisive (UF121). Taking into account the influence of diet in the presence of caries and calculus (Barroso-Barbosa *et al.* 2007), the set of oral pathologies observed in this sample, when compared to their contemporaries (see table 27 in Merino and Rissech 2022), indicates that this population, apart from poor oral hygiene, could have had a low carbohydrate (low prevalence of caries) and animal protein (low prevalence of calculus) intake (Merino and Rissech 2022). This low prevalence of calculus and caries might also be due to a certain consumption of elements with a detergent power, such as apples, or with some bacterial inhibition power, such as *Cyperus rotundus* tubers, which could contribute to the elimination of a certain degree of plaque before calculus forms and the inhibition of *Streptococcus mutans* favouring the low level of caries (Buckley *et al.* 2014).

The late 10th and early 11th centuries are known to have been a period affected by periods of famine in Barcelona (Benito i Monclús 2022). In the Early Middle Ages, access to fruit was more widely seen among the wealthy, as the poorest classes found it difficult to obtain (Glick *et al.* 2005). In the population analysed, there seems to be evidence for vitamin C deficiency (lack of fresh fruit and vegetables) mainly among growing individuals. In addition, non-adults exhibited deficiencies that indicated growth difficulties, but adults did not. In adults, the frequency of hypoplasia was low, and their height was normal for the time in which they lived (see tables 18 and 27 in Merino and Rissech 2022). In addition, 18 adults and one child suffered skull injuries from penetrating stab wounds. From this early medieval period, there is little direct information on the population (Banks 1987) in the Iberian Peninsula and much less from Barcelona.

At this time Catalonia was a crossroads between Christian (in the north) and Muslim (in the south) worlds (Banks 1980; Tischler 2019). It was a society in transition from the slave mode of production to the systems that would define the socio-economic organisation of the feudal world. In *c.* 717, Visigothic rule came to an end in this part of the Iberian Peninsula because of the Arabo-Berber conquest. Muslim rule continued in the area throughout most of the 8th century, with the city of Barcelona itself falling to Carolingian forces in the year 801. Thereafter, Catalonia became a frontier society facing the Muslim world that still ruled most of the Iberian Peninsula (Tischler 2019). In Barcelona, due to this situation, the 9th and 10th centuries correspond to a turbulent period in the city's history, with conflicts between the Franks and Muslims; eventually, the Marca Hispanica (Spanish March) was established, which led to Barcelona being besieged

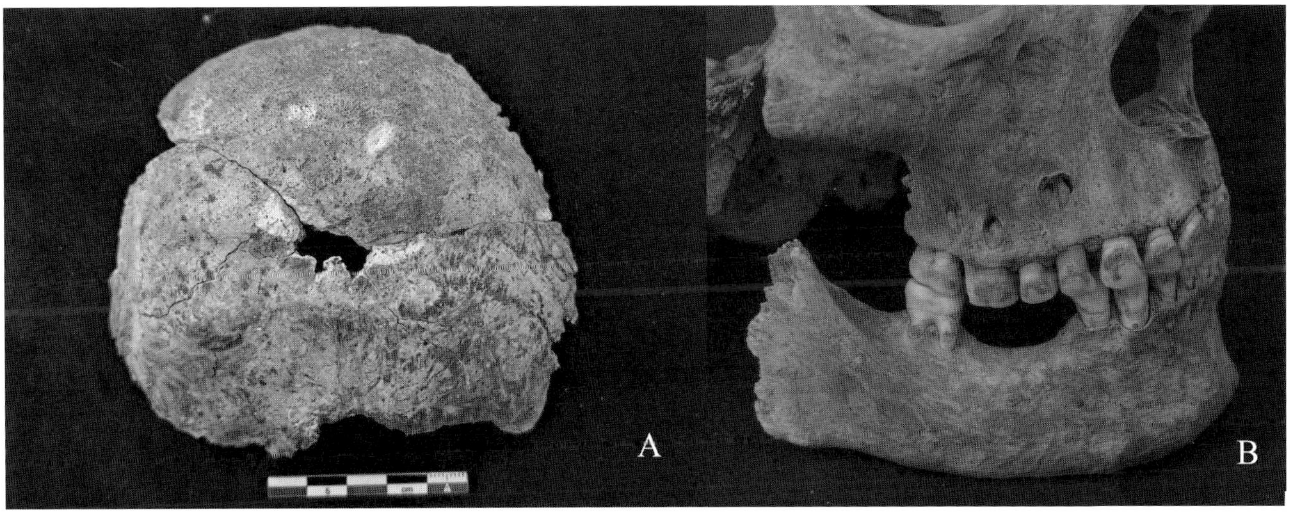

Figure 5.6. A. Occipital of an adult individual with peri-mortem *trauma, UF5; B. Individual from UF87, with numerous oral pathologies: 2 fistulas, loss of teeth and accumulation of calculus.*

several times. After the attack led by Louis the Pious in 801, the most important occasions numbered three: (i) in 826–27 during a rebellion of the nobility of the Spanish March against the Franks; (ii) during the period between 850 and 852, when the Muslims may have briefly occupied the city, possibly with the assistance of the Jewish population, and (iii) in 985 when the city was besieged and captured by Muslim forces led by al-Mansur (Chandler 2019). We believe that the individuals from the Antics Jutjats Municipals cemetery are inhabitants of Barcelona who were wounded by arrows and/or spears during these wars and sieges. Nineteen of these individuals have a cranial trauma of penetrating stab wounds, including a young child. They do not show any other stab wound in other parts of their bodies, which would indicate some type of interpersonal fighting. Although there are some authors that think that in this period internal violence was inherent in the city (Vinyoles i Vidal 1990; Sabaté i Curull 1999), from our point of view 19 from 115 individuals (17% of the population) is a high percentage to reflect only inherent urban violence.

Conclusions

To date, the archaeological site of the Antics Jutjats Municipals is unique in Barcelona. In addition to its organisation and the number of burials and type of funerary ritual, it represents a large and significant burial area in the region known as the eastern suburb (Beltrán de Heredia 2010), which appears not to have been associated with any religious centre. This characteristic distinguishes this cemetery from other burial spaces of the 9th and 10th centuries (Ripoll and Molist 2012). Starting in the 6th century, Christian burial practices were predominantly organised around urban religious buildings, basilicas or monasteries (Beltrán de Heredia 2010). In contrast, the Antics Jutjats Municipals cemetery appears to have been a planned burial area, located at a considerable distance from the Late Roman city wall.

The analysed adult individuals from the Antics Jutjats Municipals (n=44) exhibited a low prevalence of dental hypoplasia (29.4%, 13/44) and an average height (153.6 cm for women and 166.6 cm for men) comparable to their contemporaries. They had rounded long bone diaphyses and a low frequency of physical activity markers, indicating that they did not perform significant physical work. In fact, during the period when the cemetery was in use, the population of the city was gradually increasing (Banks 2003). This phenomenon could denote good access to certain foods, reflected in the sample by the low representation of dental hypoplasia, which is below the values found in other contemporary populations, as well as individual heights (Merino and Rissech 2022). Only two cases of *cribra orbitalia* were found among the 44 adult individuals. On the other hand, the non-adults (n=71) buried in this cemetery clearly faced some growth-related challenges similar to those recorded among pre-industrial individuals in Coimbra and Lisbon (Merino and Rissech 2022). In addition, these non-adults exhibited certain types of *cribra* (n = 20), such as *cribra orbitalia* (8 individuals), *cribra femoralis* (8 individuals) and *cribra humeralis* (1 individual), or a combination of *cribra orbitalia* and *femoralis* (2 individuals), and *cribra femoralis* and *humeralis* (1 individual) (Merino and Rissech 2022). However, both adults and non-adults at the Antics Jutjats site exhibited new subperiosteal bone formation (10 adults and 16 non-adults). This new bone formation may have been related to scurvy, rickets or osteitis. These pathologies indicate biological stress, which, considering previous observations in the adults, suggests that this biological stress may have occurred relatively recently among this population.

The population of the Antics Jutjats Municipals site stands out for the high number of consolidated and unconsolidated fractures observed in 30% (13/44) of adults, with more than 50% (8/13) of these individuals having more than two fractures affecting different skeletal areas. Generally, these traumatisms are complex and untreated. In addition, there are 19 traumatisms (18 adults and 1 child), in the form of penetrating stab wounds, located in the face and/or skull, and may, in some cases, be associated with the cause of death, because the lesions are perimortem and affect the inside of the head or destroy a large part of the face. Most surprising are the three young children, between 3 and 6 years of age, whose eleventh and second ribs are fractured, and who have a penetrating stab wound in the occipital bone/region, respectively. These types of lesions are extremely rare in children, and they usually indicate violent interactions (Berthold *et al.* 2018).

The high frequency and variety of types of fractures and the presence of 19 penetrating stab wounds, together with the formation of new subperiosteal bone formation and the possible cases of vitamin C and/or D deficiency, suggest some type of interpersonal violence. The fact that the cemetery of the Antics Jutjats Municipals was located at a considerable distance from the city clearly distinguishes this population from the rest of Barcelona. Most adult individuals from the Antics Jutjats Municipals seem to have had a good quality of life (they have a low frequency of hypoplasia and do not seem to have worked physically hard) when they were young; however, for some reason, at the time of their death, both they and their children faced much tougher conditions. The regular organisation of the cemetery excludes the possibility of people buried there suddenly dying from illness or a sudden armed attack. However, we believe that they could be related to one of the several sieges that the city of Barcelona experienced during the 9th and 10th centuries.

Acknowledgements

The authors thank Dr Philip Banks for his comments and linguistic corrections and the Museu de Història de Barcelona (MUHBA) and Emili Revilla for allowing this study and for giving us access to the facilities of this museum to carry it out. The authors also want to thank the Servei d' Arqueologia de Catalunya and the General Directorate of Infrastructure Management of the Department of Justice of the Generalitat of Catalonia for making it possible to study both the different individuals and the necropolis as a whole.

SGR Social, cultural and biological evolution in the Pleistocene (StEP). Directorate General of Research of the Generalitat de Catalunya. Ref.: 2021 SGR 01239.

MONBONES. Ministry of Science and Innovation (MICINN). Ref: PID2020-118194RJ-I00.

References

Alduc-Le Bagouse, A. (1988) Estimation de l'âge des non-adultes: maturation dentaire et croissance osseuse. Donnés comparatives pour deux nécropoles médiévales bas-normandes. In *Actes des 3èmes Journées Anthropologiques. Notes et Monographies Techniques* 24, 81–103. Paris, Éditions du CNRS.

Alemán, I., Botella, M.C. and Ruiz, L. (1997) Determinación del sexo en el esqueleto postcraneal. Estudio de una población mediterránea actual. *Archivo Español de Morfología* 2, 7–17.

Alesan, A. (1990) *Estudi d'una població subadulta de l'Edat del Ferro: demografia, antropometria i creixement.* Memòria del treball de recerca del Mestratge d'Especialització Professional en Biologia Humana, Universitat Autònoma de Barcelona.

Banks, P. (1980) Greeks in early medieval Barcelona? *Faventia* 2, 73–92.

Banks, P. (1987) Mensuration in early medieval Barcelona. *Medievalia* 7, 37–56.

Banks, P. (2003) El creixement físic de Barcelona, segles X-XIII. In A. Cubeles and R. Grau (eds) *El procés urbà i la identitat gòtica de Barcelona (Barcelona Quaderns d'Història 8)*, 11–33. Barcelona, Ajuntament de Barcelona.

Barroso-Barbosa, J., Guinot-Jimeno, F., Barbero-Castelblanque, V. and Bellet-Dalmau, L. (2007) La importancia de la dieta en la prevención de la caries. *Gaceta dental* 181, 116–35.

Beltrán de Heredia, J. (2010) La cristianización del suburbium de Barcino. *Monografías de Arqueología Cordobesa* 18, 363–95.

Benito i Monclús, P. (2022) Fams endèmiques, fams episòdiques. Crisis i conjuntures frumentàries als comtats de Barcelona, Osona, Manresa i Girona (990–1090). *Recerques: història, economia, cultura* 81, 5–36.

Berthold, B., Frericks, B., John, T., Clemens, V., Fegert, J.M., and von Moers, A. (2018) Abuse as a Cause of Childhood Fractures. *Deutsches Ärzteblatt International* 115, 769–75.

Black, T.K. (1978) A new method for assessing the sex of fragmentary skeletal remains: femoral shaft circumference. *American Journal of Physical Anthropology* 48(2), 227–31.

Bonnassie, P. (1975–1976) *La Catalogne du milieu du Xe à la fin du XIe siècle: croissance et mutations d'une société.* Toulouse, Publications de l'Université de Toulouse-le-Mirail.

Brickley, M., and McKinley, J.I. (2004). *Guidelines to the Standards for Recording Human Remains.* Southampton and Reading, British Association for Biological Anthropology and Osteoarchaeology. Institute of Field Archaeologists.

Brothwell, D.R. (1987) *Desenterrando huesos.* Mexico, Fondo de Cultura Económica.

Buchet, L. (1983) L'inhumation en basilique funéraire: observations, interprétations et commentaires. In L. Buchet (ed.) *Le matériel anthropologique provenant des édifices religieux*, 69–73. Paris, CNRS.

Buckley, S., Usai, D., Jakob, T., Radini, A., and Hardy, K. (2014) Dental calculus reveals unique insights into food items, cooking and plant processing in prehistoric Central Sudan. *PLoS One* 9(7), e100808.

Cameron, N. (2023) Child growth and armed conflict. *Annals of Human Biology* 50, 301–7.

Capasso, L. (1998) *Atlas of Occupational Markers on Human Remains.* Teramo (Italy), Ed. S.P.A.

Cardoso, H. and Garcia, S. (2009) The Not-so-Dark ages: ecology for human growth in medieval and early twentieth century Portugal as inferred from skeletal growth profiles. *American Journal of Physical Anthropology* 138, 136–47.

Cavalli-Sforza, L.L. and Bodmer, W.F. (1971) *The Genetics of Human Populations*. San Francisco, Freeman.

Chandler, C.J. (2019) *Carolingian Catalonia. Politics, Culture and Identity in an Imperial Province, 778-987*. Cambridge, Cambridge University Press.

Chevalier, T. (2013) La diaphyse fémorale AL 333-61: preuve d'une bipédie comme unique mode locomoteur il y a 3,2 Ma en Ethiopie?. *Bulletins et mémoires de la Société d'Anthropologie du Paris* 25, 169–89.

Cobo, S. (2022) *Estudio dental de los individuos del hipogeo Neolítico Final-Calcolítico dels Carrer París de Cerdanyola del Vallès (Barcelona)*. Barcelona, Universitat de Barcelona.

Corruccini, R.S., Handler, J.S. and Jacobi, K.P. (1985) Chronological distribution of enamel hypoplasias and weaning in a Caribbean slave population. *Human Biology* 57, 699–711.

Crétot, M. (1978) *L'arcade dentaire humaine (Morphologie)*. Paris, Julien Prélat Ed.

DeWitte, S.N. (2014) Differential survival among individuals with active and healed periosteal new bone formation. *International Journal of Paleopathology* 7, 38–44.

Ferembach, D., Schwidetzky, I. and Stloukal, M. (1980) Recommendations for age and sex diagnoses of skeletons. *Journal of Human Evolution* 9, 517–49.

Fernández, M.E. (2008) *Marcadores óseos de actividad ocupacional en las poblaciones históricas de Palat de Rey (León) y de Gormaz (Soria)*. PhD thesis, Universidad de León.

Fernández de Prado, M. I. (1978) *Antropología de restos humanos medievales en el País Vasco*. 194. Unpublished.

Glick, T., Livesey, S.J. and Wallis, F. (2005) *Medieval Science, Technology, and Medicine: an Encyclopedia*. New York, Routledge.

Guglielmino-Matessi, C.R., Gluman, P. and Cavalli-Sforza, L.L. (1979) Climate and evolution of skull metrics in man. *American Journal of Physical Anthropology* 50, 549–64.

Hillson, S. (2001) Recording dental caries in archaeological human remains. *International Journal of Osteoarchaeology* 11, 249–89.

Humphrey, L.T. (2010) Weaning behaviour in human evolution. *Seminars in Cell & Developmental Biology* 21, 453–61.

Jordana, X. (2007) *Caracterització i evolució d'una comunitat medieval catalana. Estudi bioantropològic de les inhumacions de les esglésies de Sant Pere*. PhD thesis, Universitat Autònoma de Barcelona.

Jordana, X., Alesan, A., Alfonso, J., Safont, S. and Malgosa, A. (2003) *Els enterraments de la Plaça Vila de Madrid. Informe Antropològic de la intervenció arqueològica a la Plaça Vila de Madrid (Barcelona) durant la campanya 2001–2002*. Unpublished.

Jordana, X. and Malgosa, A. (2002) Terrassa, una Villa Medieval en transición a la época moderna. Estudio bioantropológico de la necrópolis de la Plaça Vella. *Revista Española de Antropología Biológica* 23, 1–25.

Kent, J.C. (2007) How breastfeeding works. *Journal of Midwifery & Women's Health* 52, 564–70.

Krenz-Niedbala, M. (2001) Biological and cultural consequences of the transition to agriculture in human populations on Polish territories. *Variability and Evolution* 9, 89–99.

Krogman, W.M. and Işcan, Y.M. (1986) *The Human Skeleton in Forensic Medicine*. Springfield, Illinois, Charles C. Thomas Publisher.

Lewis, M. (2018) *Paleopathology of Children. Identification of Pathological Conditions in the Human Skeletal Remains of Non-Adults*. London, Academic Press.

Luna, L., Aranda, C., and Amorim-Alves, A. (2017) Reflexiones sobre el relevamiento y análisis comparativo de patologías osteoarticulares en restos esqueletales humanos. *Revista Argentina de Antropología Biológica* 19, 1–8.

Martin, R. and Saller, K. (1957) *Lehrbuch der Anthropologie*. Stuttgart, Ed. G. Fischer.

Merino, M. and Rissech, C. (2022) *Intervenció Arqueològica Passeig Lluís Companys, 1-5. Antics Jutjats Municipals de Barcelona. Districte de Ciutat Vella*. Codi 069/18. Unpublished report, Servei d'Arqueologia de la Generalitat de Catalunya.

Milner, G.R., Humpf, D.A., Harpending, H.C. (1989) Pattern matching of age at death distributions in paleodemographic analysis. *American Journal of Physical Anthropology* 80, 49–58.

Olivier, G. (1960) *Pratique Anthropologique*. Paris, Vigot Frères Eds.

Oms Llohis, J.I. (2016) *Testimonio de los dientes. Antropología dentaria de la transición neolítica hasta el presente*. Barcelona, Edicions Bellaterra.

O.N.U. (1984) *Manuel X. Techniques indirectes d'estimation démographique*. (*Études démographiques, 81*). New York, Nations Unies.

Ortner, D.J. (2003) *Identification of Pathological Conditions in Human Skeletal Remains*. 2nd ed. London, Academic Press.

Pearson, K. (1899) Mathematical contributions to the Theory of Evolution. V. On the reconstruction of the stature of prehistoric races. *Philosophical Transactions of the Royal Society* 192, 169–244.

Pujol-Bayona, A., Rissech, C., Prida, D. and Turbón, D. (2011) Estudio bioantropológico del conjunto funerario de época tardorromana Carrer Ample 1 de Barcelona. *Revista Española de Antropología Física* 32, 603–10.

Ripoll, G. and Molist, N. (2012) Arqueología funerària a Catalunya de l'Antiguitat tardana al món medieval. *Monografies d'Olèrdola* 3.1, 17–32. Barcelona, Museu d'Arqueologia de Catalunya.

Rissech, C. and Black, S. (2007) Scapular development from neonatal period to skeletal maturity. A preliminary study. *International Journal of Osteoarchaeology* 17, 451–64.

Rissech, C., Estabrook, E.F., Cunha, E. and Malgosa, A. (2006) Using the acetabulum to estimate age at death in adult males. *Journal of Forensic Sciences* 51, 213–29.

Rissech, C., Garcia, M. and Malgosa, A. (2003) Sex and age diagnosis by ischium morphometric analysis. *Forensic Science International* 153(3), 188–96.

Rissech, C. and Malgosa, A. (1991) Importancia relativa de la longitud del pubis y la anchura del ilion en el estudio del dimorfismo sexual de los coxales. *Boletín de la Sociedad Española de Antropología Biológica* 12, 29–43.

Rissech, C. and Malgosa, A. (1997) Sex prediction by discriminant function with central portion measures of innominate bones. *Homo* 48, 22–32.

Rissech, C. and Malgosa, A. (2005). Ilium growth study: applicability in sexual and age diagnostic. *Forensic Science International* 147, 165–74.

Rissech, C. and Malgosa, A. (2007) Pubis growth study: applicability in sexual and age diagnostic. *Forensic Science International* 173, 137–45.

Rissech, C., Sañudo, J.R. and Malgosa, A. (2001) The acetabular point: a morphological and ontogenetic study. *Journal of Anatomy* 198(6), 743–48.

Rissech, C., Schaefer, M. and Malgosa, A. (2008) Development of the femur – Implications for age and sex determination *Forensic Science International* 180, 1–9.

Rocha, M.A. (1995) Les collections ostéologiques humaines identifiées du Musée Anthropologique de l'Université de Coimbra. *Antropologia Portuguesa* 13, 17–38.

Ruff, C.B. (1987) Sexual dimorphism in human lower limb bone structure: relationship to subsistence strategy and sexual division of labor. *Journal of Human Evolution* 16, 396–416.

Sabaté i Curull, F. (1999) La violencia en la cotidianidad bajomedieval catalana. *Aragón en la Edad Média* 14–15, 1389–408.

Safont, S., Malgosa, A. and Subirà, E. (2000) Sex assessment on the basis of long bone circumference. *American Journal of Physical Anthropology* 113, 317–28.

Samaras, T.T. and Elrick, H. (2002). Height, body size, and longevity: is smaller better for the human body. *The Western Journal of Medicine* 176, 206–8.

Serra-Molinos, J. (2023) *Intervenció arqueològica al Passeig Lluís Companys, 1-5. Antics jutjats municipals de Barcelona. Districte de Ciutat Vella. Codi 069/18.* Ajuntament de Barcelona. SABCD-Q100-Q104-Q158-Q160-069/18 (h.1).

Stloukal, M. and Hanáková, H. (1978) Die Länge der Längsknochen altslawischer Belvökerungen, unter besonderer Berücksichtigung von Wachstumsfragen. *Homo* 29, 53–69.

Souich, PH. (1979) Estudio anthropológico de la necrópolis medieval de La Torrecilla (Arenas del Rey, Granada). *Antropología y Paleoecología humana* 1, 27–33.

Tischler, M.M. (2019) Using the Carolingian past in a society of transformation: the case of early medieval Septimania/ Catalonia in the long tenth century (900–1050). *Medieval Worlds* 10, 72–86.

Trinkaus, E., Ruff, C.B., Churchill, S.E. and Vandermeersch B. (1998) Locomotion and body proportions of the Saint-Césaire 1 Châtelperronian Neandertal. *Proceedings of the National Academy of Sciences* USA 95, 5836–40.

Turbón, D. and Rissech, C. (2024) Biología del Desarrollo en las sociedades modernas. Impactos de los conflictos armados. In *Retos Vitales*, Chap. VIII, *Desafíos y realidades en la investigación moderna*, 1472–85.

Ubelaker, D.H. (1989) *Human Skeletal Remains. Excavation, Analysis, Interpretation.* 2nd ed. Washington, Taraxacum.

Vinyoles i Vidal, T.M. (1990) La violència marginal a les ciutats medievals (exemples a la Barcelona dels volts del 1400). *Revista d'Història Medieval* (Universitat de València) 1, 155–77.

Weaver, D.S. (1979) Application of the likelihood ratio test to age estimation using the infant and child temporal bone. *American Journal of Physical Anthropology* 50, 263–70.

6

Bioanthropological study of human remains from funerary unit UF228 (14th to 15th centuries) at the archaeological site of the friary of Santa Caterina (Barcelona)

Carme Rissech, Marina Gascà and Anna Llauradó

Santa Caterina was one of the most important friaries in Barcelona (Spain) during the medieval and post-medieval periods. Founded in 1223, Santa Caterina belonged to the Order of Preachers (Dominicans). The main goal of this study is to provide information about the individuals buried in this convent, particularly those interred in Funerary Unit UF228, datable to the 14th and 15th centuries. Funerary Unit UF228 was a rectangular collective tomb (2.25 m by 0.82 m), outside the nave of the church and located between two buttresses against the external wall of the apse. According to the archaeological information, the individuals were buried successively and occupied the entire space; the adults were placed in the centre and the immature individuals at the sides of the tomb or in the spaces not occupied by adults. Standard bioanthropological methods were applied to carry out the study. Results indicate a total of 18 individuals, of whom 5 are adults (4 men and 1 woman) and 13 immature individuals (2 babies, 5 infants, 4 children and 2 juveniles). Adult individuals exhibited hypoplasia, caries and a mean stature of 165 cm in males and 150.3 cm in females. In addition, the oldest individual displayed lesions compatible with neoplasia. The sex and age profiles suggest they were a secular familial group. The biological profile and pathological conditions together with the location of this funerary unit in the friary indicate that they correspond to a moderately prosperous family, wealthy enough to be able to pay the corresponding fee to be buried in this area. According to previous studies and the available archaeological data, the friars and people closely related to the community were buried in the cloister, the people from the highest social classes were buried inside the church as close as possible to the high altar, moderately well-off people against the outer part of the church wall and less well-off people were buried in the external area, beyond the apse.

Keywords: Bioanthropology, palaeopathology, burials, medieval period, Santa Caterina friary.

Introduction

The friary of Santa Caterina in Barcelona (Spain) belonged to the Order of Preachers (Dominicans), who devoted their lives largely to preaching, teaching and catechism. The Order of Preachers was founded in Toulouse and four years later, in 1219, they arrived in Barcelona at the invitation of Bishop Berenguer de Palou (Ortoll 1996). The friary was located in what is now Avinguda Francesc Cambó, in the Ciutat Vella district of Barcelona (Fig. 6.1). During the medieval and post-medieval periods, the friary of Santa Caterina was one of the most important religious houses in Barcelona. It was a key centre of Dominican training and Catholic devotion in the city, the architecture and dimensions of the friary being similar to those of the Cathedral of Barcelona or the largest of the parish churches, Santa Maria del Mar (Huertas and Aguelo 2006). After the confiscation of Mendizábal, it was demolished in 1836 and replaced by the market of Santa Caterina. Between 1997 and 2004, the market was reformed, and it was during the course of this refurbishment that the archaeological artefacts and skeletal remains appeared. The archaeological excavations were carried out by the company CODEX- Arqueologia i Patrimoni, S.C.C.L between the years 1999 and 2003, under the direction of Josefa Huertas and Jordi Aguelo (Huertas and Aguelo 2006).

The main goal of this study is to provide information about the lifestyle, health and socio-economic status of the individuals buried in this friary, particularly those interred in Funerary Unit (UF) 228 (Fig. 6.1), datable to the 14th and 15th centuries. This will be done through the bioanthropological analysis of the skeletal remains found in this funerary unit, located against the external area of the nave wall of the church between two buttresses of the polygonal apse (Fig. 6.2).

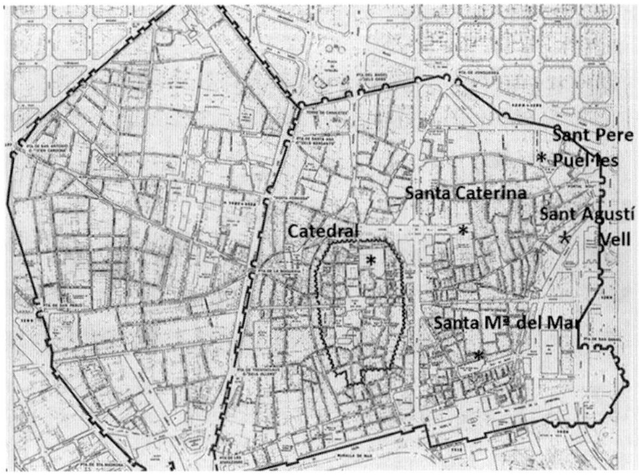

Figure 6.1. Plan of Barcelona with the Roman and medieval walls showing the location of the friary of Santa Caterina.

Material and methods

Material

The material analysed in the present study corresponds to the skeletal remains exhumed from UF228 (14th and 15th centuries) of the friary of Santa Caterina during the rescue excavations carried out between 1999 and 2003. This material is housed in the store in Zona Franca (Barcelona) at the Museum of the History of Barcelona (Collections Centre – MUHBA) and was transported and analysed in the Laboratory of the Unit of Human Anatomy in the Department of Basic Medical Sciences of the Faculty of Medicine and Health Sciences at the Rovira i Virigili University of Reus (Tarragona, Spain).

UF228 (or chapel H) is a rectangular collective tomb (2.25 m × 0.82 m) outside the nave of the church, located between two external buttresses of the apse (Fig. 6.2). According to the archaeological information, the bodies were oriented east–west, with the head to the west. They were buried successively and occupying the entire space,

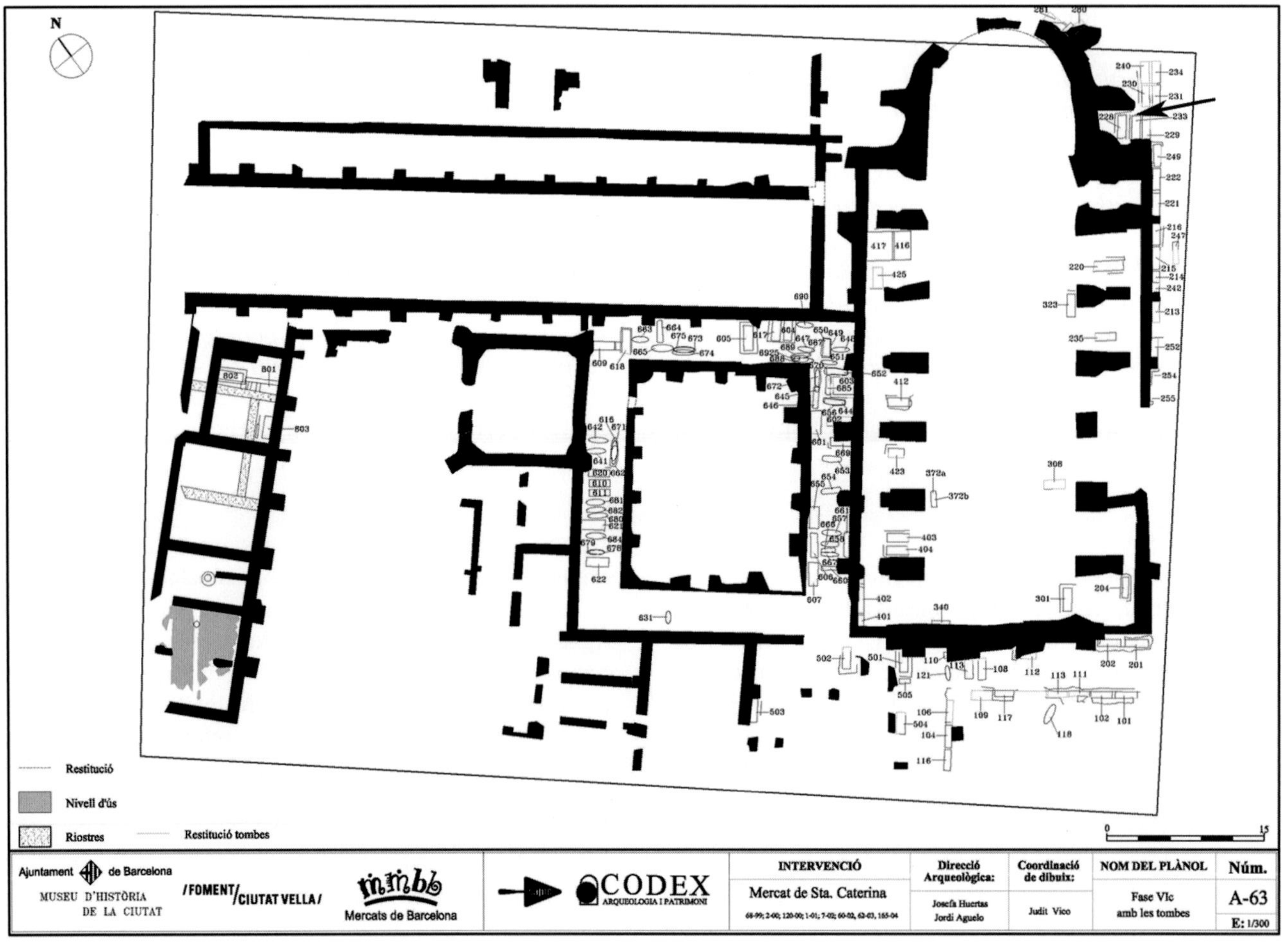

Figure 6.2. Santa Caterina friary in Barcelona site plan, indicating the location of UF 228 (arrow). Image modified from Huertas and Aguelo (2006).

with the adults being placed in the centre and the immature individuals at the sides or in the spaces not occupied by adults (Huertas and Aguelo 2006).

Methods

The bioanthropological analysis of this material was carried out at the human bone laboratory of the Human Anatomy Unit in the Departament de Ciències Mèdiques Bàsiques de la Facultat de Medicina i Ciències de la Salut de la Universitat Rovira i Virigili de Reus (Tarragona, Spain). The material arrived in boxes containing individualised skeletons. However, if isolated skeletal elements that could not be attributed to the individual buried were found, they were grouped (considering the area of origin) according to anatomical categories and by matching them by size, degree of bone maturation, sex, colour, shape and anomalies (Brothwell 1987). We also took into account joint congruence, pathological group and the symmetry of paired bones (Brothwell 1987).

The skeletal material was cleaned and reconstructed, when possible, with adhesive tape, which was removed after study. For each individual, this bioanthropological study included: the skeletal description, the degree of preservation, the estimation of age and sex, and the metrical analysis of the skeletal elements based on standard measurements. The different procedures will be described below.

To establish the skeletal preservation of each individual, the three preservation indices (IP1, IP2, IP3) of Walker *et al.* (1988), modified by Alesan (1990), were calculated. They indicate the percentage of bones preserved in relation to the total number of bones present in the skeleton, allowing us to know the quality and quantity of information that can be obtained from the sample. The first of these indices (IP_1) is calculated in relation to the long bones of the upper and lower limbs [IP_1= (Preserved long bones of the upper and lower limbs *100) / 12)]. The second of these indices (IP_2) is calculated in relation to the previous bones plus the bones of the pelvic and scapular girdles [IP_2= (Previous long bones plus coxal bones, scapulae, clavicles and sacrum preserved *100) / 19)]. The third of these indices considers the previous bones plus the cranium and mandible [IP_3= (Previous bones plus cranium, maxilla and mandible preserved *100) / 22)].

Sex estimation is the first step in any human osteological analysis study whether forensic or archaeological. To estimate sex, in immature individuals we utilised the Schutkowski (1993) method based on mandibular and ilium morphology. In adults, we considered the morphological characteristics of the cranium, mandible and particularly those of the os coxae (Ferembach *et al.* 1980; Rissech and Malgosa 1991; 1997; 2007). In addition, we also used the sacrum (Krogman and İşcan 1986), the metrical characteristics of the long bones (Olivier 1960; Alemán *et al.* 1997) and classic observations on the size, robusticity and muscular insertions of postcranial bones (Martin and Saller 1957; Olivier 1960).

To estimate the age at death, all the skeletal elements useful for this diagnosis were used, prioritising the most reliable criteria according to the characteristics of the developmental stage of the individual. We consider as immature those individuals who are growing. The characteristics used to determine an individual as adult are the closure of the spheno-occipital synchondroses, the presence of the third molar, and the closure of the epiphyses of the main long bones.

For the age estimation of immature individuals, the recommendations of Ferembach *et al.* (1980) and the criteria described by Krogman and İşcan (1986), Scheuer and Black (2000) and Rissech (2016) were followed. Specifically, in children up to 12 years of age, the criterion of dental development was prioritised (Crétot 1978; Ubelaker 2007). We also considered the maximum diaphyseal length of the long bones (Rissech *et al.* 2013a), the growth of the coxal bone and scapula (Stloukal and Hanáková 1978; Alduc-Le Bagouse 1988; Rissech and Malgosa 1997; Rissech and Black 2007; Rissech *et al.* 2013b) and temporal tympanic plate development (Weaver 1979).

As regards the skeletons in the juvenile stage (aged over 12), the degree of ossification and epiphyseal fusion of the bones (Brothwell 1987; Ferembach *et al.* 1980; Krogman and İşcan 1986; Rissech *et al.* 2013a; 2013b) were mainly considered, but also the maximum diaphyseal length of the long bones, the growth of the coxal bone and the scapula, as previously cited (Rissech 2016).

In mature individuals the morphological changes to the auricular surface of the ilium and those of the acetabulum were prioritised, as described by Lovejoy *et al.* (1985) and Rissech *et al.* (2006; 2007) respectively. Additionally, we also used the morphological changes in the rib sternal end (İşcan and Loth 1984; Krogman and İşcan 1986), dental attrition (Brothwell 1987) and the degree of cranial suture fusion (Masset 1982).

For the anthropometrical analysis, we used the standard measurements (Martin and Saller 1957; Olivier 1960; Fazekas and Kósa 1985; Krogman and İşcan 1986; Rissech *et al.* 2013a; 2013b) used in biological anthropology. Due to the poor preservation of the samples, both the right and left sides were considered, depending on the individual. In the rare event that both values existed (we say 'rare' because of the poor preservation), an average was calculated. From these measurements we calculated skeletal indices to describe the shape and size of the bones. Bearing in mind the period in which the individuals analysed lived, their stature was estimated by the Pearson formulae (Pearson 1899), which are based on a 19th-century French population. This method was the best to apply to these individuals because the reference sample on which the method is based is biologically closer to the individuals analysed than other methods in terms of time and space.

During the analysis, the non-metrical traits of the cranium (Hauser and DeStefano 1989) and post-cranium (Brothwell 1987) were documented. The non-metrical traits have a genetic component that makes them both important and useful for analysing the familial and populational relationships of the sample. In addition, using the presence/absence criteria, we collected data on taphonomic alterations (Lloveras *et al.* 2016; Pokines *et al.* 2021) and occupational stress markers (Capasso *et al.* 1999; Fernández 2008). The occupational stress markers are useful as a means to gain an approximate idea of the individual's lifestyle.

Palaeopathological analysis was undertaken by macroscopic observation of the skeletal sample following the indications of Ortner (2003). The most frequent pathological conditions observed in past populations are congenital, inflammatory, traumatic, tumoral and degenerative lesions. Lesions caused by nutritional deficiencies and infections are also very frequent (Aufderheie and Rodríguez-Martin 1998; Campillo 2001; Lewis 2018). For trauma analysis, the fracture pattern of bones was taken into account following the instructions of Spitz and Fisher (1993), Etxeberria *et al.* (1993) and Etxeberria (2003).

For oral pathological conditions, the lesions analysed were the following: caries, alveolar recession, fistulas, maxillary and mandibular exostoses, diastemas, periodontitis, wear and tear. Those alterations that may be the result of an oral pathology with antemortem tooth loss and postmortem tooth loss were also considered. Finally, environmental stress indicators such as hypoplasia, the presence or absence of dental calculus and dental wear were also evaluated. In order to facilitate the collection of data related to dental parts, the dental file developed by Cobo (2022) was used.

Results

Results indicate a total of 18 individuals: 5 adults and 13 immature individuals (Table 6.1). The adults comprised 4 men and 1 woman. Three of these men were 16–25, 33–45 and 40–44 years old, and the age of the fourth man could not be estimated. The woman is also of indeterminate age (Table 6.1).

As regards the immature individuals, they comprise: two babies (one of them aged 2 months and the other between

Table 6.1. Summary of the individuals found in UF 228 of the Santa Caterina archaeological site in Barcelona

ID	*Sex*	*Age*	*Stature*	*IP_3*	*Pathology*	*Oral pathology*
UE 2070						
Immature	Indet	0.5–2 yr	–	1%	–	–
UE 2071					–	–
Immature	Indet	0.57–1 yr	–	1%	–	–
Immature	Indet	~ 5 yr	–	1%	–	–
Immature	Indet	10–11 yr	–	1%	–	–
Adult	Male	33–45 yr	–	77.27%	Neoplasia?	Hypoplasia, periodontitis, attrition
Adult	Male	Indet.	–	1%	–	
UE 2072						
Immature	Indet	10–11 yr	–	1%	–	Hypoplasia
Adult	Male	16–25 yr	165 cm	55%	–	Hypoplasia, caries
Adult	Female	Indet.	152.1 cm	18%	–	–
UE2073						
Immature	Indet	1–3 mnth	–	14%	–	–
Immature	Indet	2 mnth	–	14%	–	–
Immature	Indet	3 yr	–	50%	–	Hypoplasia
Immature	Indet	7–9 yr	–	1%	–	–
Immature	Indet	7–9 yr	–	1%	–	–
UE2074						
Immature	Female	~ 13 yr	148.5 cm	55%	–	Hypoplasia
UE2076					–	–
Immature	Indet	3–6 yr	–	36%	–	–
UE2077					–	–
Adult	Male	40–44 yr	165 cm	46%	–	–
Immature	Indet	8–14 yr	–	1%		–

ID: individual; ♂: masculine; ♀: feminine

1 and 3 months), 5 infants (aged between 6 months and 1 year, between 6 months and 2 years, 3 years, 5 years and between 3 and 6 years), 4 children (two aged between 7 and 9 years and two aged between 10 and 11) and 2 juveniles (1 female aged about 13 and the second, whose sex could not be determined, aged between 8 and 14) (Table 6.1). The immature individuals are the most abundant in the sample, corresponding to 72% (13/18) of the total sample. The sex and age distribution of the sample leads us to consider that UF 228 corresponds to a family group burial.

In general, the individuals from UF 228 are poorly preserved (Table 6.1). Some of them are represented by only one or two skeletal elements (IP3 of 1%), mainly immature individuals (Table 6.1). This means many measurements could not be taken on them and part of the information has been lost. Figure 6.3 shows three of the best-preserved skeletons (UE 2071, a male aged between 33 and 45 years; UE 2073, aged around 3 years; UE 2074, a female aged around 13 years). Tables 6.2 to 6.4 contain the absolute values and indices from the cranial and postcranial skeleton; Table 6.2 shows the cranial skeleton, and Tables 6.3 and 6.4 present the upper limb and lower limb respectively. Results indicate that the crania display the typical Mediterranean morphology (Table 6.3). This means long crania and long faces. Particularly, the male UE2071, aged between 33 and 45 (Figure 6.3), displays hyperdolichocrany (cephalic index between 65 and 69.9), medium skull height (metrocranium), with a vertico-transversal index between 92 and

Table 6.2. Absolute values and cranial indices of the individuals found in UF 228 of the archaeological site of Santa Caterina in Barcelona

Id	*UE2071*	*UE2072*	*UE2072*	*UE2073*	*UE2073*	*UE2073*	*UE 2074*	*UE 2076*	*UE 2077*
Sex	♂	♂	♀	Indet.	Indet.	Indet.	♀	Indet.	♂
Age	33–45 yr	16–25 yr	Indet.	2 mnth	~ 3 yr	7–9 yr	13 yr	3–6 yr	40 yr
Cranial max. breadth	131.0	–	–	–	–	–	–	–	–
Cranial max. length	190.0	–	–	–	–	–	–	–	–
Cranial vault height	124.0	–	–	–	–	–	–	–	–
Max. frontal breadth	109.0	–	–	–	–	–	–	–	–
Min. frontal breadth	91.0	–	–	–	–	–	–	–	–
Frontal sagittal chord	115.0	–	–	–	–	–	–	–	–
Parietal sagittal chord	122.0	–	–	–	–	–	–	–	–
Occipital sagittal chord	65.0	–	–	–	–	–	–	–	–
Total sagittal chord	173.0	–	–	–	–	–	–	–	–
Occipital squama sagittal chord	50.0	–	–	–	–	–	–	–	–
Frontal sagittal arc	133.0	–	–	–	–	–	–	–	–
Parietal sagittal arc	140.0	–	–	–	–	–	–	–	–
Occipital sagittal arc	70.0	–	–	–	–	–	–	–	–
Total sagittal arc	393.0	–	–	–	–	–	–	–	–
Occipital squama sagittal arc	50.0	–	–	–	–	–	–	–	–
Foramen magnum breadth	24.0	–	–	–	–	–	–	–	–
Foramen magnum height	38.0	–	–	–	–	–	–	–	–
Facial breadth	123.0	–	–	–	–	–	–	–	–
Facial height	122.0	–	–	–	–	–	–	–	–
Upper facial height	73.0	–	–	–	–	–	–	–	–
Bigoniac breadth	98.0	–	–	–	–	–	–	–	–
Orbital breadth	42.0	–	–	–	–	–	–	–	–
Orbital height	35.5	–	–	–	–	–	–	–	–
Nasal breadth	27.0	–	–	–	–	–	–	–	–
Nasal height	53.0	–	–	–	–	–	–	–	–
Palatal breadth	37.0	–	–	–	–	–	–	–	–
Palatal length	54.0	–	–	–	–	–	–	–	–
Ascending ramus breadth	33.0	–	–	–	–	–	–	–	–

(Continued)

Table 6.2. Absolute values and cranial indices of the individuals found in UF 228 of the archaeological site of Santa Caterina in Barcelona (Continued)

Id	*UE2071*	*UE2072*	*UE2072*	*UE2073*	*UE2073*	*UE2073*	*UE 2074*	*UE 2076*	*UE 2077*
Ascending ramus height	72.0	–	–	–	–	–	–	–	–
Mandibular length	–	–	–	–	–	–	100.0	–	–
Mandibular arch width	–	–	–	–	–	–	36.0	–	–
Cranial index	69.0	–	–	–	–	–	–	–	–
Vertico-longitudinal I.	65.3	–	–	–	–	–	–	–	–
Vertico-transversal I.	94.5	–	–	–	–	–	–	–	–
Transvers cranial-facial I.	64.7	–	–	–	–	–	–	–	–
Fronto sagittal I.	86.5	–	–	–	–	–	–	–	–
Parietal sagittal I.	87.1	–	–	–	–	–	–	–	–
Frontal transversal I.	83.5	–	–	–	–	–	–	–	–
Fronto-parietal transv. I.	69.5	–	–	–	–	–	–	–	–
Foramen magnum I.	63.2	–	–	–	–	–	–	–	–
Occipital sagittal I.	92.9	–	–	–	–	–	–	–	–
Total sagittal I.	44.0	–	–	–	–	–	–	–	–
Occipit. squama sagtital I.	100.0	–	–	–	–	–	–	–	–
Total facial I	99.2	–	–	–	–	–	–	–	–
Upper facial I	59.4	–	–	–	–	–	–	–	–
Iugomandibular I	79.7	–	–	–	–	–	–	–	–
Orbital I	84.5	–	–	–	–	–	–	–	–
Nasal I	50.9	–	–	–	–	–	–	–	–
Palatal I	68.5	–	–	–	–	–	–	–	–
Mandibular ramus I	44.4	–	–	–	–	–	–	–	–
Mandibular robusticity I	54.6	–	–	–	–	–	–	–	–

97.9, and a very narrow (hyper-leptoprosopic) face with a total facial index greater than 95.

Regarding the postcranial measurements of the upper limbs (Table 6.3), in general, the humerus exhibits euribrachy (round diaphyses), with a mean diaphyseal index greater than 76.5; however, the clavicles are robust (clavicle robusticity index >25.5). Regarding the lower limb (Table 6.4), the individuals exhibit a marked linea aspera, showing a medium or high pilaster (Table 6.4). The pilastric index refers to the degree of protrusion of the linea aspera, which indicates the use of the lower limb (insertions of the vastus internus, vastus externus and biceps femoris). On the other hand, these individuals display mesocnemic and euricnemic tibiae, indicating a slight degree of flattening of the tibia (Table 6.4). These adult individuals exhibit a mean stature of 165 cm in males and 150.3 cm in females, considering also the female individual aged around 13 years (Table 6.1). These means correspond to the mean stature observed among their contemporaries (Table 6.5).

In addition, in the individuals from UF 228, several physical activity markers were observed in the upper and lower limbs, especially in the adult individual UE2071, between 33 and 45 years of age. In the upper limb, the musculoskeletal indicators of occupational activity were observed in the insertion of the costoclavicular ligament and in the deltoid muscle in the left clavicle. They are related to repetitive movements of abduction and elevation of the shoulder and internal rotation of the arm, associated with the movements in trades such as tailors, blacksmiths or farmers (Capasso *et al.* 1999; Fernández 2008). In the lower limbs, particularly on the diaphysis of the left femur, a very marked linia aspera was observed; this musculoskeletal indicator of occupational activity is typical of weavers or individuals who ride horses (Capasso *et al.* 1999; Fernández 2008). On the other hand, the insertion site of the internal obturator muscle (external rotator of the femur) located medial to the greater trochanter, was marked. This physical activity marker is usually present in loom-weavers (Fernández 2008). In the left tibia, the insertion of the soleus muscle is marked. This musculoskeletal indicator of occupational activity usually appears in individuals who walk on a regular basis (Capasso *et al.* 1999; Fernández 2008). Regarding pathological conditions, the male individual UE 2071, aged between 33 and 45, displayed multiple small osteolytic

Table 6.3. Absolute values and indices of the upper limbs of the individuals found in UF 228 of the archaeological site of Santa Caterina in Barcelona

Id	*UE 2071*	*UE 2072*	*UE 2072*	*UE 2073*	*UE 2073*	*UE 2073*	*UE 2074*	*UE 2076*	*UE 2077*
Sex	♂	♂	♀	Indet.	Indet.	Indet.	♀	Indet.	♂
Age	33–45 yr	16–25 yr	Indet.	2 mnth	~ 3 yr	7–9 yr	13 yr	3–6 yr	40 yr
Clavicular length	–	–	–	–	–	–	110.0	–	–
Clavicular perim. at midshaft	42.0	–	–	–	–	–	29.5	–	–
Scapular length	–	–	–	–	–	–	119.0	80.0	–
Scapular width	–	–	–	–	–	–	85.0	–	–
Scapular spine length	–	–	–	–	–	–	99.0	–	–
Glenoidal surface max. width Ø	28.0	–	–	–	–	–	–	–	–
Glenoidal surface max. height Ø	45.0	–	–	–	–			–	–
Humeral max. length	–	–	–	–	103.5	–	–	102.0	343.0
Humeral max. physiological length	–	–	–	–	–	–	–	–	336.0
Humeral min. perim.	–	59.0	–	–	34.5	–	51.5	38.0	66.0
Humeral perim. at midshaft	71.0	–	–	–	–	–	–	–	67.0
Humeral min. Ø at midshaft	21.0	–	–	–	10.7	–	15.5	11.0	19.0
Humeral max. Ø at midshaft	23.0	–	–	–	11.5	–	16.0	12.0	23.0
Humeral distal epiphysis width	–	–	–	–	28.8	–	–	33.0	67.0
Humeral max. Ø proximal epiphysis	–	–	–	–	–	–	–	28.0	49.0
Humeral min. Ø proximal epiphysis	–	–	–	–	–	–	–	20.0	46.0
Ulnar max. length	–	–	–	–	103.0	–	221.5	97.0	–
Ulnar physiological length	–	–	–	–	–	–	–	–	242.0
Ulnar min. perim.	–	51.0	–	–	20.0	–	32.0	21.0	40.0
Ulnar perim.. in middle	–	–	–	–	–	–	–	–	57.0
Ulnar ante.-post. Ø radial notch	–	33.0	–	–	16.8	–	22.0	11.0	29.0
Ulnar transv. Ø radial notch	–	20.0	–	–	12.3	–	17.0	11.0	23.0
Ulnar max. Ø in middle	–	–	–	–	7.7	–	10.0	7.0	20.0
Ulnar min. Ø in middle	–	–	–	–	6.7	–	9.0	6.0	16.0
Ulnar max. distal epiphysis width	–	–	–	–	8.2	–	14.0	–	–
Ulnar min. distal epiphysis width	–	–	–	–	–	–	10.0	–	–
Radial max. length	–	–	–	–	93.0	–	198.0	–	253.5
Radial physiological length	–	–	–	–	–	–	–	–	245.0
Radial min. perim.	–	40.0	–	–	22.0	–	38.0	25.0	47.5
Radial perim. at midshaft	–	–	–	–	–	–	–	–	51.0
Radial tuberos. Perim.	–	53.0	–	–	28.0	–	44.5	30.0	61.0
Radial max. Ø in middle	–	–	–	–	7.2	–	5.0	8.0	17–5
Radial min. Ø in middle	–	–	–	–	6.4	–	4.0	8.0	13.5
Radial max.m proximal epiphysis width	–	–	–	–	10.6	–	15.0	–	–

(Continued)

Table 6.3. Absolute values and indices of the upper limbs of the individuals found in UF 228 of the archaeological site of Santa Caterina in Barcelona (Continued)

Id	*UE 2071*	*UE 2072*	*UE 2072*	*UE 2073*	*UE 2073*	*UE 2073*	*UE 2074*	*UE 2076*	*UE 2077*
Radial max. distal epiphysis width	–	–	–	–	15.0	–	19	–	36.5
Radial head Ø	–	–	–	–	–	–	–	–	63.0
Humeral diaphyseal index at midshaft	91.3	–	–	–	–	–	–	–	82.6
Humeral robusticity index	–	–	–	–	–	–	–	–	19,53
Glenoidal index	62.2	–	–	–	–	–	–	–	–
Clavicular robusticity index	26.0	–	–	–	–	–	–	–	–
Clavicular Terry index	19.5	–	–	–	–	–	–	–	–
Ulnar diaphyseal index at midshaft	85.7	–	–	–	–	–	–	–	–
Ulnar robusticity index	–	–	–	–	–	–	–	–	23.6
Radial robusticity index	–	–	–	–	–	–	–	–	19.39
Radial diaphyseal index in middle	–	–	–	–	–	–	–	–	77.1
Platolenia index (Verneau)	–	–	–	–	–	–	–	–	79.4

lesions, mostly circular, with clearly defined margins in the skull, vertebrae, sacrum, ribs, clavicle, scapula and pelvis. The lesions are of various sizes that range between 1 and 11 mm in diameter. New bone formation or evidence for the healing of these lesions were not observed. All this suggests a possible neoplasia; however, all these pathological conditions will be analysed in greater depth in another paper. In addition, this individual also revealed an antemortem stab injury, probably made with a knife-like utensil, without the presence of bone remodelling. This knife-like lesion was located on the lateral epicondyle in the distal epiphysis of the left femur. Furthermore, and because of the age, in some vertebrae and articular surfaces of this individual, osteophytes can be observed.

Regarding oral palaeopathology, all (or almost all) teeth are preserved in only two adults (EU 2071, a male aged between 33 and 45; and UE 2072, a male aged between 16 and 25) and some of the teeth in three immature individuals (UE 2072, aged between 10 and 11; UE 2073, around 3 years of age; and UE 2074, aged between 10 and 13). In the two adults, particularly in the male individual UE 2071 aged between 33 and 45 (Fig. 6.3), all the maxillary teeth are preserved except for the second right incisor and the first incisor and left canine, which were lost postmortem. In the right hemimandible, the first incisor was lost postmortem, and the second molar had been lost in life, observable by the reabsorbed alveolus. The left hemimandible is missing the first incisor, also lost postmortem. In this individual a very marked alveolar setback in the majority of the mandibular and maxillary teeth should be highlighted. Most of this alveolar setback is between 3 and 6 mm and in some dental elements it is greater than 6 mm. This setback is so pronounced that it implies that some teeth were about to be lost. A large amount of calculus can be observed in all the dentition, especially on the labial area, although in some dental pieces it can also be seen on the lingual and distal area. A large presence of band-type hypoplasia stands out, especially in the maxilla. The left hemimandible has two fistulas below the first and second molars; however, in spite of these fistulas, the lack of caries stands out in this individual. In all his dental pieces, a very significant amount of wear can be detected, so there is practically no enamel on the occlusal area. There is evidence for periodontitis in both mandibular canines. The presence of dental calculus is considered the main cause of periodontal disease (Gómez 2012). Therefore, the periodontitis of this individual could be due to poor oral hygiene and to eating less abrasive foods. The presence of abundant calculus is usually related to a diet rich in animal protein.

In the adult male UE 2072, between 16 and 25 years of age, all the teeth are preserved, with the exception of the upper left second premolar, the upper right first incisor and the lower left third molar. The main characteristic of the dentition of this individual is the presence of caries and enamel hypoplasia in all the preserved teeth. It also displays dental attrition (Fig. 6.4).

In the immature individuals (UE 2072, between 10 and 11 years of age, UE 2073, aged around 3 years, and UE 2074, between 10 and 13 years of age), despite retaining only a few dental pieces (UE 2072: the right hemimandible: canine, first premolar and molar. UE 2073: the right hemimaxilla: canine and first and second molars; the left hemimaxilla first and

Table 6.4. Absolute values and indices of the lower limbs of the individuals found in UF 228 of the archaeological site of Santa Caterina in Barcelona

Individual	*UE 2071*	*UE 2072*	*UE 2072*	*UE 2073*	*UE 2073*	*UE 2073*	*UE 2074*	*UE 2076*	*UE 2077*
Sex	♂	♂	♀	Indet.	Indet	Indet	♀	Indet.	♂
Age	33–45 yr	16–25 yr	Indet.	2 mnth	~ 3 yr	7–9 yr	13 yr	3–6 y.r	40 yr
Tibial max. length	–	–	329.0	72.0	–	–	–	–	
Tibial min. perim.	–	73.0	67.0	30.0	–	47.0		–	83.0
Tibial perim. at midshaft	–	79.0	72.0	31.0	–	–	–	–	99.0
Tibial ante.-post. Ø at nutrient foramen	–	27.0	28.0	11.2	–	16.8	–	–	37.0
Tibial transv. Ø at nutrient foramen	–	32.0	22.0	8.7	–	17.8	–	–	27.0
Tibial ante.-post. Ø at midshaft.	35.0	22.0	21.0	10.2	–	–	–	–	34.0
Tibial transv. Ø at midshaft.	–	27.0	19.0	7.6	–	–	–	–	25.0
Tibial perim. nutrient foramen	–	91.0	81.0	33.5	–	56.0	–	–	107.0
Tibial distal epiphysis width	–	54.0	44.0	14.5	–	–	–	–	41.0
Tibial proximal epiphysis width	–	–	–	–	–	41.3	–	–	66.0
Fibular min. perimeter	41.5	34.0	–	–	–	–	–	–	47.0
Fibular distal epiphysis width	–	–	–	–	–	–	–	–	27.0
Fibular min. Ø at midshaft.	12.0	–	–	–	–	–	–	–	14.0
Fibular max. Ø at midshaft	14.0	–	–	–	–	–	–	–	18.0
Femoral max. length	–	442.0	–	–	–	–	–	–	450.0
Femoral physiological length	–	442.0	–	–	–	–	–	–	445.0
Femoral perim. in middle	–	86.0	–			–	–	–	99.0
Femoral Subtrochan. perim.	–	102.0	–	–	–	–	–	–	107.0
Femoral vertical Ø of head	–	42.0	–	–	–	–	–	–	44.0
Femoral transv. Ø of head	–	–	–	–	–	–	–	–	47.0
Femo. ante-post. Ø subtrochan.	–	29.0	–	–	–	–	–	–	30.0
Femo. transver. Ø subtrochan.	–	33.0	–	–	–	–	–	–	34.0
Femoral ante.-post. Ø at midshaft	29.0	28.0	–	–	–	–	–	–	33.0
Femoral transv. Ø at midshaft	24.0	26.0	–	–	–	–	–	–	30.0
Femo. distal epiphysis width	–	87.0	–	–	–	–	–	–	84.0
Femoral head perim.	–	–	–	–	–	–	–	–	150.0
Hip bone height	–	–	–	–	–	–	–	–	219.0
Ilium length	–	–	–	–	–	–	–	91.0	135.0
Ilium width	–	–	–	–	–	–	123.0	76.0	164.0
Ischium length	–	–	–	–	–	–	–	49.0	91
Ischium width	–	–	–	–	–	–	–	31.0	
Coctylosciatic width	–	46.5	–	–	–	–	–	–	38.0
Acetabular Ø	–	51.5	–	–	–	–	–	–	55.0
Tibial mid-diaphyseal index	–	122.7	90.5	–	–	–	–	–	–
Tibial ncnemic index	68.6	118.6	78.6	–	–	–	–	–	–
Tibial length width index	–	–	20.4	–	–	–	–	–	–
Femur robusticiy index	–	–	–	–	–	–	–	–	22.0
Pilastric index	120.8	–	–	–	–	–	–	–	110.0
Platymeric index	–	–	–	–	–	–	–	–	88.2

Table 6.5. Mean male and female statures from different funerary areas of Santa Caterina and other series of different periods from the Iberian Peninsula

Population	*Period*	*Males*	*Females*	*References*
Empúries	Roman	165.87	147.32	Pons 1949
Plaça Vila de Madrid	Roman	164.89	148.72	Jordana *et al.* 2003
Romans de Tarragona	Roman	163.18	151.82	Pons 1949
Can Solà del Racó	Roman	160.49	152.16	Alesan *et al.* 1998
La Solana s.VI-VIII	Late Roman	166.14	154.12	Safont *et al.* 1997
Vilassar de Mar	Late Roman	164.22	153.61	Safont *et al.* 2001
Visigots castellans	Late Roman	167.61	154.41	Varela 1974
Ample 1	Late Roman	159.17	153.34	Pujol *et al.* 2011
La Olmeda 12th and 13th century	Medieval	163.34	155.43	Hernández and Turbón 1991
Sant Pere (Terrassa)	Medieval	166.97	154.93	Jordana and Malgosa 2002
Sta. María de Hito 6th–13th century	Medieval	165.36	155.33	Galera and Garralda 1992
Castilla y León	Medieval	164.17	154.37	López-Martínez 2000
Porta dels Fillols 12th century	Medieval	166	153.4	Ferré 2013
Santa Caterina UF 221 13th and 14th century	Medieval	169.02	154.53	Cevallos *et al.* 2023
Santa Caterina cloister courtyard 13th century	Medieval	166	–	Trujillo *et al.* 2023
Santa Caterina apse/cloister corridor 13th–15th century	Medieval	167	–	Trujillo *et al.* 2023
Santa Caterina UF 228, 14th and 15th centuries	Medieval	165	150.3	Present study
Jews of Montjuïc	Medieval	161.95	149.1	Prevosti and Prevosti 1951
Medievals from Catalunya	Medieval	165.8	154.01	Vives 1990
Muslims of la Torrecilla	Medieval	164.12	150.65	Souich 1978
Sant Agustí, church yard 14th–16th century	Medieval	161.41	154	Rissech *et al.* 2023
Sant Agustí, cloister 14th–16th century	Medieval	162.02	–	Rissech *et al.* 2023
Santa Caterina UF 221 beginning 16th century	Modern	163.02	156.75	Cevallos *et al.* 2023
Castilla y León	Early Modern	164.56	152.93	Lopez-Martínez 2000
Plaça Vella (Terrassa)	Early Modern	164.62	152.46	Jordana and Malgosa 2002

second molars; the right hemimandible: two incisors, canine and both molars; left hemimandible: the second incisive. UE 2074: the right hemimaxilla: first and second molars; the left hemimaxilla: the canine and the first molar; the right hemimandible: the canine, two premolars and the first and second molars; the left hemimandible: the second molar), the main characteristic of their dentition is the presence of enamel hypoplasia in all the dental pieces retained. Enamel hypoplasia is a defect in the formation of enamel during the development of the tooth. The causes which produce this hypoplasia are varied; they include, among others: congenital and hereditary anaemia, rickets, avitaminosis, perithyroidal alterations, local infections, poisoning and traumatisms. For this reason, enamel hypoplasia has been considered as an environmental stress marker occurring during the growth stage of an individual. In this case, the considerable number of teeth revealing enamel hypoplasia in these immature individuals (and also in adults) indicate that these individuals had some difficulties during their growth.

Discussion

This study presents the bioanthropological analysis of the individuals buried in Funerary Unit UF 228 (14th to 15th century) at the archaeological site of the friary of Santa Caterina in Barcelona. The results indicate a total of 18 individuals: 5 adults and 13 immature individuals (Table 6.1). The adult individuals comprise 4 men and a woman. Three of these men are 16–25, 33–45 and 40–44 years old, and the fourth is of undetermined age. The woman is also of undetermined age (Table 6.1). The immature individuals comprise: two babies (one aged 2 months and the other between 1 and 3 months old), 5 infants (one aged between 6 months and 1 year, another aged between 6 months and 2 years, the third around 3 years of age, the fourth aged 5 years and the last infant between

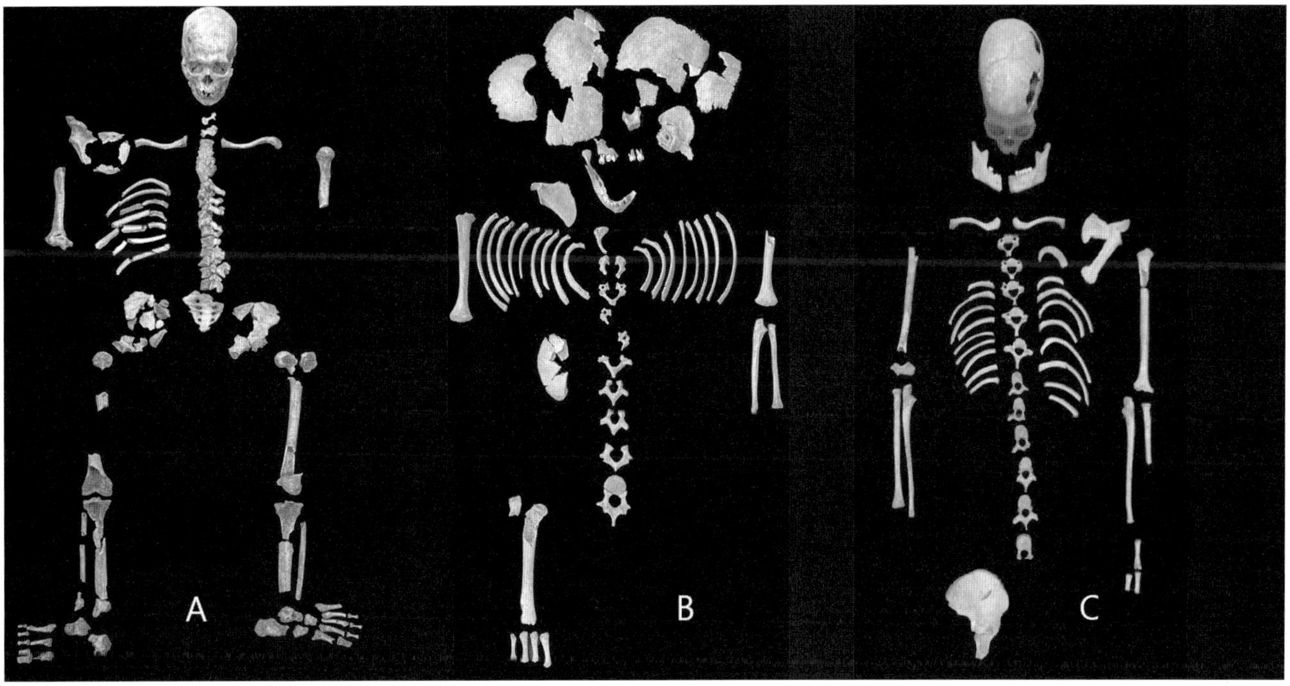

Figure 6.3. The three best-preserved skeletons in the sample. They correspond to UE 2071 (male aged between 33 and 45) (A); UE 2073 (around 3 years of age) (B); UE 2074 (a female aged around 13) (C).

3 and 6 years of age), 4 children (two between 7 and 9 years old and two aged between 10 and 11) and 2 youths (1 female of approximately 13 years of age and an individual of undetermined sex aged between 8 and 14).

The biological profile, considering previous results derived from the analysis of the other funerary units at this friary (Cevallos *et al.* 2023; Trujillo *et al.* 2023; Sanz-Boscà 2023) and the pathological conditions observed in the individuals buried in UF 228, indicate that this funerary unit may have corresponded to a secular family group of weavers. However, the possibility of them having been a family group needs to be confirmed by DNA analysis and the assertion that they were weavers should be confirmed through more information. The documentary sources available make it clear that as from *c.* 1250 Santa Caterina started to become a favoured burial place for some of the leading patricians of Barcelona (Zaldívar 2012). Unfortunately, the extant 13th-century wills referring to burial at the Dominican house are related to these upper ranks of urban society, and there are few, if any, that refer to artisans, even successful ones (Andrés 1985, 130–32). However, it seems clear that many of the parishioners of the adjoining Sant Pere neighbourhood, definitely not the wealthiest district in the city, chose to be buried in the friary (Fig. 6.1). An agreement between the nuns of the Benedictine house of Sant Pere de les Puel·les, which gave its name to the district, and the Dominican friars of Santa Caterina, signed in 1250, at the same time as the earliest documentary evidence for burials there, specifies the proportion – one-sixth – of the bequests to the Dominicans that the nuns were to receive for each

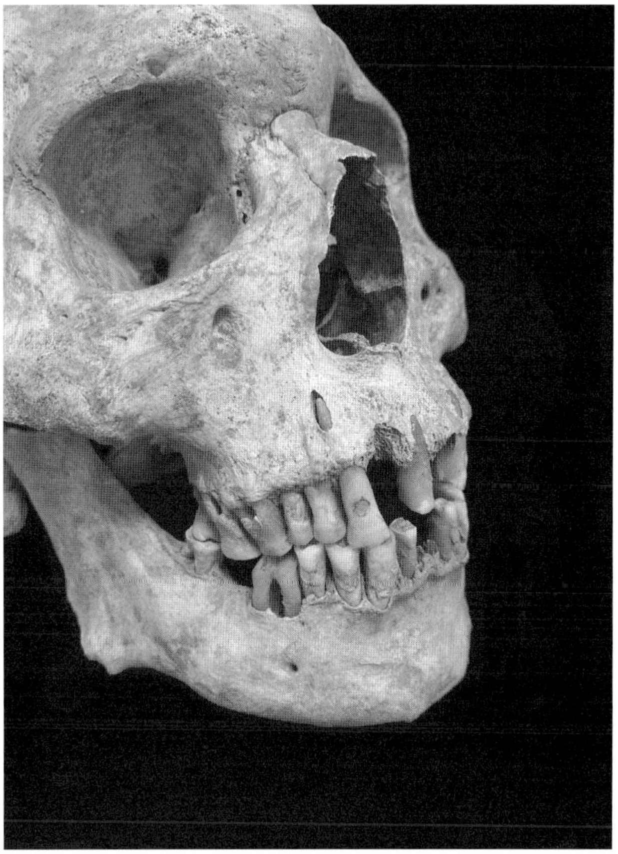

Figure 6.4. Male skeleton UE 2071, aged between 33 and 45 years, displaying dental attrition, alveolar setback, calculus and hypoplasia.

individual from the parish of Sant Pere buried at Santa Caterina (Banks in press). Moreover, there is documentary evidence for many weavers residing in the parish of Sant Pere, particularly between carrer de Sant Pere Més Alt and carrer de Sant Pere Més Baix and for this number having increased noticeably towards the end of the 13th century (Banks in press). These two circumstances lead one to suppose that families associated with cloth production may well have been buried at Santa Caterina in the later 13th and early 14th centuries.

The dental characteristics, apart from poor oral hygiene, the frequency of caries and calculus, indicate a diet rich in animal protein and low in carbohydrates. However, if we take into account the prevalence of enamel hypoplasia and the stature observed in these individuals, they do not seem to have come from the noble or wealthy sectors of Barcelona society in this time. This opinion is also reinforced by the location of this funerary unit in the friary of Santa Caterina, which is outside the nave of the church. This family was probably moderately prosperous, sufficiently well-off to afford the corresponding fees to be buried in this Gothic chapel (chapel H) outside the nave of the church, located between two external buttresses of the apse. According to previous studies and to the archaeological data available from the Santa Caterina site, friars and people closely related to the community were buried in the cloister (friars in the central courtyard of the cloister or garth, and friars together with people closely related to the community in the cloister galleries), the people from the highest social classes were buried inside the church (the higher their social rank, the closer to the high altar they wanted to be), and moderately well-off people against the outer part of the church wall, while the less well-off were buried in the external area, beyond the apse (Cevallos *et al.* 2023; Trujillo *et al.* 2023; Sanz-Boscà 2023).

Medieval monasteries, friaries and nunneries display spatial hierarchies in relation to the location of burials. This is due to the medieval tendency to classify individuals according to their social status (defined by religious role, lineage, sex, age or pathology) and the medieval belief that the holiest point was physically located at the high altar of the monastery church. These factors led lay individuals to value burial within the church itself very highly, and particularly near the high altar or side altars rather than in the churchyard (Harding 1992; Gilchrist and Sloane 2005).

The burial of friars and people closely related to the community in the cloister is in accordance with the concept of the cloister as the spiritual heart of the friary (Gilchrist and Sloane 2005). The cloister consisted of an open space or quadrangle in the centre (sometimes occupied by a garden or garth), surrounded by the cloister galleries. The friary church and the conventual buildings were distributed around the cloister. Because of the significance of the cloister, and although the monastic superiors initially favoured burial in the chapter house, the cloister courtyard could be used as a burial ground for the priors (Butler 1993; Gilchrist and Sloane 2005), and both friars and lay benefactors could commonly be buried in the cloister galleries (Barber and Thomas 2002), with patterns relating to both sex and age, especially in relation to friaries (Gilchrist and Sloane, 2005).

In monasteries, the burials of founding families, patrons and wealthy benefactors might also be found, as can probably be detected in the cloister gallery of Santa Caterina. In this area, the burials are mostly male; however, there is a small group of female burials, suggesting that they held a particular relationship to the friary community, through patronage, a religious role, or family ties. In fact, because of the high frequency of physical activity markers displayed by the women found in the cloister gallery of the Santa Caterina friary site, it was suggested they could have had family ties with or have been domestic staff for the friary (Sanz-Boscà 2023).

Conclusions

In conclusion, the individuals buried in UF 228 possibly comprised a family group of weavers, who did not come from the noble or wealthy sectors of Barcelona society of that time. They probably belonged to a moderately prosperous family, wealthy enough to be able to afford the cost of being buried in the Gothic chapel outside the nave of the church. The highest medieval class used to be buried inside the church near the high altar, moderately well-off people against the outer part of the church wall whereas the less well-off were buried in the external area, beyond the apse. The friars and people closely related to the community were usually buried in the cloister itself.

Acknowledgements

Museu d'Història de la Ciutat de Barcelona (MUHBA), particularly Emili Revilla. Philip Banks for his interesting comments. MonBones: PID2020-118194RJ-100 AEI/MINECO. SGR Evolució social, cultural i biològica al Pleistocè (StEP), Ref: 2021 SGR 01239.

References

Alduc-Le Bagouse, A. (1988) Estimation de l'âge des non-adultes: maturation dentaire et croissance osseuse. Donnés comparatives pour deux nécropoles médiévales bas-normandes. In *Actes des 3èmes Journées Anthropologiques. Notes et Monographies Techniques* 24, 81–103. Paris, Centre National de la Recherche Scientifique.

Alemán, I., Botella M.C. and Ruiz, L. (1997) Determinación del sexo en el esqueleto postcraneal. Estudio de una población mediterrània actual. *Archivo Español de Morfología* 2, 7–17.

Alesan, A. (1990) *Estudi d'una població subadulta de l'Edat del Ferro: demografia, antropometria i creixement.* Master's thesis, Universitat Autònoma de Barcelona.

Alesan, A., Safont, S., Alfonso, J., Malgosa, A. and Egea, R. (1998) *Informe antropològic de les inhumacions trobades durant la campanya de l'any 1997 al jaciment de Can Solà del Racó (Matadepera, Barcelona)*. Unpublished report.

Andrés, R.M. (1985) *El Convent de Santa Caterina de Barcelona: s.*XIII. Undergraduate dissertation, Universitat de Barcelona. Available at: Dipòsit Digital de la Universitat de Barcelona: El Convent de Santa Caterina de Barcelona: S. XIII (ub.edu)

Aufderheide, A.C. and Rodriguez-Martin, C. (1998) *The Cambridge Encyclopedia of Human Paleopathology*. Cambridge, Cambridge University Press.

Banks, P. (in press), Construint el barri: la parròquia de Sant Pere de les Puel·les i el patrimoni urbà del Monestir (segles XII i XIII). In I. Brugués and X. Costa (eds) *Construint la ciutat en femení Sant Pere de les Puel·les a la Barcelona medieval*. Barcelona, Publicacions de l'Abadia de Montserrat.

Barber, B. and Thomas, C. (2002) *The London Charterhouse*. MoLAS Monograph Series 10. London, Museum of London Archaeology Service.

Brothwell, D. (1987) *Desterrando huesos*. Mexico, Fondo de cultura económica.

Butler, L. (1993) Cistercian abbots' tombs and abbey seals. In M.P. Lillich (ed.) *Studies in Cistercian art and architecture*, 78–89. Cistercian Studies Series 4.

Campillo, D. (2001) *Introducción a la paleopatología*. Barcelona, Edicions Bellaterra.

Capasso, L., Kennedy, K. and Wilczak, C. (1999) *Atlas of Occupational Markers on Human Remains*. Teramo, Edigrafital.

Cevallos, A., Rissech, C., Huertas, J. and Lloveras, L. (2023) Estudio bioantropológico de los restos humanos procedentes de la unidad funeraria (UF) 221 del antiguo convento de Santa Caterina (1243–1836), Barcelona. *Rodis, Journal of Medieval and Post-Medieval Archaeology* 6, 171–216.

Cobo, S. (2022) *Estudio Dental de los Individuos del Hipogeo Neolítico Final-Calcolítico del Carrer París de Cerdanyola del Vallès (Barcelona)*. Master's thesis, Universidad de Barcelona.

Ciétot, M. (1978) *L'arcade dentaire humaine (Morphologie)*. Paris, Julien Prélat.

Etxeberria, F. (2003) Patología traumática. In A. Isidro and A. Malgosa (eds) *Introducción a la Paleopatología. La enfermedad no escrita*, 195–207. Barcelona, Masson.

Etxeberria, F., Baraibar, J.P., De La Rua, C. and Vegas, J.I. (1993) Identificación postmortem y mecanismos de producción de las fracturas diafisarias del cúbito,. In *Actas del II Congreso Nacional de Paleopatología*, 219–323. Valencia, Ministerio de Educación y Ciencia, Asociación Española de Paleopatología, Universidad de Valencia.

Fazekas, I.G. and Kósa, F. (1978) *Forensic Fetal Osteology*. Budapest, Akadémiai Kiadó.

Ferembach, D., Schwidetzky, I. and Stloukal, M. (1980) Recommendations for age and sex diagnoses of skeletons. *Journal of Human Evolution* 9, 517–49.

Fernández, M. (2008). *Marcadores óseos de actividad ocupacional en las poblaciones históricas de Palat de Rey (León) y de Gormaz (Soria)*. PhD thesis, Universidad de León.

Ferré, M. 2013, *Estudi bioantropològic de les restes humanes trobades en les tombes en cista de la necròpolis posta dels Fillols de la Seu Vella (Lleida) datada del s. XII*. Master's thesis, Universitat de Barcelona.

Galera, V. and Garralda, M.D. (1992) La población medieval cántabra de Santa María de Hito. Aspectos morfológicos y etnohistóricos. *Boletín de la Sociedad Española de Antropología Biológica* 13, 69–87.

Gilchrist, R. and Sloane, B. (2005) *Requiem. The Medieval Monastic Cemetery in Britain*, 56–77. London, Museum of London Archaeology Service.

Gómez González, S. (2012) *Paleopatología dental de poblaciones históricas (siglos III–XIII) en la provincia de Alicante: estudio de la variabilidad como respuesta a factores de hábitat y dieta*. PhD thesis, Universitat d'Alacant.

Harding, V. (1992) Burial choice and location in later medieval London. In S.R. Basset (ed.) *Death in Towns: Urban Responses to the Dying and the Dead, 100–1600*, 119–35. Leicester, Leicester University Press.

Hauser, G. and DeStefano, G.F. (1989) *Epigenetic Variants of the Human Skull*. Stuttgart, Schweizerbart.

Hernández, M. and Turbón, D. (1991) Parámetros del esqueleto postcraneal en la población medieval de 'La Olmeda'. *Boletín de la Sociedad Española de Antropología Biológica* 12, 61–80.

Huertas, J and Aguelo, J. (2006) *Memòria de la intervenció arqueològica al solar del mercat de Santa Caterina*. Barcelona. Unpublished report.

İşcan, M.Y. and Loth, S.R. (1986) Determination of age from the sternal rib in white females. A test of the phase method. *Journal of Forensic Sciences* 31, 990–9.

Jordana, X. (2003) Els enterraments de la Plaça Vila de Madrid, Informe Antropològic de la intervenció arqueològica a la Plaça Vila de Madrid (Barcelona). Unpublished report.

Jordana, X. and Malgosa, A. (2002) Terrassa, una Villa Medieval en transición a la Época Moderna: Estudio bioantropológico de la necrópolis de la Plaça Vella. *Revista Española de Antropología Biológica*. 23, 1–25.

Krogman, W.M. and Işcan, Y.M. (1986) *The Human Skeleton in Forensic Medicine*. Springfield, Illinois, Charles C. Thomas Ltd.

Lewis, M. (2018) *Paleopathology of Children. Identification of Pathological Conditions in the Human Skeletal Remains of Non-Adults*. London, Academic Press.

López-Martínez, B. 2000, Estudio Antropológico de poblaciones históricas de Castilla y León. PhD thesis, Universidad de León.

Lloveras, L., Rissech, C. and Rosado Mendez, N.Y. (2016) Tafonomia forense. In C. Sanabria Medina (ed.) *Patología y antropología forense de la muerte: la investigación científico-judicial de la muerte y la tortura, desde las fosas clandestinas, hasta la audiencia pública*, 453–523. Bogotá, Forensic Publisher.

Lovejoy, C.O., Meindl, R.S., Pryzbeck, T.R. and Mensforth, R.P. (1985) Chronical metamorphosis of the auricular surface of the ilium. A new method for the determination of adult skeletal age at death. *American Journal of Physical Anthropology* 68, 15–28.

Martin, R. and Saller, K. (1957) *Lehrbuch der anthropologie*. Stuttgart, G. Fischer.

Masset, C. (1982) *Estimation de l'âge au décès par les sutures crâniennes*. Thèse de Sciences Naturelles, Multigraphée. Université Paris VII.

Olivier, G. (1960) *Pratique Antropologique*. Paris, Ed. Vigot Frères.

Ortner, D. (2003) *Identification of Pathological Conditions in Human Skeletal Remains*. 2nd ed. San Diego, Academic Press.

Ortoll, E. (1996) Algunas consideraciones sobre la iglesia de Santa Caterina de Barcelona. *Locus amoenus* 2, 47–63.

Pearson, K. (1899) Mathematical contributions to the Theory of evolution. V. On the reconstruction of the stature of prehistoric races. *Philosophical Transactions of the Royal Society* 192, 169–244.

Pokines, J.T., L'Abbé, E.N. and Symes, S.A. (eds) (2021) *Manual of Forensic Taphonomy.* 2nd edition. Boca Raton, CRC Press.

Pons, J. (1949) Restos humanos procedentes de la necrópolis de época romana de Tarragona y Ampurias (Gerona). *Trabajos del Instituto 'Bernardino de Sahagún'* 7, 19–206.

Prevosti, M. and Prevosti, A. (1951) Restos humanos procedentes de una necrópolis judaica de Montjuic (Barcelona). *Trabajos del Instituto 'Bernardino de Sahagún'* 12, 69–148.

Pujol, A., Rissech, C., Prida, D. and Turbón, D. (2011) Estudio bioantropológico del conjunto funerario de época tardorromano Carrer Ample 1 de Barcelona. *Revista Española de Antropología Física* 32, 1–19.

Rissech, C. (2016) La estimación de la edad de individuos adultos y subadultos a partir de material esquelético. In C. Sanabria Medina (ed.) *Patología y antropología forense de la muerte: la investigación científico-judicial de la muerte y la tortura, desde las fosas clandestinas, hasta la audiencia pública*, 237–92. Bogotá, Forensic Publisher.

Rissech, C. and Black, S. (2007) Scapular development from neonatal period to skeletal maturity. A preliminary study. *International Journal of Osteoarchaeology* 17, 451–64.

Rissech, C., Estabrook, G.F., Cunha, E. and Malgosa, A. (2006) Using the acetabulum to estimate age of death in adult males. *Journal of Forensic Sciences* 51, 213–29.

Rissech, C., Estabrook, G.F., Cunha, E. and Malgosa, A. (2007) Estimation of age at death for adult males using the acetabulum, applied to four Western European collections. *Journal of Forensic Sciences* 52, 774–8.

Rissech, C. and Malgosa, A. (1991) Importancia relativa de la longitud del pubis y la anchura del ilion en el estudio del dimorfismo sexual de los coxales. *Boletín de la Sociedad Española de Antropología Biológica* 12, 29–43.

Rissech, C. and Malgosa, A. (1997) Sex prediction by discriminant function with central portion measures of innominate bones. *Homo* 48, 22–32.

Rissech, C. and Malgosa, A. (2007) Pubic growth study: applicability in sexual and age diagnostic. *Forensic Science International* 173, 137–45.

Rissech, C., Márquez-Grant, N. and Turbón, D. (2013a) A collation of recently published Western European formulae for age estimation of subadult skeletal remains: recommendations for forensic anthropology and osteoarchaeology. *Journal of Osteoarchaeology* 58, S163–S168.

Rissech, C., López-Costas, O. and Turbón, D. (2013b) Humeral development from neonatal period to skeletal maturity—application in age and sex assessment. *International Journal of Legal Medicine* 127, 201–12.

Rissech, C., Revuelta, B., Creo, O., Llauradó, A., Jové, M. and Lloveras, L. (2023). Estudi bioantropològic de les restes humanes exhumades al Convent de Sant Agustí Vell (S. XIV–XVI) de Barcelona. *Rodis, Journal of Medieval and Post-Medieval Archaeology* 6, 143–70.

Safont, S., Alesan, A. and Malgosa, A. (1997) Informe antropològic de les restes del jaciment de La Solana (Cubelles, Barcelona). Unpublished report.

Safont, S., Armentano, N. and Malgosa, A. (2001) Informe antropològic de les inhumacions trobades durant la campanya de l'any 1998 en el jaciment de la Piscina Municipal de Vilassar de Mar, Maresme, Barcelona. Unpublished report.

Sanz-Boscà, A. (2023) *Vida i mort al convent dels predicadors de Santa Caterina, Barcelona.* Master's thesis, University of Barcelona.

Scheuer, L. and Black, S. (2000) *Developmental Juvenile Osteology.* London, Academic Press.

Schutkowski, H. (1993) Sex determination of infant and juvenile skeletons. I. Morphognostic features. *American Journal of Physical Anthropology* 90, 199–205.

Souich, P. (1978) Estudio antropológico de la necrópolis medieval de la Torrecilla (Arenas del Rey, Granada). PhD tesis, Facultad de Filosofía y Letras de la Universidad de Granada.

Spitz, U. and Fisher, W.U. (1993) *Medicolegal Investigation of Death: Guidelines for the Application of Pathology to Crime Investigation.* 3rd edition. Springfield, Illinois, Charles Thomas publisher.

Stloukal, M. and Hanáková, H. (1978) Die Länge der Längsknochen altslawischer Belvökerungen, unter besonderer Berücksichtigung von wachstumsfragen. *Homo* 29, 53–69.

Trujillo, J., Rissech, C., Huertas, J. and Lloveras, L. (2023). El convento de Santa Caterina: estudio bioantropológico de una muestra de las inhumaciones del claustro gótico y la cabecera del ábside. *Rodis, Journal of Medieval and Post-Medieval Archaeology* 6, 217–42.

Ubelaker, D. (2007) *Enterramientos humanos: excavación, análisis, interpretación. Desterrando huesos.* 7th edition. Donostia, Sociedad de Ciencias Aranzadi Zientzi Elkartea.

Varela, T. (1974) Estudio antropológico de los restos óseos procedentes de necrópolis visigodas de la Península Ibérica, Tesis Doctoral. *Trabajos de Antropología* 17(2-3-4), 7–157.

Vives, E. (1990) *La població catalana medieval. Origen i evolució.* Vic, Eumo Editorial.

Walker, P., Johnson, J.R. and Lambert, P.M. (1988) Age and sex biases in the preservation of human skeletal remains. *American Journal of Physical Anthropology* 76(2), 183–8.

Weaver, D.S. (1979). Application of the likelihood ratio test to age estimation using the infant and child temporal bone. *American Journal of Physical Anthropology* 50, 263–69.

Zaldívar, A.M. (2012) Patricians' embrace of the Dominican Convent of St. Catherine in thirteenth-century Barcelona. *Medieval Encounters* 18(2), 174–206.

7

The cloister of the monastery of Santa Maria de Roses in the early modern period: study of a secular funerary space in a religious centre

Marc Bouzas, Neus Coromina, Lluís Palahí and Jordi Vivo

Founded in the 10th century, the monastery of Santa Maria de Roses (Roses, Girona, Spain) has played a major role in the town's history, first as the focal point from which the medieval town of Roses itself developed, and later as a religious and secular power. From the 16th century onwards, the monastery and the town of Roses were enclosed within the walls of a new fortress traditionally known as la Ciutadella. A century later, the civilian population moved to a new space located outside these defences. The monastery, in full decline, continued to exist as a religious house until the end of the 18th century, and its church fulfilled, until the end, the functions of a parish church. In the 17th and 18th centuries, the interior of the church and a small part of the cloister were used as a burial space for the town's population. The written sources reveal that the interior of the church was mainly for the (economically and socially) most outstanding families, while the graves of more modest inhabitants were located in the surrounding areas.

This study presents the results of the work carried out in the years 2022 and 2023 in the cloister area, a space intensively used in the early modern period for burials. From an anthropological point of view, our research has enabled us to delve deeper into a type of population context that has been little studied both in terms of the type of population and, especially, in terms of its chronology. This analysis is complemented by an analysis of the adaptation of the monastic architecture to the new situation and the difficulties involved in making military, monastic and funerary uses compatible in a very small space, particularly when profoundly affected by a series of armed conflicts.

Keywords: Monastery, anthropology, funerary practices, lay people, early modern period, Roses.

Introduction

The town of Roses is located in the extreme north-east of the Iberian Peninsula, dominating the bay of the same name. This is an exceptional space due to its location – on the main maritime route in the north of the western Mediterranean – and its protection from winds, which made it a natural harbour (Fig. 7.1). From a commercial and military perspective, this strategic location has long marked the history of Roses. These characteristics were the main reason for a Greek colony being founded there in the 4th century BC (*Rhode*) (Puig and Martin 2006, 614) and explain the existence of a *vicus* in the Roman period and Late Antiquity. It is this strategic value that made Roses the main port of the County of Empúries in the medieval era, culminating in the 16th century with the construction of a bastioned fortification, which, from that moment onwards, emphasised the military significance of the bay.

The Benedictine monastery of Santa Maria was founded in the 10th century (Palahí *et al.* 2022) and was the nucleus around which the town of Roses was built in the 11th century. It is not clear whether the new town had its own church at some point or whether the monastery also acted as a parish church throughout its history. Whatever the case, the monastery was clearly established in the early modern era.

In the middle of the 16th century, the town became a true frontier fortress with the construction of new bastioned

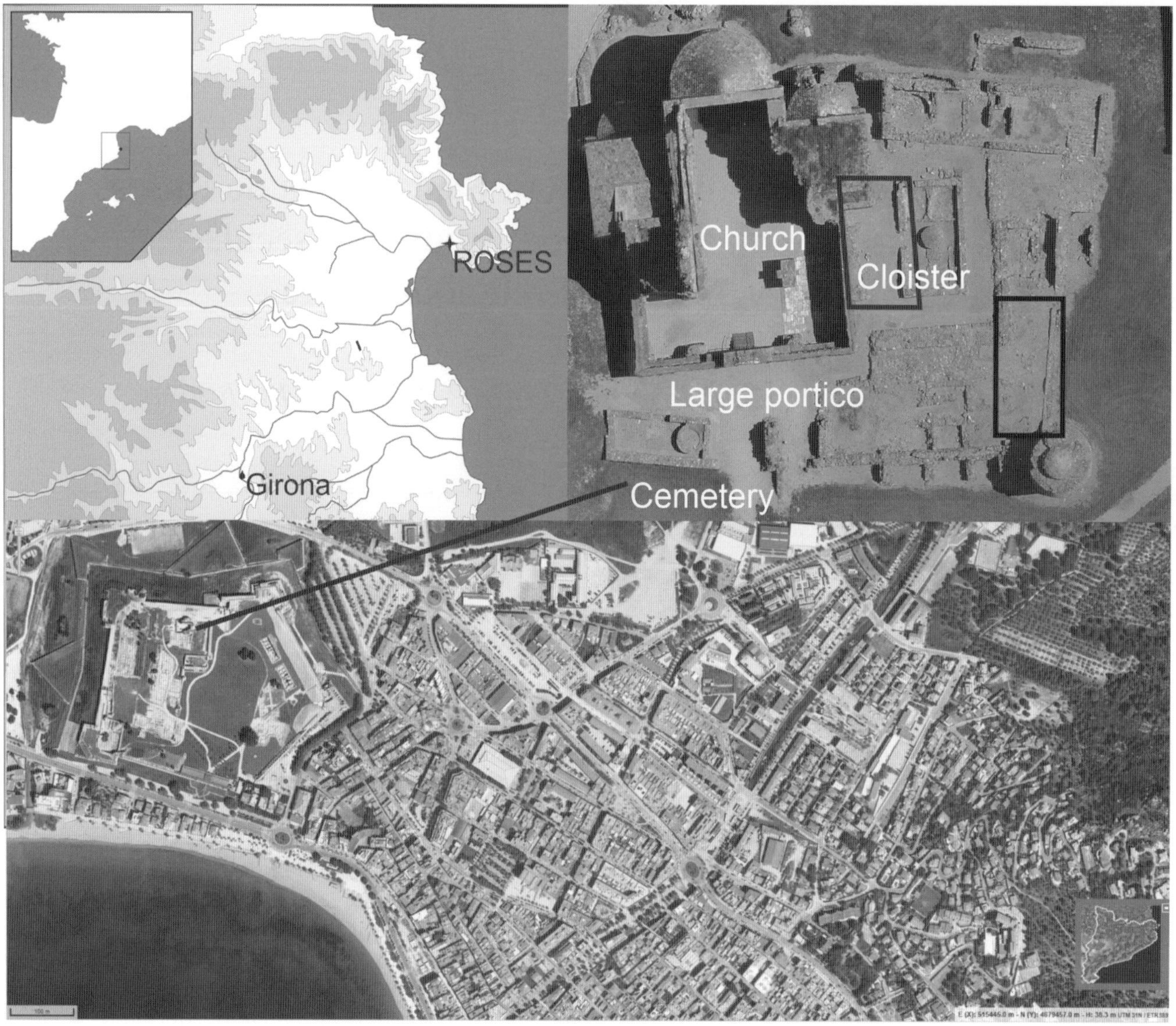

Figure 7.1. Location map for Roses and the archaeological site of the monastery of Santa Maria de Roses.

defences and the presence of a permanent military garrison (De la Fuente 1998). A century later, the civil population left the interior of the fortress and created a new urban centre a few hundred metres to the north of the fortified enclosure (De la Fuente 1998, 179–84). From then on, the monastery, which was already in full decline, was surrounded by an eminently military area. However, the church continued to function as a place of worship and parish cemetery until the end of the 18th century. The area of the parish cemetery was originally located to the west of the monastery, but in the early modern period (17th–18th century) it expanded and occupied spaces within the monastery. These included the area in front of the church entrance, the interior of the church and part of the cloister. The parish death registers note that on 31 January 1794, the governor of the stronghold transmitted to the rector an order prohibiting further burials in the cemetery situated around the church of Santa Maria. At that time, the monastic community no longer existed. A few months later, the church was destroyed during bombardments of the stronghold by French troops during the War of the Convention (1793–1795) (Solà and Puig 2012, 196).

The latest archaeological excavations carried out in the cloister of the monastery enabled part of the parish cemetery to be recorded. The written evidence, particularly the death registers (CAT ADG 3/257 1 4 1 01 and 02), can be used to date the period in which this space was used as a funerary area with great accuracy (September 1696 to January 1794). The importance of this source is that it indicates the sector of the cemetery where each of the deceased was buried, a fact that allows a direct link to be established with the anthropological study carried out. The anthropological study of a sample of the burials recovered, associated with data from the death registers, provides valuable information on the social configuration of the town of Roses in the 18th century.

Materials and methods

The materials come from two essential sources that are not always available to researchers. The first of these is the archaeological excavations carried out in the monastic cloister, which led to the recovery of a significant number of individuals buried in this space. However, as will be seen, the lack of space meant that considerable disturbance took place, involving the destruction of a significant number of graves to generate new spaces for the newly deceased. The anthropological study of these remains constitutes the central focus of the study.

The second source of material is the study of the death register entries from the period preserved in the Diocesan Archive. These enable us to identify the people who were buried in the cloister area, which is where the excavations took place. In addition, this volume often provides data such as the age or profession of the deceased (or of their family). The same source also describes the people buried in other nearby spaces such as the church itself. Consequently, a comparison can be made between the types of population that were buried in one space or the other.

The precisely defined period when the first and last individuals were buried in this space provides a portrait of a substantial part of the population of Roses in the 18th century.

Written sources

The Girona Diocesan Archive houses the death registers for this period (CAT ADG 3/257 1 4 1 01 and 02). These documents are relevant, since from the time of Council of Trent (1545–1563) it was obligatory to keep an adequate record of deaths. Although the first volume preserved has 1666 as its starting date, in fact the first entries date from 1660. These are documents of unequal value, given that the data that are included on each deceased person vary depending on who was responsible for registering the event. For example, some of the priests included the place of burial and the family background or, particularly in the first years of the volume, details such as the donation made for the funeral rites. Other priests only entered the name of the deceased without further details. In some cases, the profession of the deceased was given or, especially in the case of child burials, the father's occupation, which gives us an idea of the family's socio-economic level. For this reason, the data that are gathered are a good indication of the number of people buried in the spaces that constituted the parish cemetery. However, it should be borne in mind that when certain aspects, such as professions, are analysed, the data are not exhaustive.

A controversial element is the definition of the age group that is referred to in the documents as 'albats'. Some authors consider that 'albats' generally refers to unbaptised children, although they accept that the term could have a wider meaning in the documents (Puig *et al.* 2002, 304). Many researchers consider that in fact the documentation refers to 'albats' up to the time of communion, which would put their age in a range from 9 (Nadal 1992, 3–48) to 13 years of age (Gual and Millàs 1999, 71–2). The documents do not specify more about the term, which is why it is not possible to narrow down these age ranges, although most are likely to refer to very young children (0–3 years).

An important fact is that from 1784 the 'albats', that is, young children, are not registered individually and are listed at the end of the annual register without specifying their burial place, which is given the generic name of 'cemetery'. From the notes on previous years, we know that the cloister or the church were also common burial places for these infants.

According to the topographical indications provided by the documentation, we can divide the funerary space of the parish cemetery into four main zones (Fig. 7.1). The most frequently mentioned place name is the cemetery itself, a space that was situated to the north-west of the church, in the former monastery square. The second and most privileged space was the interior of the church and the third was the cloister. A fourth space corresponded to what was called in the documentation 'davant el portal Major' (in front of the main entrance), an area situated to the west of the church, between the façade and the boundary of the monastic enclosure.

The first reference to burials in the cloister is dated September 1696 and the last corresponds to 25 January 1794, only five days before the military authorities prohibited interments within the Citadel, as described in the register of deaths. From that time, the deceased were buried in the 'new' cemetery, situated in the surroundings of the new parish church that was being constructed in the centre of the town outside the fortress. Usually, the written sources refer to the 'cloisters' but, on some occasions, they mention the 'portal petit' (small doorway) to refer to the entrance that, situated in the southern wall of the church's nave, connected it to the cloistered courtyard.

According to the burial registers, in these 98 years, at least 231 individuals were buried in this space. This represents an average of little more than two people per year, which might seem a small number but in fact represents a very high density if it is considered that the available space was barely 50 m^2. Excavations undertaken in the east and south galleries of the cloister show that only the north gallery and the space in front of the main façade of the church (the main entrance) were used as funerary spaces. This largely explains the high level of destruction of most of the identified skeletons, given that the interment of each newly deceased person implied the total or partial destruction of previous remains.

As will be seen, a comparative analysis of the area of study inside the church – a privileged space par excellence – reveals a certain distribution of the deceased depending on their social status.

Archaeological studies

The archaeological work that started in 2022 focused on two spaces in the monastery's cloister: the north and south galleries (Palahí *et al.* 2024) (Fig. 7.1). The north gallery is the closest to the church, with which it is connected through a door described in the documents as the 'porta petita' (small door) to distinguish it from the 'porta gran' (large door) situated in the west façade. This seems to be the area where the cemetery zone named in the documentation as the 'cloisters' seems to be concentrated, given that no burials have been recovered from the south gallery for this period, and the same is true of the east gallery, which was excavated in the years 1945–1946 (Nolla 1997, 122). The west gallery still remains to be excavated, but currently this space has pebble paving that is dated to the 18th century, with no appreciable repairs. Consequently, funerary use of this area can be ruled out.

The north gallery is a space of barely 50 m^2. During the archaeological excavation, a total of 63 individuals could be identified. However, in many cases, unfortunately, the remains are partial as they were affected by other subsequent graves. In some cases, the decomposition of the coffins caused the remains of a deceased person to become mixed with those of the person immediately below. In other cases, when new graves were dug, the remains of previously deceased people were partially destroyed. For this reason, the study focused on 56 individuals, the best preserved (the ones from which it proved possible to recover more than three articulated skeletal elements). According to the death registers, more than half of those buried in the cloister (53.25%) were 'albats': infants or children, within a range that could reach up to 13 years old (a broad concept that includes in a single group the perinatal, children I and children II of the anthropological study). However, the excavations only made it possible for seven burials that could be situated within this age range (12.5%) to be recovered.

The space was barely 5 metres wide, but not all the surface area could be used as a funerary space, given that against the church's façade were features such as stone steps. Other elements that also reduced the useful space, such as the steps to access the place of worship or a structure that occupied all the north-east corner, which was probably a notable family tomb, or an ossuary, were distributed along the length of the façade.

The bodies were laid out in accordance with two orientations. Most were positioned from north to south, aligned with the south wall that separated the pathway of the central courtyard from the cloister. In contrast, the remains closest to the church's façade tended to be laid out from east to west. To bury them, a simple grave was dug in the subsoil and the body was put in it in a shroud or a wooden coffin. The only grave goods that were found in some of the tombs were small rosaries.

The inside of the church was the subject of two archaeological surveys (1999–2000). In these surveys, it was determined that the use of the church's subsoil as a funerary space began in the 15th century and continued until the 18th century (Puig *et al.* 2002). In the latter century, various family tombs were constructed. These are quadrangular pits lined with stone and mortar walls and covered with a vault that had an opening so that several bodies could be placed inside. To build these tombs, existing graves had to be destroyed, as well as much of the historical stratigraphy of this space; an area that had been inhabited since the 4th century BC. It was these large local families, as well as soldiers and ecclesiastics, who mainly had the privilege of being buried in the church in the 18th century. Some of these families are mentioned in the death registers, particularly in the latest entries (especially from 1790). For example, we could mention the Llach, Marés, Coll, Ferré or Gelabert families.

Anthropological material

Between 2022 and 2023, a total of 63 individuals were identified in the north gallery of the cloister of Santa Maria de Roses. Of these, 56 were suitable for anthropological study. The state of preservation was not particularly good as the space, which was quite small, had been used intensively, which led to the superimposition of many of the graves. Consequently, many individuals were only partially preserved.

The anthropological analysis of the individuals consisted of estimating their age, sex and the diagnosis of bone and dental diseases. Based on metric data, the stature of some of the adult individuals was also estimated.

Broad parameters were used to determine age (Buikstra and Ubelaker 1994). To establish the age of subadult individuals, four groups of age ranges were used: perinatal (up to 3 months), child I (0–3 years), child II (3–12) and adolescent (12–20). To determine the age in all these subadult cases, the main markers such as dental development and the state of maturity and growth of the skeleton were considered (Schaefer *et al.* 2009). For adult individuals, we differentiated between young adults (20–35 years), middle adults (35–50) and old adults (over 50 years). To classify adult individuals, dental wear was mainly used as a marker (Brothwell 1981), as in many cases, even if the cranium had not survived, teeth were recovered from the jaw or maxillary fragments. In addition, in the best-preserved individuals, classical methods were used, such as observation of the auricular surface, the pubic symphysis and cranial sutures (Lovejoy *et al.* 1985; Meindl and Lovejoy 1985; Brooks and Suchey 1990). If a skeleton could not be classified into any of these categories, but it was recognised as that of an adult individual, the concept of 'adult' without further determination was used, particularly in cases of partial or badly preserved bodies in which age markers could not be observed, but all the (observable) long bone epiphyses are fused (+20 years).

The determination of sex was carried out using morphological criteria of the pelvis, cranium and jaw (Ferembach *et al.* 1980). The poor preservation of long bones does not allow metrical parameters to be used for sex. However, as with the determination of age, some of the worst preserved individuals could not be classified and had to be considered indeterminate. For the determination of stature, the formulae of Trotter and Gleser (1958) and Trotter (1970) were used. Although the state of preservation of the remains is not very good, some biometric measurements were taken in the field, to avoid the partial destruction of the long bones during their extraction.

Finally, to gain an overview of the study population, the wear of the main joints was quantified numerically to determine whether this wear could be correlated with the professional activities described in the death registers. It should be considered that rheumatoid or inflammatory diseases and traumas, among other factors, could alter the course of joint wear. Consequently, this quantification is only intended to give an idea of the general trend in the population. The categories are: not observable (the skeletal part in question has not been preserved or the degree of wear cannot be observed due to taphonomic agents); 0 (no wear is evident); 1 (the first signs of wear can be observed on the joint surface); 2 (wear of the joint surface is evident but there are no acetabular rim osteophytes; 3 (notable joint wear and formation of osteophytes); 4 (the joint has been adversely affected).

In addition, the same quantitative exercise was undertaken with some muscle insertions. Several factors were considered to select which insertions to analyse, such as whether there was skeletal diversity, or the insertions were in areas that were considered in these individuals (a simplification of Mariotti *et al.* 2007). All the quantification was undertaken by the same person to avoid subjective bias (Davis *et al.* 2013). As with the joint wear, the results could provide an overview of this population, without trying to individualise specific markers of work and instead focus on 'entheseal changes' (Villotte and Knüsel 2013).

The selection consisted of deltoid insertions in the humerus, the brachial biceps insertions in the radius, the muscle insertions of the phalanges of the hand, the rectus femoris tendon insertion in the patella, the patellar ligament insertion in the tibia, the origin of the soleus on the tibia and the Achilles tendon insertion. In this case, the categorisation was as follows: not observable (the insertion in question has not survived or cannot be observed); 0 (there is no morphological change in the enthesis); 1 (there are slight changes or thickening of the enthesis); 2 (there are moderate changes in the entheses but no osteophytes); and 3 (there are moderate changes in the entheses and the presence of osteophytes).

The analysis of dental disorders, arthropathies and entheses was carried out in young individuals and adults following macroscopic observation (Chimenos 2003; Ortner 2003). Children were excluded.

Results

Written source results

One of the most interesting results arising from the study of the death registers are the social differences established between the area of the cloister and the interior of the church, a space that was reserved for certain social statuses and families. Notably, in the 18th century, a series of small family crypts was built inside the church; these were owned by the main families in Roses, many of whom had collaborated and financed the numerous altars that were distributed around the church.

An analysis of the death registers revealed some information that complements the anthropological studies. Of the 231 burials in the cloister, 118 were male (of whom 73 were 'albats') and 105 were female (58 'albats') (Table 7.1). This implies that infants or children of a young age comprised 53.25% of the people buried in this space. This percentage is similar to that of the area in front of the main entrance (54%), but much higher than that found inside the church (28.19%).

In addition, interesting nuances emerge from the analysis of the occupations that were carried out. In the case of the numerous 'albats', the profession refers to that of the father, but it is a certain indicator of the economic and social level of the family. The most common occupations were those of peasant farmer and fisherman, which constituted the basis of the socio-economic structure in Roses in the 18th century (Table 7.2). Peasant farmers represented 37.74% of those buried in the cloister, 14.29% in the area of the main entrance and 21.59% inside the church (Given the length of the death registers and the aim of this article, the space called the 'cemetery' in the registers is not analysed here. This was the main burial area, whose study is pending for future research).

Table 7.1. Distribution of sex and age according to the death book

Sex	*Cloister*		*Main entrance*		*Church*	
Male	118		63		220	
Female	105		58		139	
Not specified	8		3		17	
Total	213		124		376	
ALBAT		*%*		*%*		*%*
Male	73	61.86	35	55.56	62	28.18
Female	50	47.62	32	55.17	44	31.65
Total	123	53.25	67	54.03	106	28.19

Table 7.2. Summary of the different professions and their distribution according to the death register (in italics. original citation)

Profession	*Cloister*	*% Cloister*	*Main entrance*	*% Main entrance*	*Church*	*Total*	*% Total*
Grocer/*Adroguer*	4	2.52		0	6	10	2.07
Apothecary/*Apotecari*	10	6.29		0		10	2.07
Cooper/*Barriler*	3	1.89	6	6.12	4	13	2.69
Barrel maker/*Boter*	4	2.52		0	4	8	1.65
Shopkeeper/*Botiguer*		0	1	1.02		1	0.21
Field hand/*Bracer*	2	1.26	2	2.04		4	0.83
Surgeon and hospital workers/ *Cirurgià i treb. Hospital*	13	8.18	5	5.1	15	33	6.82
Blacksmith/*Ferrer*	2	1.26		0		2	0.42
Carpenter/*Fuster*	1	0.63		0		1	0.21
Merchant/*Mercader*		0	1	1.02	2	3	0.62
Builder/*Mestre de cases*	1	0.63		0		1	0.21
Military/*Militars*	5	3.14	4	4.08	67	76	15.7
Dealer/*Negociant*	7	4.4	8	8.16	12	27	5.58
Notary/*Notari*		0		0	1	1	0.21
Farmer/*Pagès*	60	37.74	14	14.29	49	123	25.41
Shepherd/*Pastor*	1	0.63		0		1	0.21
Ship's master/*Patró vaixell*	1	0.63	1	1.02	3	5	1.03
Fisherman/*Pescador*	43	27.04	55	56.12	42	140	28.93
Clergy/*Religiós*		0		0	21	21	4.34
Worker/*Treballador*	2	1.26	1	1.02	1	4	0.83
Total	159		98		227	484	

Fishermen represented 27.04%, 56.12% and 18.5% respectively. In terms of the peasant farmers and fishermen buried inside the church, it should be considered that some of them were within the family crypts discussed above.

In contrast, two social classes were buried almost exclusively within the church. One of these comprised ecclesiastics, whether they were monks from the monastery or priests. All of them were interred in privileged places within the church, generally in the area of the chancel. The other professional group was composed of soldiers. Of the 78 soldiers registered, 67 were buried in the church, including some high-ranking officers and governors of the stronghold. Notably, the few soldiers documented in the area of the cloister or the main entrance date to a very specific period that is situated between 1710 and 1724. At this time, it seems that the area of the monastery was undergoing various remodelling works after the military events that occurred during the War of the Spanish Succession (1701–1714) (De la Fuente 1998, 362–7).

Most professions from the secondary sector such as coopers, blacksmiths, apothecaries or labourers were buried outside the church. Again, the exception seems to correspond to those families that had a family tomb inside the church. That is, although some professions were found inside and outside the church, families were generally not found in both places, which indicates a certain difference of status within the same profession.

Anthropological results (result of the study of the population buried in the north gallery of the cloister)

Age and sex

As shown in Figure 7.2 and Table 7.3, a total of 56 individuals were analysed from the north gallery of the cloister of Santa Maria de Roses. Regarding the distribution of age in this population segment according to the skeletal remains, there was only a very small representation of child remains ('albats'): no perinatal individuals; three child I individuals (5%) and four child II individuals, which correspond to 7% of the sample. Regarding adult individuals, only 5% of the sample are young people while individuals classified as young adults (13 individuals) comprise 23% of the total. The best represented population segment is that of mature adults with 19 individuals (34%). In contrast, only three individuals are classified as old adults. The rest of the sample is made up of indeterminate adults, with 11 individuals representing 20% of the entire population.

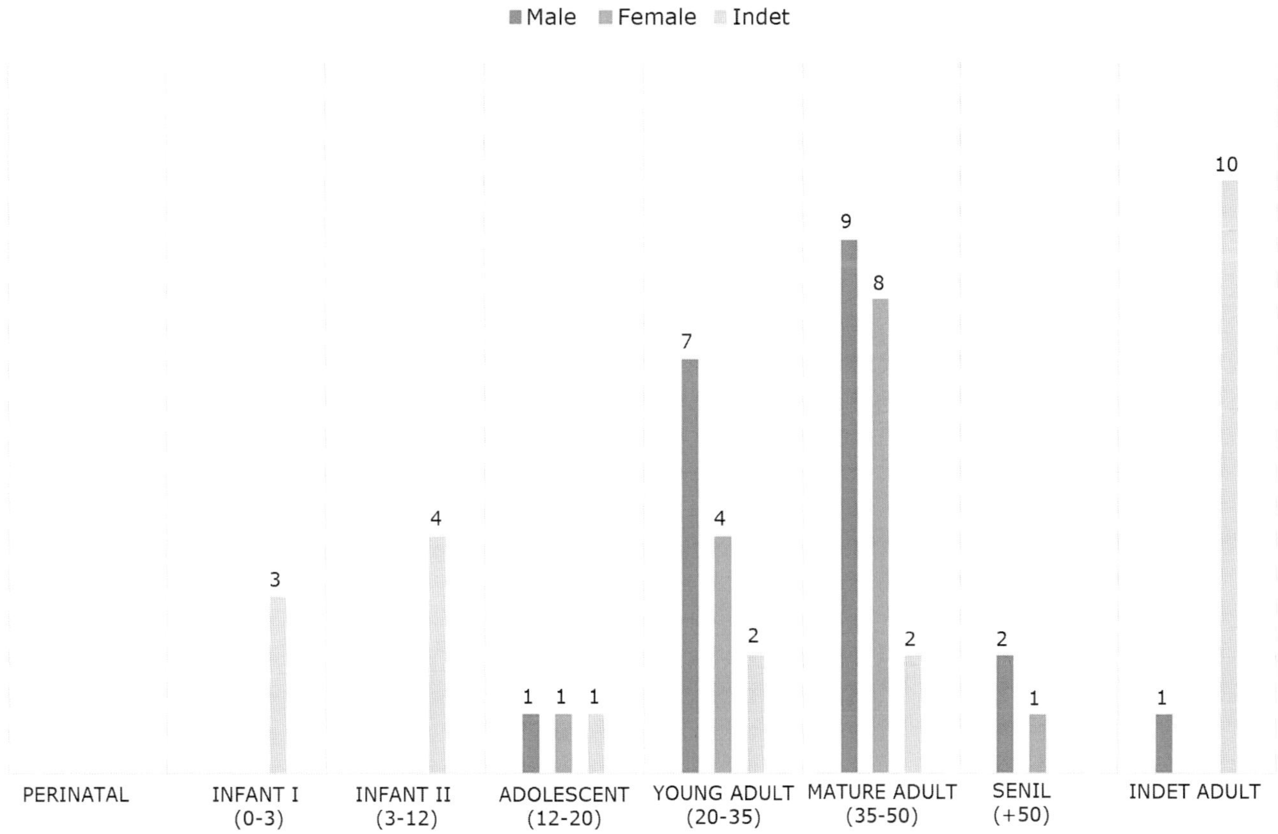

Figure 7.2. Graph of distribution of sex and age according to anthropological study.

Regarding the distribution of sex in the sample, unfortunately a high percentage of individuals are indeterminate: 22 individuals, representing 39% of the sample. Male individuals (36%) predominate over the 25% of individuals identified as female.

If the distribution of age by sex is analysed, the first observation is that the high number of individuals of indeterminate sex is partly due to the fact that the sex of the seven children (and one of the adolescent individuals) could not be determined. Of the individuals that could be sexed, it can be observed that, both in men (9=16.07%) and women (8=14.29%), the most represented group is the mature adult range. In percentage terms, there are more young men (7=12.5%) than women (4=7.14%). In contrast, representatives of old individuals include two men (3.57%) and only one woman (1.8%), although this result is not very significant, as there are only a total of three individuals.

Among the pathological results, we will mainly describe the dental disorders (Fig. 7.3; Table 7.4). The population analysed in this study had a high incidence of dental hypoplasia (a pathology that could be a result of periods of nutritional stress or disease during infancy and childhood). Of the total number of individuals in whom this 'disease' could be observed (17), 94% suffered from it and only 6% (just one individual) did not. The incidence of caries is much lower: of the 21 individuals in whom this dental pathology could potentially be observed, it was identified in 10 (48%). In the other 11 individuals, no evidence could be found (52%). On the other hand, in the 20 observable individuals, 12 had alveolar bone resorption in some teeth and only 8 did not. Finally, dental calculus was present in 16 of the 22 observable individuals (around 67%).

Table 7.3. Distribution of sex and age according to anthropological study

	Male	*Female*	*Indeterminate*	*Total*
Perinatal	0	0	0	0
Child I (0–3)	0	0	3	3
Child II (3–12)	0	0	4	4
Adolescent (12–20)	1	1	1	3
Middle Adult (20–35)	7	4	2	13
Mature Adult (35–50)	9	8	2	19
Old Adult (+50)	2	1	0	3
Indeterminate Adult	1	0	10	11
Total	20	14	22	56

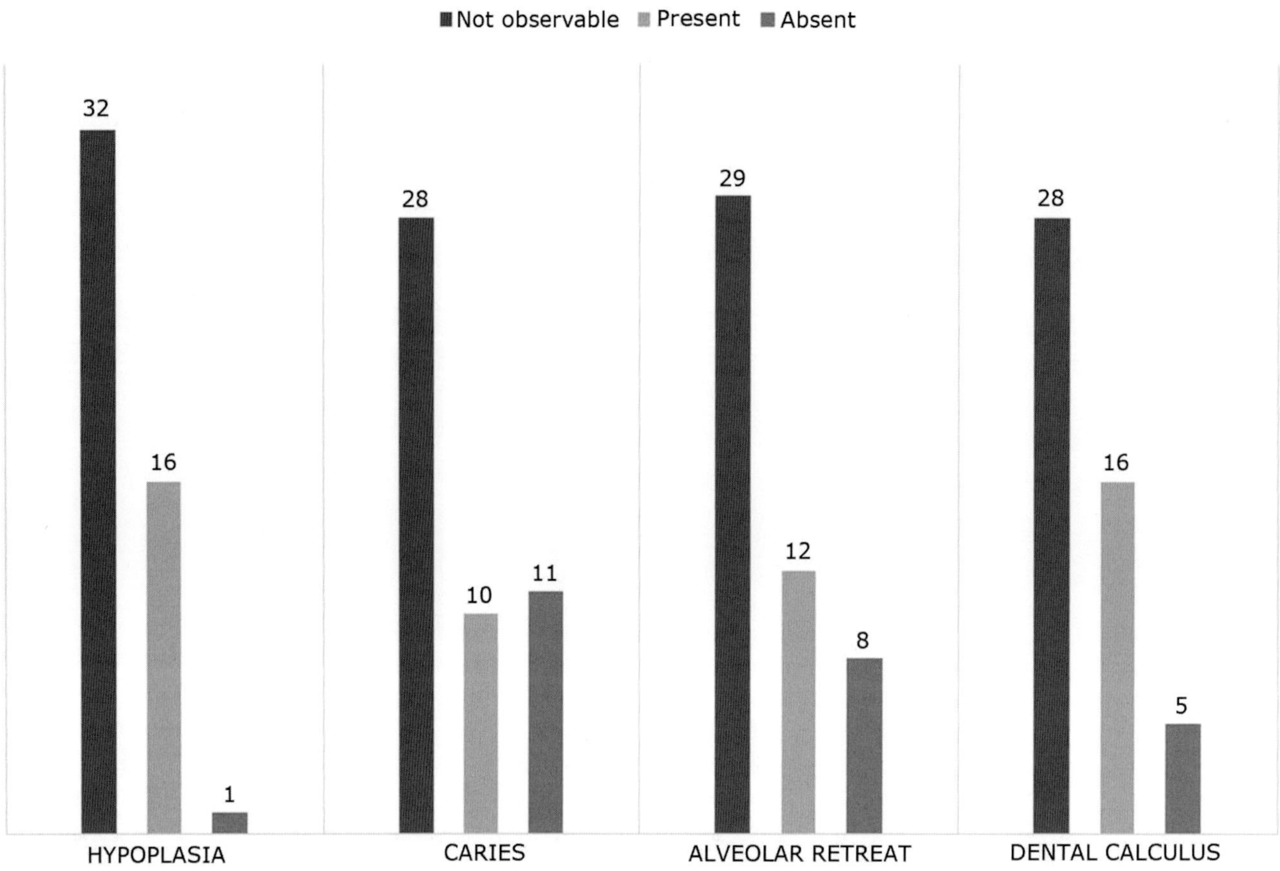

Figure 7.3. Graph of dental diseases.

Table 7.4. Summary of dental diseases

	Not observable	*Present*	*Absent*
Hypoplasia	32	16	1
Caries	28	10	11
Alveolar resorption	29	12	8
Dental calculus	28	16	5

Height

Of the total sample, the height of 23 individuals could be calculated: 11 women and 12 men (Fig. 7.4). The maximum height of a man is 176.90 cm and the minimum is 160.52 cm, with a mean of 168.84 cm (median 169.58 cm). For women, the highest value is 169.29 cm and the lowest is 149.93 cm, with a mean of 159.91 cm (median 160.92 cm). As shown in Figure 7.4, there is slightly more variation in women than in men, although they follow a similar trend.

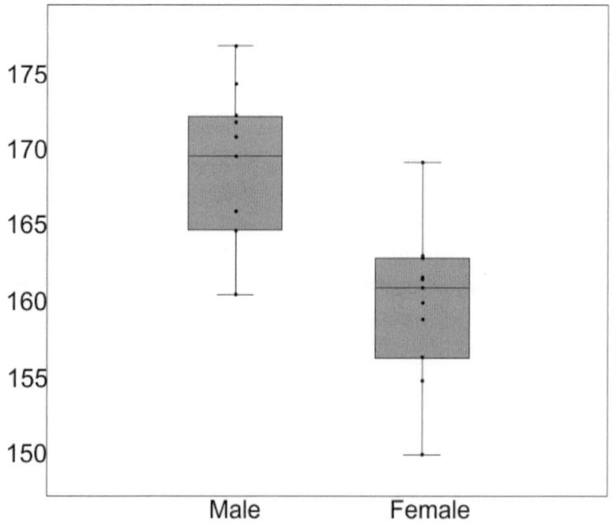

Figure 7.4. Height boxplot of the population studied.

Joint wear

The general picture offered by the quantification of joint wear is far from exhaustive, as the poor state of preservation of some of these joints means that we can only obtain a partial view. However, the results do indicate trends in the population, as shown in Figure 7.5 and Table 7.5. The shoulder and pelvis are the most poorly preserved parts, as they can only be observed in 49% and 43% of the sample respectively. In contrast, the elbow is the best represented part, as it can be observed in 59%. The results for the knee and the ankle are similar, at 57%

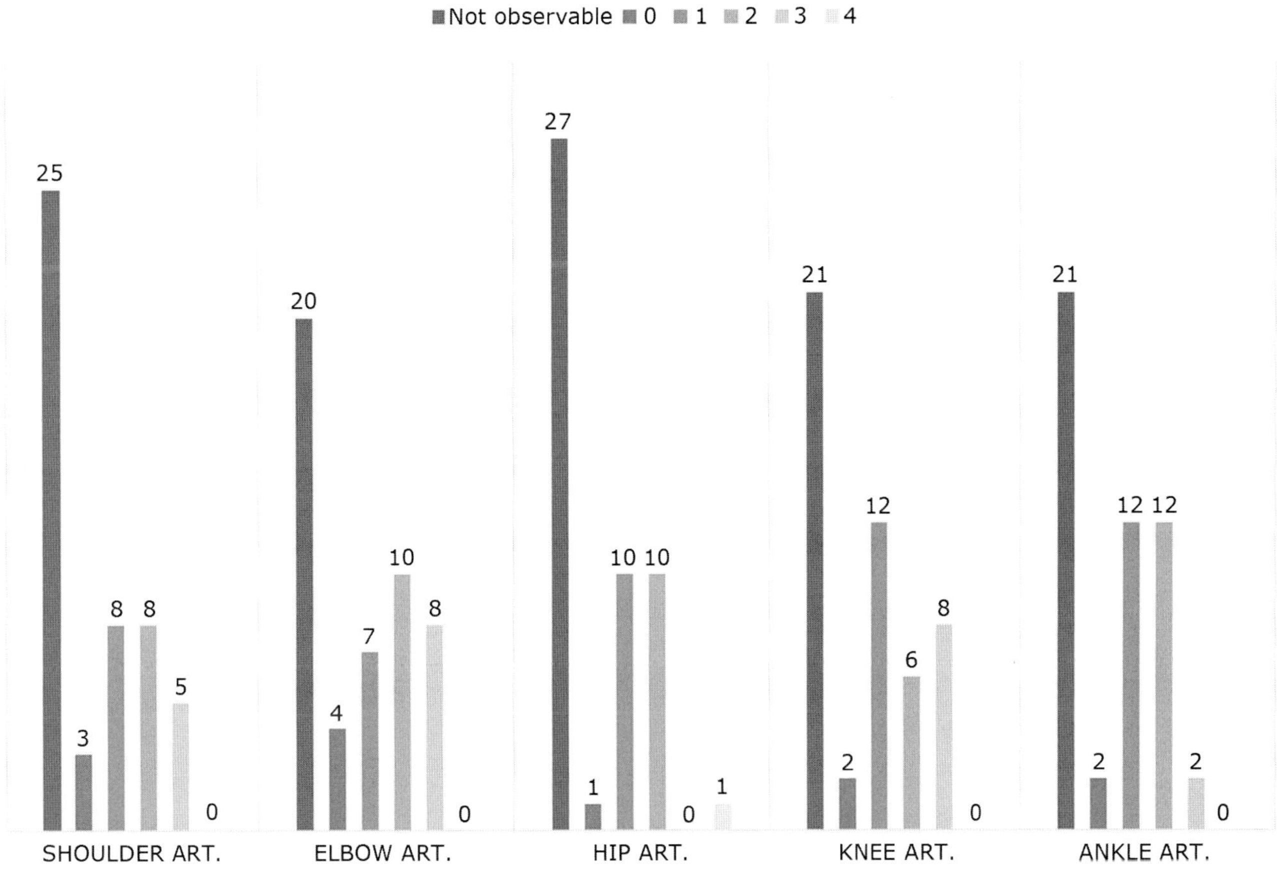

Figure 7.5. Graph of the different degrees of joint wear.

of the observable sample. From the results obtained in this process, only in one case can an individual with joint wear be observed with the highest rating. This was for one of the most poorly preserved parts of the skeleton: the pelvis.

The general trends in the study population are that the joints with the least wear are the pelvis and the ankle, in contrast to the shoulder, elbow and knee, which are the most affected zones. The knee joint is notable, as 29% of individuals have considerable wear of this joint, in many cases with the presence of osteophytes.

Table 7.5. Summary of the different degrees of joint wear

	Not observable	*0*	*1*	*2*	*3*	*4*
Shoulder art.	25	3	8	8	5	0
Elbow art.	20	4	7	10	8	0
Hip art.	27	1	10	10	0	1
Knee art.	21	2	12	6	8	0
Ankle art.	21	2	12	12	2	0

Muscle insertions

As with joint wear, the information generated from the observation of muscle insertions is also partial, due to the state of preservation of skeletal remains (Fig. 7.6; Table 7.6). The two insertions that are most poorly represented are the insertion of the rectus femoris in the patella, which could only be analysed in 16 individuals (around 33%) and the patellar ligament in around 35%. The insertion of the brachial biceps has similar results, as it could only be observed in approximately 39% of the sample. In contrast, the insertion of the deltoids was the best represented at 61%, followed by the insertion of the Achilles tendon (55%) and the insertion of the soleus (51%).

The insertions with a higher percentage of incidence of slight changes (1) in the zone of the enthesis are the rectus femoris of the patella (62%) and the Achilles tendon of the calcaneus (48%). In contrast, in the insertion of the deltoids in the humerus (73%) and the phalanges of the hand (67%) there is a predominance of moderate changes (2). Finally, there is a higher percentage of individuals with pathological insertions (3) in the soleus at 32% and the brachial biceps at 26%. The soleus insertion has 52% of moderate changes (2) and is the muscle insertion with highest incidence of changes.

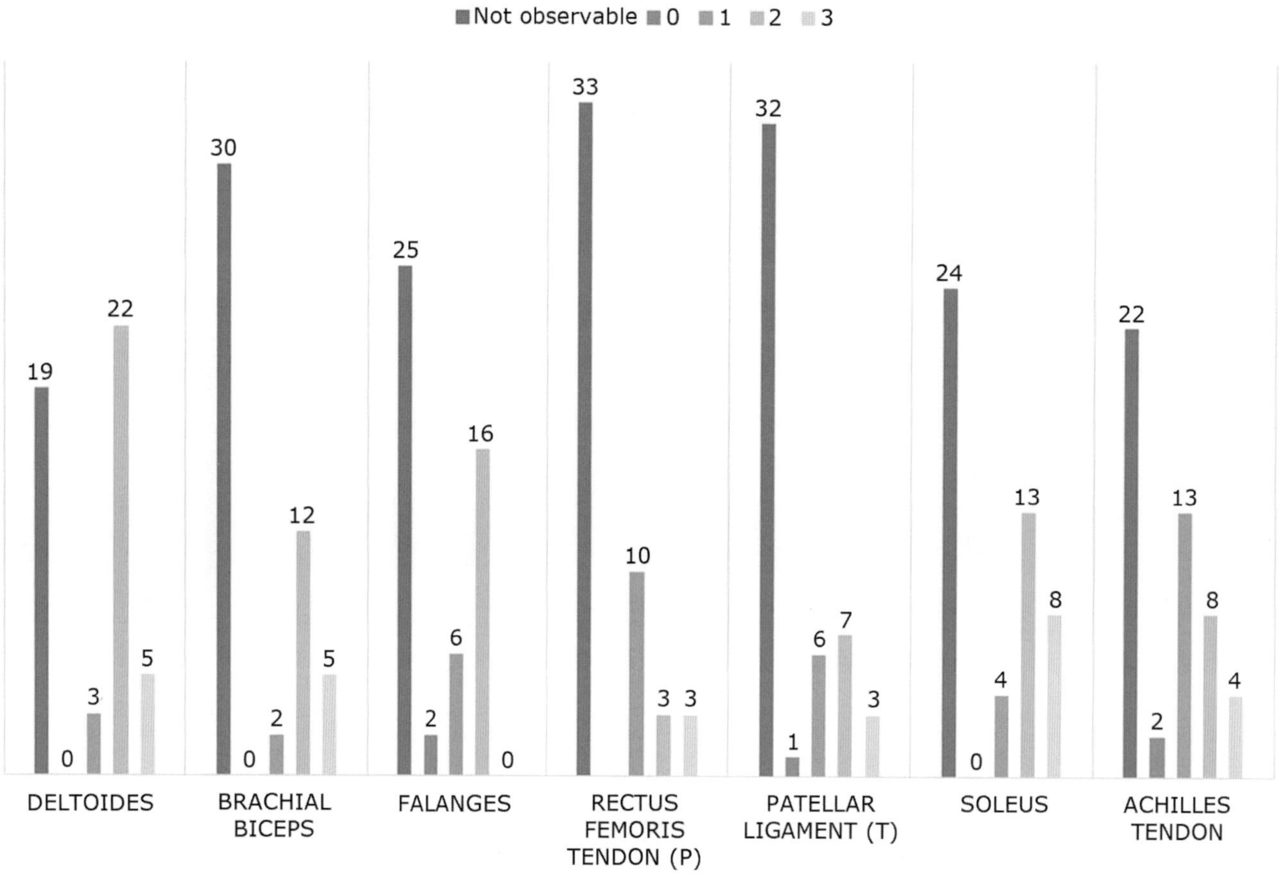

Figure 7.6. Graph of the different degrees of entheseal changes.

Table 7.6. Summary of the different degrees of entheseal changes

Total	*Not observable*	*0*	*1*	*2*	*3*
Deltoids	19	0	3	22	5
Brachial biceps	30	0	2	12	5
Phalanges	25	2	6	16	0
Rectus femoris tendon (p)	33	0	10	3	3
Patellar ligament (t)	32	1	6	7	3
Soleus	24	0	4	13	8
Achilles tendon	22	2	13	8	4

Discussion

One of the most important points in this study is the difference between the number of individuals entered in the death registers and the number of individuals recovered during the archaeological excavations. A total of 63 individuals were recovered (of whom 56 were in a suitable condition to be studied) out of 231 individuals in the registers (27%).

The archaeological sample only includes seven child individuals and no perinatal individuals. Consequently, we recorded 49 individuals, counting adults and young people. The death registers include 108 individuals, if young people and adults are considered. Therefore, the anthropological sample represents 45.37% of this population. Part of this difference is explained by the superimposition of tombs, which led to the constant destruction of the oldest graves. The loss of a significant part of the anthropological sample is not just a problem in our cemetery. It is an element that should be considered when populations recovered in old necropolises are analysed.

The death registers, excluding the 'albats', mention 45 men (41.67%) and 55 women (50.93%) in addition to at least eight entries in which the text is damaged and does not enable identification (7.41%). In contrast, in the anthropological sample, the trend is the opposite: out of 49 individuals, 20 are male (40.81%), 14 are female (28.57%) and 15 are not identified (30.61%). These data lead to a clear reflection: the problem of statistically analysing populations with a high number of individuals whose sex is not clearly determined. This can distort the overview of the population analysis if other sources of information are not available, as in our case.

These data reveal another important factor: the small number of 'albats' present in the archaeological record.

'Albats' made up the majority of deaths in the early modern era, and not just in Roses. An example can be found in a study undertaken in the Miravet district in the 17th to 18th centuries. This area covers five municipalities, in which the mortality of 'albats' between 1700 and 1821 was generally above 50% and in no period below 43% (Cañabate 2018, 105). However, 'albats' account for only 12.5% of the individuals identified archaeologically. Inside the church, of the 63 individuals studied, 13 (21%) are in the category of 'albats' (3 between 0 and 1 year; 8 between 1 and 5 years and 2 between 5 and 10 years) (Puig *et al.* 2002) Various factors may have influenced the preservation problems. The first derives from the constitution of the deceased, since perinatal and infant bones are weaker. Another factor could be dampness, given that the burials in the cloister were deposited in sand levels that filtered a lot of water. There is no doubt that the bodies deposited inside the church were better protected, placed under a flagstone pavement, protected from damp. The tombs in the church appear to have been marked on the floor and the destruction of graves was much less. In addition, the construction of family tombs in the 18th century also facilitated the good preservation of the remains (Puig *et al.* 2002, 305). According to the death registers, in the period from 1660 to 1800, this age group comprised 28% of the total. This figure is much more accurate but notably much lower than that found in other spaces (in front of the main entrance, 'albats' make up 54% of the individuals recovered). In the north gallery of the cloister, the total lack of perinatal remains and the very low proportion of child individuals could hypothetically be explained by the fact that the burial space for this population segment – particularly children of an age younger than between 0 and 3 – was within a structure, such as a communal ossuary, that has not survived to the present. Unfortunately, the death registers include under the concept of 'albat', children belonging to a wide age range (between 0 and 13 years). The registers do not specify how many of these were in the earliest stages of life, in a historical period in which the death of the newborn and children under 3 years of age was very high.

This is not a new problem as it has already been discussed on various occasions, particularly in studies of rural necropolises in Late Antiquity, where the burial regulations were much laxer. In these places, it has been proposed that some child burials that were not found in the excavated spaces may have been deposited in a different place to the rest of the burials (Chavarría and Marinato 2015; Roig 2015; Prat *et al.* 2019). In addition to the problems arising from the state of preservation of remains that have already been discussed, the customs in force in each period regarding the treatment of children's bodies or, in the case of Christian cemeteries, the impossibility of burying the unbaptised in sacred ground, must be added. At this point it should be noted that the death registers indicate that practically all the deceased mentioned in them were baptised.

The information about professions taken from the death registers reflects the difference between the population buried in the cloister and that within the church. It suggests that the population within the cloister was of lower status. However, the position of this cloister and that of the area known as the 'Portal Major' was close to the church and therefore near the altar and the most sacred area of the church building. This meant that they were spaces that were more desired by the faithful than the cemetery itself. The people buried in these spaces were probably not the most privileged members of the population, but they were also not the poorest, who would have ended up being buried in the cemetery.

Anthropologically, we have hardly any comparative data with the population buried within the church. The available information corresponds to an excavation on which only some basic data were published. One of these pieces of information is that the average height was 165.8 cm for men and 153.8 cm for women (Puig *et al.* 2002, 304). If we compare this with the average height in the north gallery of the cloister, which was 168.84 cm for men and 159.91 cm for women, it can be seen that in general the population was slightly taller. This is a relatively surprising fact since a higher status in the same population (same geographical space and same climatic conditions) would mean that the higher status group (better nutrition and quality of life, among other factors) should have a higher average height (Komlos 2009). The population buried in the church and in the cloister had a different social status on paper (for example, according to the number of ecclesiastics in the church and their absence from the cloister), but this does not necessarily translate into different food or quality of life. It has already been mentioned that the population buried in the cloister was probably not the most privileged, but neither did they constitute one of the poorest population segments.

Another way of measuring the social status of these two populations with anthropological data would be by observing the dental diseases. The problem is that we do not have data on the population buried in the church to make the comparison. However, the data on the population in the cloister are revealing: the incidence of dental hypoplasia is very high (94%), which indicates potential nutritional deficiencies during childhood (Chimenos 2003). Although it must not be forgotten that hypoplasia can also be caused by some other factors, the high incidence helps to give strength to our hypothesis. In contrast, the presence of caries is not as high: fewer than half of the individuals who had retained teeth had caries (48%). However, this fact is complementary to the high presence of alveolar bone resorption and dental calculus, which indicates widespread bad dental hygiene among this population segment.

Finally, one of the aspects that is well reflected in the death registers is the principal activity of the population in the north gallery of the cloister. A total of 37.74% worked in peasant farming and 27.04% in fishing (64.78% of the population). Other occupations recorded in this part of the necropolis are those with a high degree of physical labour, such as blacksmiths, coopers and carpenters. If we observe the degree of joint wear together with the record of changes in the entheses described above (Figs 7.5 and 7.6; Tables 7.5 and 7.6), the data are consistent with the professions described in the death registers. However, no differences between sexes were found in muscle and joint disorders. Consequently, we could consider that although the profession entered in the death registers is usually that of the husband, women who are registered as peasant farmers or fisherwomen probably collaborated in some of the physical tasks.

Conclusions

The results of this combined study of written sources and anthropology provide deeper knowledge of the social and economic structure of the town of Roses in the 18th century. They also reveal the importance of working in a multidisciplinary manner and of being able to compare, whenever possible, data from different sources that are seldom related. As we have seen, the comparison between documentary and anthropological data enables us to qualify and correct some of the data significantly.

The study also highlights the importance of contextualising the space that is analysed. Thus, for example, the death registers show that the number of children buried in the church was, for different reasons, much lower than that established considering the usual demography of the period. There were various reasons for this. First, the church was a privileged space where the unbaptised (newborns) could not be buried. Other factors could also have an impact. Some of the people buried in the church (monks) did not have any descendants and others (soldiers) may not have had their relatives at the barracks. If the characteristics of the space are not considered, the results obtained could be completely biased.

The study again brings up the problem of some necropolises with perinatal and child deaths and their archaeological record. Although it is not the subject of this study, it also reveals the difficulty presented by some of the documents. Concepts such as that of 'albat', which do not have a clear definition and could vary depending on the period and even depending on the population, can make it harder to understand certain aspects of demography in the early modern era.

Archive: Arxiu Diocesà de Girona

CAT ADG 3/257 1 4 1 01 Libre de Difunchs [de la Iglésia parroquial] de Santa [Maria] de la vila de Rosas, Bisbat de Girona (1666–1752)

CAT ADG 3/257 1 4 1 02 amb O Hosp. Llibre de Òbits de la Iglésia parroquial del Monastir de Santa Maria de Rosas, Bisbat de Gerona (1752-1808, OH 1719–1742)

References

Brooks, S. and Suchey, J. (1990) Skeletal age determination based on the os pubis: a comparison of the Acsádi-Nemeskéri and Suchey-Brooks methods. *Human Evolution* 5 (3), 227–38.

Brothwell, D. (1981) *Digging Up Bones: The Excavation, Treatment, and Study of Human Skeletal Remains*. Ithaca, Cornell University Press.

Buikstra, J.E. and Ubelaker, D. (1994). *Standards for Data Collection from Human Skeletal Remains*. Fayetteville, Arkansas, Arkansas archeological survey research series no 44.

Cañabate, J. (2018) *La batllia de Miravet (segles XVII i XVIII). Demografia, societat i economia*. PhD thesis, Universitat de Barcelona. https://diposit.ub.edu/dspace/handle/2445/132871

Chavarria, A. and Marinato, M. (2015) Frammentazione e complessità nelle pratiche funerarie altomedievali in Italia settentrionale. In P. Arthur (ed.) *VI Congresso Nazionale degli Archeologi Medievisti Italiani (Lecce 2015)*, 61–68. Firenze, All'Insegna del Giglio.

Chimenos, E. (2003) Perspectiva odontoestomatológica en Paleopatologíaa. In A. Isidro and A. Malgosa (eds) *Paleopatología: la enfermedad no escrita*, 151–61. Barcelona, Masson.

Davis, C.B., Shuler, K.A., Danforth, M.E. and Herndon, K.E. (2013) Patterns of interobserver error in the scoring of entheseal changes. *International Journal of Osteoarchaeology* 23, 147–51. https://doi.org/10.1002/oa.2277

De la Fuente, P. (1998) *Les fortificacions reials del golf de Roses en l'època moderna*. Col·lecció Papers de Recerca, 3. Figueres, El Brau.

Ferembach, D., Schwidetzky, I. and Stloukal, M. (1980) Recommendations for age and sex diagnoses of skeleton. *Journal of Human Evolution* 9, 517–49.

Gual, F.X. and Millàs, A.C. (1999) *La població del Baix Llobregat a l'època dels Àustria*. Barcelona, Fundació Salvador Vives i Casajuana.

Komlos, J. (2009) Anthropometric history: an overview of a quarter century of research. *Anthropologischer Anzeiger* 67(4), 341–56.

Lovejoy, C.O., Meindl, R.S., Pryzbeck, T.R. and Mensforth, R.P. (1985) Chronical metamorphosis of the auricular surface of the ilium: a new method for the determination of adult skeletal age at death. *American Journal of Physical Anthropology* 68, 15–28.

Mariotti, V., Facchini, F. and Giovanna Belcastro, M. (2007) The study of entheses: proposal of a standardised scoring method for twenty-three entheses of the postcranial skeleton. *Collegium Antropologicum*. Mar. 31(1), 291–313. PMID: 17598416

Meindl, R.S. and Lovejoy, C.O. (1985) Ectocranial suture closure: a revised method for the determination of skeletal age at death based on the lateral-anterior sutures. *American Journal of Physical Anthropology* 68, 57–66. doi: 10.1002/ajpa.1330680106

Nadal, J. (1992) *Bautismos, desposorios y entierros. Estudios de historia demográfica*. Barcelona, Ariel España.

Nolla, J.M. (1997) Roses a l'antiguitat tardana. El cementiri de Santa Maria. *Annals de l'Institut d'Estudis Empordanesos* 30, 107–46.

Ortner, D. (2003) *Identification of Pathological Conditions in Human Skeletal Remains*. San Diego, Academic Press.

Palahí, Ll., Pujol, M. and Aguelo, X. (2022) Les muralles del monestir de Santa Maria i la vila de Roses a l'Edat Mitjana. *RODIS. Journal of Medieval and Post-medieval Archaeology* 5, 125–50.

Palahí, Ll., Vivo, J., Coromina, N. and Bouzas, M. (2024) Ciutadella de Roses. Excavacions a la vila medieval i al claustre del monestir de Santa Maria. In *XVII Jornades d'Arqueologia de les Comarques de Girona*. Blanes.

Prat, M., Coromina, N. and Nolla, J.M. (2019) Les necròpolis tardoantigues del nord-est de la provincia Tarraconensis. In J. López Vilar (ed.) *Tarraco Biennal. Actes 4t Congrés Internacional d'Arqueologia i Món Antic. VII Reunió d'Arqueologia Cristiana Hispànica: el cristianisme en l'Antiguitat Tardana. Noves perspectives,* 139–46. Tarragona, Universitat Rovira i Virgili–Institut d'Estudis Catalans.

Puig, A.M. and Martin, A. (2006) *La colònia grega de Rhode (Roses, Alt Empordà)*. Serie Monogràfica, 23. Girona, Museu d'Arqueologia de Catalunya.

Puig, A.M., Agustí, B., Teixidor, M. and Martin, A. (2002) Segona campanya d'excavacions a l'església de Santa Maria de la Ciutadella de Roses (Alt Emporda). In *VI Jornades d'Arqueologia de les Comarques de Girona*, 301–6. Sant Joan de les Abadesses.

Roig, J. (2015) Necròpolis, aixovars, dipòsits humans anòmals en estructures no funeràries als *territoria* de *Barcino* (Barcelona) i *Egara* (Terrassa) entre el baix imperi romà i l'època visigòtica (s. IV-V al VIII). In *Necropolis and Funerary World in Rural Areas*, 33–86. Studies on the Rural World in the Roman Period, 9. Girona, Publicacions del Laboratori d'Arqueologia, Història Antiga i Prehistòria de la Universitat de Girona.

Schaefer, M., Black, S. and Scheuer, L. (2009) *Juvenile Osteology: A Laboratory and Field Manual.* Elsevier, Academic Press.

Solà, X. and Puig, A.M. (2012) El monestir de Santa Maria de Roses (1592–1835): l'annexió a Amer, la militarització i l'extinció. *Annals de l'Institut d'Estudis Empordanesos* 43, 177–98.

Trotter, M. (1970) *Estimation of Stature from Intact Long Bones.* In T. D. Stewart (ed.) *Personal Identification in Mass Disasters*, 71–83. Washington DC, Smithsonian Institution Press.

Trotter, M. and Gleser, G.C. (1958) A re-evaluation of estimation of stature based on measurements of stature taken during life and of long bones after death. *American Journal of Physical Anthropology* 16(1), 79–123. doi: 10.1002/ajpa.1330160106

Villotte, S. and Knüsel, C.J. (2013), Understanding entheseal changes: definition and life course changes. *International Journal of Osteoarchaeology* 23, 135–46. https://doi.org/10.1002/oa.2289

8

The life and death of the Poor Clare nuns of the Holy Trinity in Monte Sant'Angelo (Puglia, Southern Italy): archaeological, anthropological, pathological, botanical, entomological, textile, chemical and documentary data

Ginevra Panzarino, Elena Varotto, Francesco Maria Galassi, Agustín Pastor, Gianni Gallello, Stefano Vanin, Giuseppina Carta, Francesco Breglia, Annalisa Biselli, Donatella Pian and Elena Dellù

This study presents the first multidisciplinary results of the archaeological investigations conducted in the crypt of the Poor Clares of the Holy Trinity in Monte Sant'Angelo (Foggia, Puglia, Italy). This monastery was constructed in 1404 and rebuilt in around 1630. Beneath the nave of the church, there is an underground crypt, which comprises four rooms in which the members of a community of nuns were buried between the early 17th century and the late 19th century. The skeletal remains were laid on wooden boards crafted from white fir, wearing monastic habits and holding rosaries bearing medals and crucifixes. The climatic conditions of the crypt have enabled histological, parasitic, wood and textile materials to be preserved. The skeletal sample consists of approximately 20 bodies, and the remains of other individuals were moved into corners to accommodate new burials. All of them were covered with quicklime, as confirmed by spectroscopic analysis. The assemblage is composed of the remains of elderly women – with the exception of one young nun – who devoted their lives to cooking, sewing, painting and silversmithing. From osteological analysis, it is evident that these individuals were accustomed to kneeling for praying and lifting weights during their cloistered life. Numerous pathologies have been recorded among the skeletons, including severe osteoarthritis on the hands, back, knees and feet, but also traumas from accidents such as leg fractures. All the data were cross-referenced with documentary sources, particularly the diary of a nun named Donna Constantia, which provides insights into the community's lifestyle, health, clothing and rituals as well as the composition of the community and the social relations between the nuns and the outside world. The excavation is part of an ongoing project aimed not only at making the crypt accessible to visitors, but also at recovering the oldest church dating back to the 15th century, where the remains of other nuns may still be preserved.

Keywords: Poor Clare nuns, crypt, bioarchaeology, multidisciplinary, female monasticism.

Introduction

From the Middle Ages to the present day, Monte Sant'Angelo (Puglia, Southern Italy, Fig. 8.1) has been one of the most important pilgrimage destinations, primarily due to its Sanctuary of Saint Michael, which forms part of a pilgrimage route connecting locations from Ireland to the Holy Land.

In the old town, immediately beside the Sanctuary of Saint Michael and other medieval churches, numerous monasteries can also be found. The Monastery of the Holy Trinity was founded in 1404 for the Poor Clares, whose monasticism was initially a prerogative of the feudal aristocracy, allowing families to preserve their heritage and enhance their social prestige (Tancredi 1940, 254–6;

Figure 8.1. Location of Monte Sant'Angelo (Southern Italy).

Cavallini 2012). One notable figure is Donna Constantia Jordana (1595–1652), who became a nun in 1603, at the tender age of eight. She wrote a diary found in the State Archives of Naples (Cavallini 2015). Donna Constantia lived during the tumultuous years of the Council of Trent (1545–1563), a period marked by significant changes to the monastery, including the rebuilding of the church around 1630, thanks to the noble Gambadoro family. The new church comprises a single nave with a large presbytery, a sacristy and a refectory that leads to the bell tower. The main altar is characterised by an artistic tabernacle decorated with a marble bas-relief and is flanked by two side altars with marble and plaster decorations. On both sides of the church there is a 'matroneo' (women's gallery), which include windows and a choir. Additionally, there is a historic organ, numerous canvases and silverwork made by the Poor Clares themselves. In fact, in this period an important school of painting and sacred craftsmanship also developed within the monastery. In 1888, the monastic community was dissolved, forcing the nuns to leave the monastery. In 1905, the Confraternity of Holy Trinity was established to manage the church while the remainder of the monastery was transferred to municipal ownership and repurposed as a library and other facilities.

The excavation, study and display project of the crypts, where the bodies of the Poor Clare nuns are preserved, commenced in 2020 with the support of the Confraternity and is overseen by the Soprintendenza of the Ministry of Culture. The project involves a significant number of researchers and aims to make the crypt accessible to the public.

Study area

The Holy Trinity stands in the historic medieval district of 'Junno', which features a number of significant architectural monuments (1504 m^2 distributed over four levels with a total height of approximately 11 m). At the centre of the church's nave there is an underground room, the funerary crypt of the monastery, which was used for the burial of the Poor Clare nuns and other deceased, probably between the first half of the 17th century and the late 19th century (Fig. 8.2).

The crypt is composed of four rooms, three of which are continuous and oriented north–south, with a fourth room connected by a corridor. The majority of the walls are coated with multiple layers of whitewash.

The first room (Room 1), directly accessible from the trapdoor, is a small quadrangular chamber that houses the remains of 'the Abbess'. The skeleton is presented in an almost complete state, with the bones linked by a thin copper alloy wire while the vertebrae are stacked on an iron bar nailed to a wooden pole with a round base support. According to local tradition, the skeleton used to be displayed outside the church during Easter (oral tradition) and on the Day of the Dead (All Sous' Day) to collect donations and as a *memento mori* (Tancredi 1940, 64). Next to the skeleton, close to other disarticulated bones, a coin dating back to the 15th century provides a reference to the monastery's foundation.[1]

The second room (Room 2) is a quadrangular chamber in which the majority of the bodies and finds are concentrated so that it can be defined as the actual 'burial chamber' of the crypt (Fig. 8.3). To the right of the entrance, a wooden

Figure 8.2. Aerial view of the Monastery of the Holy Trinity (A), the façade of the church (B) and the entrance to the crypt of the Poor Clares (C).

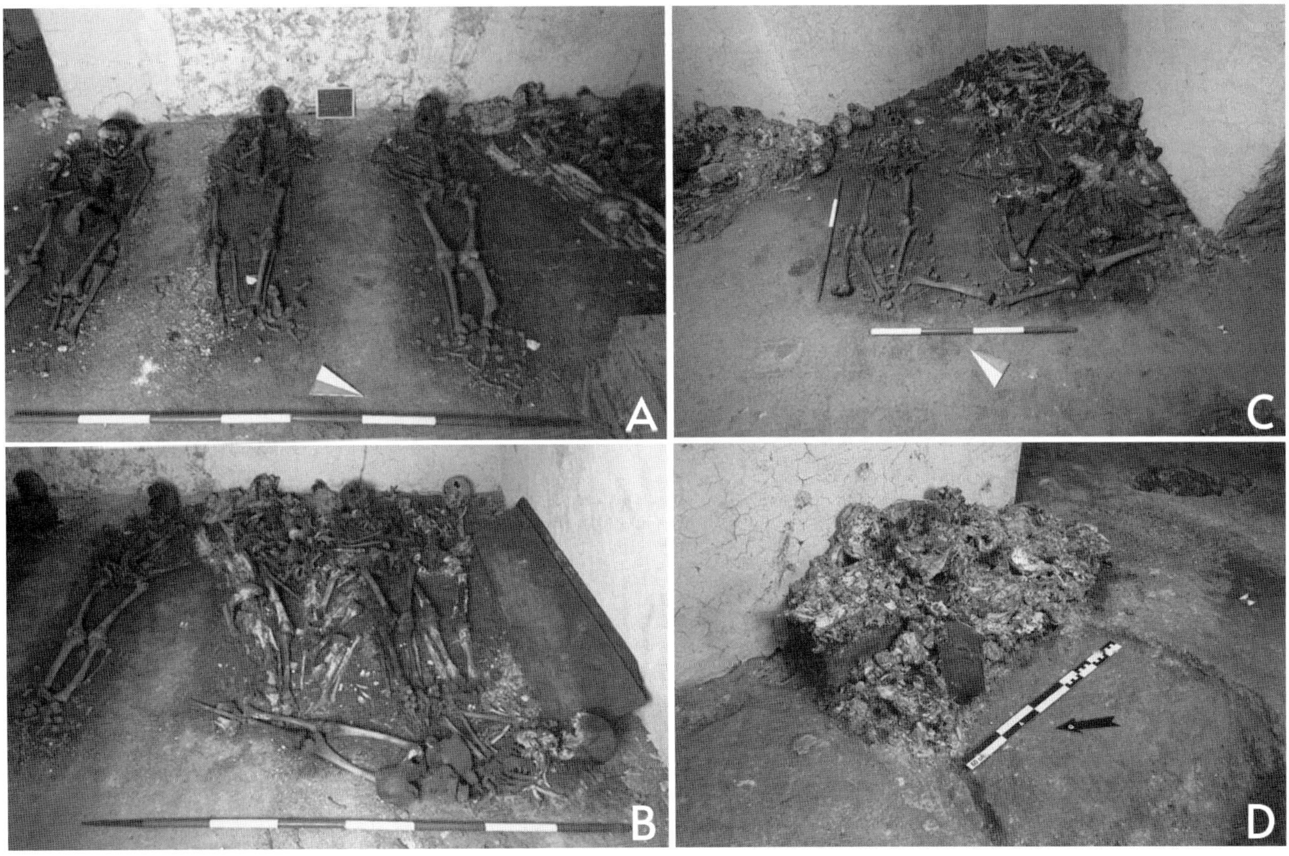

Figure 8.3. The bodies placed on the wooden boards (A–B), the piles of disarticulated bones (C) and the shelf with the skulls (D).

base with a cross has been preserved. The origin of the latter is unknown, but it probably served as a marker for the underground funerary spaces. The room contains 15 bodies in primary position. On the western side, 9 individuals are aligned east–west with their heads facing the wall, while one body is oriented north–south; on the eastern side, 5 individuals are similarly oriented east–west, with their heads turned towards the wall. In addition, there are two large accumulations of disarticulated bones, concentrated along the eastern wall. The first group of bones is located at the centre of the eastern wall. Beneath this, a large wooden feature was found, probably used as a support for accumulating skulls both above and in front, which can be interpreted as a 'shelf' for displaying the most significant bones. Furthermore, the pile and the shelf appear to rest over a circular impression in the floor, which may represent the imprint of a lost object and/or a kind of strainer for the 'double burial' (Pezzini 2006; Fornaciari 2013; Papa *et al.* 2021). This kind of ritual entails the facilitation of the discharge of bodily fluids within a designated space with tables and channels, prior to the final deposition. The second accumulation of disarticulated bones, more substantial in volume, was located in the south-eastern corner of the room. It comprised post-cranial bones and overlay two incomplete individuals. The presence of these accumulations is clearly linked to the need to free up space for new burials.

Room 3 is empty, and Room 4 is a small trapezoidal chamber decorated with a polychrome fresco adorned with plant motifs and letters, which was unfortunately damaged during restoration work in the 1990s. Only four bodies were buried in this room: two bodies positioned side-by-side occupy the central space, one of which is accompanied by a tile used as a funerary cushion. The other bodies were placed along the southern wall, one of them disarticulated. This room was probably a privileged space, possibly dedicated to the most important members of the monastic community such as the abbess, the vicar, the novice mistress and the counsellor.

Materials and methodology

The assemblage of bones comprises 22 articulated bodies, 2 accumulations, 2 groups of mixed bones representing several individuals and 4 layers of sparse bone fragments scattered across the floor. Overall, the remains are in a good state of preservation, although some bodies and bones (especially those placed in the corners of Room 2) are covered with white powder and partially destroyed. Additionally, a large number of bones exhibit green discolouration due to the presence of metallic objects placed in their vicinity. The bones are complete and measurable (Martin and Saller 1956–1959). Each body was examined to determine sex and age (Buikstra and Ubelaker 1994), stature (Trotter and Gleser 1977), pathological conditions (Ortner 2003), biomechanical stress (Mariotti *et al.* 2007), new articular surfaces, degenerative joint diseases and dental pathologies.

The botanical samples examined (numbering 35) come from a variety of sources: the wooden boards on which the nuns were laid (6 samples, 3 of which from the most complete pieces), wooden parts of rosaries, sediment samples and wooden furniture components. The wooden elements appear dehydrated, with no permanent conservation phenomena affecting the organic tissues (*i.e.* combustion, mineralisation, etc.). Thus, the decomposition of wood is still ongoing, although slowed down by the crypt's environmental conditions. The residues were dry-sieved and examined under a stereomicroscope in order to identify any plant remains. When environmental finds were identified, seeds were distinguished from wood and charcoal.[2] The seeds were therefore analysed under a stereomicroscope, while the wood fragments were analysed under a metallographic microscope, after preparing the sample. In both cases, the anatomical characteristics of the sampled remains were compared to determine the genus and, where possible, the species. With regard to the wooden fragments that were found to be dehydrated and in a state of semi-mummification (making them particularly spongy), it was decided to take sub-samples of the material and subject them to controlled combustion inside a muffle furnace. The objective of this process was to ascertain the state of conservation of the anatomical structures and to facilitate observation of fractures along useful planes. Entomological specimens were extracted after sieving residue samples (numbering 22, total weight 10.1 kg) using sieves with different meshes (1 cm to 250 um). The sieved insect fragments were manually collected with sterile tweezers and paintbrushes and stored in sterile plastic vials. The specimens were later observed and photographed using a S9i Leica stereomicroscope. Subsequently, the samples were identified using specific keys (Porta 1923–1934; 1949; Skidmore 1985; Giordani *et al.* 2019) and compared with an existing reference collection of identified specimens.

The microclimatic conditions of the crypt also allowed the partial natural mummification of the bodies, preserving human tissues (such as muscles, tendons, brains and intestines)[3] as well as fragments of the nuns' clothing worn at the time of burial.

The textile fragments (a total of 20 from 32 samples) investigated were generally in a mediocre to poor state of preservation, resembling paper due to the pronounced dehydration of the fibres, which appear rigid and discoloured. An initial dusting was conducted on these samples to remove the most substantial deposits, thereby enabling a preliminary analysis of the fabric while awaiting further research.

The osteological samples covered with white powder were analysed using a non-destructive portable X-ray fluorescence (pXRF) analyser from Bruker (Titan S1).[4]

Finally, all the data collected were compared with the information recorded in the diary of Donna Constantia Jordana (Cavallini 2015), who recounts details not only of religious life, but also of daily life including the number of sisters, food and typical treats produced, illnesses they suffered from, clothes worn, but also an intriguing episode regarding the movement of the bodies of the nuns buried in the monastery.

Results

With the exception of the body of known as 'the Abbess', the excavation and study of the skeletal remains have documented four chambers containing a total of 21 deposits (*infra*), corresponding to the most recent phase of burial.

The previously buried individuals were instead rearranged close to the walls of the crypt in two large accumulations, to create space for newer bodies.[5] Additionally, some bones were scattered across the floor within Rooms 1 and 2, as well as the corridor, reflecting the general use of the crypt over time.

The funerary and taphonomic study has yielded intriguing insights into the funerary practices and methods of occupying the underground environment. The nuns were placed directly on the floor of the crypt, which was made of beaten soil, or alternatively, on studded white fir boards with small circular iron studs. All the bodies were placed supine. The upper limbs are generally flexed over the abdomen and pelvis, while the lower limbs are extended and parallel. In most cases the feet are disarticulated as well as some lower limbs. The nuns were buried wearing monastic clothing (*infra*), of which numerous fragments are preserved, including the veil, still adhering to the skull, as well as hooks and buttons of the clothing. Each body was accompanied by several rosaries made from grains and seeds, typically tied with copper alloy hooks and cords (with one instance of a rosary made entirely of metal wire). These rosaries were often decorated with medals of varying sizes and iconography (particularly depicting Saint Michael), along with metal and wooden crucifixes, generally dating to the 17th–18th century.[6] The rosaries were mostly found between the clasped hands, fallen into the pelvis, tied to the forearm or probably hung from the belt of the habit, thus found between the femurs. An elongated metallic element was discovered on the chest of one body and among the piles of bones; this may represent a penitent object, possibly part of a cilice belt.

Apart from the Abbess, who was identified as a male aged 35–45 years, the biological sample consists of 21 bodies belonging to adult female individuals (9 aged between 50 and 60 years or older, 11 between 40 and 50 years and only 1 individual younger, of 25 to 35 years) (Table 8.1).

The 'Abbess' is 1.70 m tall while the nuns are between 1.46 and 1.65 m tall. Their bones are generally slender and exhibit muscular markings associated with strenuous activities and significant loads on the back and lower limbs.

In terms of pathologies, osteoarthritis of the upper limbs is recorded, particularly affecting the humeri and glenoid cavities, with some cases of humeral head compression and eburnation of the joints. At least four nuns show signs of rhizarthrosis in the first finger of the hands due to repetitive manipulation activities. The spinal column exhibits osteoarthritis, characterised by osteophytic spurs, fusions, disc wedging and hernias. Additionally, there is a possible case of rib cage compression. The pelvic bones often present osteophytes, with two cases of sacroiliac ankylosis. The lower limbs exhibit severe arthritis in the knees and ankles, resulting from the daily activities involving genuflection for prayer, penance and everyday tasks. One nun suffered a tibia and fibula fracture that healed with callus formation but

Table 8.1 Age and sex estimation of the anatomically connected bodies

Individual	*Age*	*Sex*
1 (The 'Abbess')	35–45	M
2	60	F
3	45–55	F
4	40 55	F
5	50–60	F
6	45–50	F
7	40–45	F
8	50–60	F
9	45–55	F
10	50–60	F
11	45–50	F
12	60	F
13	60	F
14	50–60	F
15	50–60	F
16	60	F
17	45–55	F
18	45–55	F
19	40–50	F
20	25–35	F
21	40	F
22	40	F

without alignment, leading to limb shortening and resulting in a limp. There are four cases of individuals who were completely edentulous, which is to be expected with age, along with numerous instances of antemortem tooth loss. Among the dentoalveolar pathologies noted, dental wear, even destructive caries, and abscesses are noteworthy. Benign tumours were also observed, such as classic button-shaped excrescences (button osteoma) on several cranial vaults. Additionally, one case of proximal synostosis of the tibia and fibula was identified, probably due to osteochondroma (known as 'kissing bone').

The state of conservation of the osteological material is generally excellent, most likely due to the micro-climatic conditions of the crypt. This underground space, with no ventilation except for the access hatch in the floor of the nave, maintains a constant humidity of 98.8% and an average temperature of 13°C.[7]

A spectroscopic analysis of the bones and the white traces left on their surface revealed that, once cleaned, the calcium/phosphorus ratio falls within the normal range of living bone. In contrast, bones affected by lime exhibited a calcium/phosphorus ratio that was significantly higher in favour of calcium, while phosphorus levels were lower (Alapont *et al.* 2023).

Figure 8.4. Layers of lime on the skulls (A–B) and diaphyses of long bones (C).

The data are corroborated by analyses of the pure lime used on the walls, which also shows a calcium/phosphorus ratio clearly favourable to calcium and very low for phosphorus. The application of lime to the walls and bodies served a hygienic purpose, addressing the high concentration of decomposing corpses, the associated odours, and the proliferation of bacteria. The use of quicklime on the cadavers may have caused their partial explosion and pulverisation (Fig. 8.4). However, the effect of lime on the bodies should not be underestimated, as it has also been correlated to their partial desiccation (Schotsmans *et al.* 2012; 2014a; 2014b; 2016; Gutiérrez Galiacho *et al.* 2023). Thus, the state of conservation of the osteological material may have been further enhanced by the presence of lime.

Thanks to the environmental and micro-climatic conditions of the crypt, numerous fragments of light-coloured and brown cloth have been preserved on the bones, which are attributable to the nuns' typical religious garments. Many of these fragments can be identified, including veils still secured with pins on the skulls, made of two different types of lightweight vegetable fibre canvas. One type features a gauze-like effect and is composed of multiple layers without seams (Fig. 8.5). Additionally, bandages for injured hands were recovered, differing in size, material and weave (five types). At the chest level, numerous elements, probably belonging to shirts, petticoats and stockings, were identified, including copper alloy hooks with circular eyelets and rounded buttons, made of metal and bone. One textile features an accentuated twist of the warp and weft, resulting in a finished fabric that is characterised by an 'embossed' effect, lightness and transparency. Fragments of leather or hide, cords and various parts of knitted products were also discovered, including nets and parts of fabric that probably belonged to stockings. Notably, at least one case suggests the presence of shoes, based on a metal buckle with a wooden loop found at the feet of a nun.

The botanical data set comprises 265 seeds and fruit macro-remains, as well as 113 identified charred wood remains. (Fig. 8.6). Among the seeds, only three olive pits were found in a deposit and two peach stones in one of the bone piles. Over 98% of the samples analysed come from two contexts that can be interpreted as waste deposits. These mainly consist of woody seeds of fruit plants, including olive pits, cherries, plums, grape pips, as well as fragments of shells of walnuts, almonds and chestnuts. Interestingly, even if sporadic, the presence of fragments

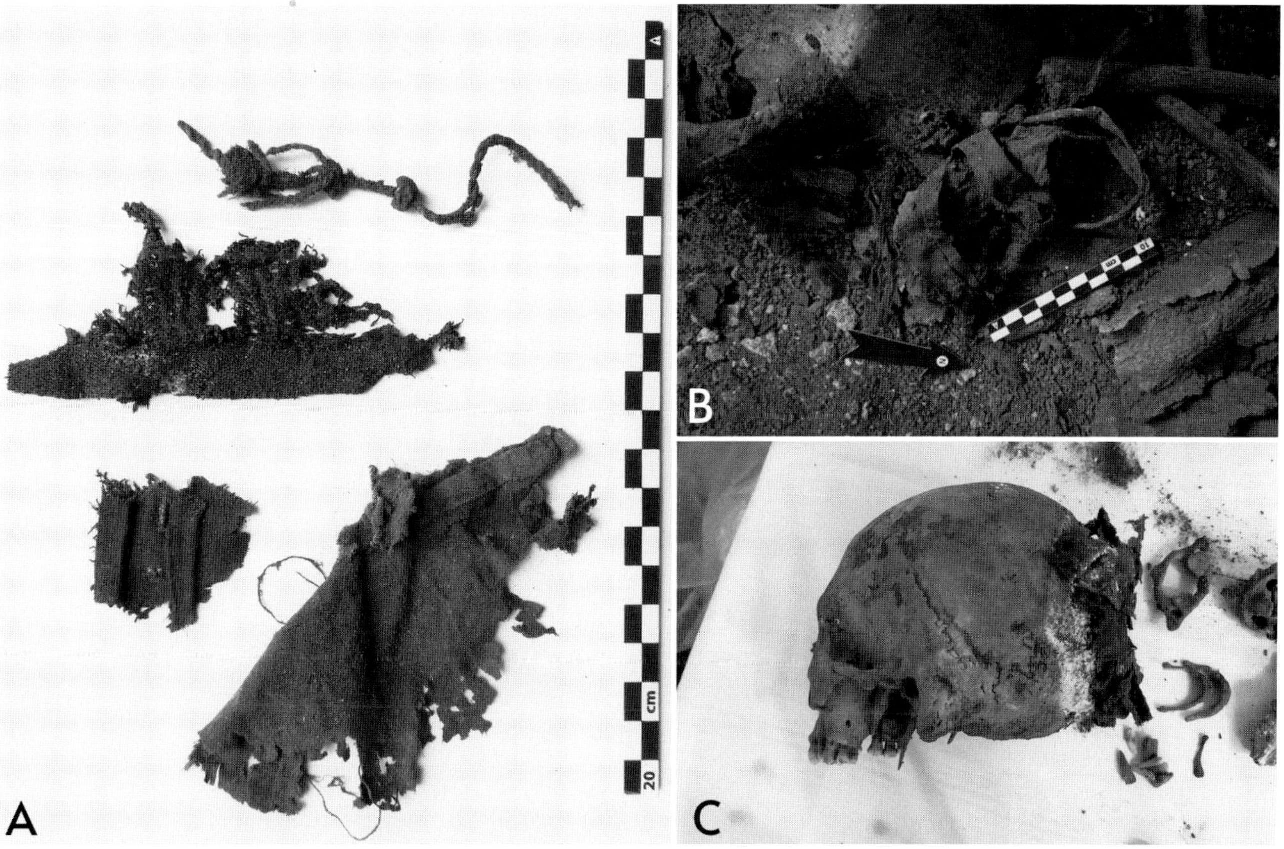

Figure 8.5. Fragments of clothing (A), bandages (B) and veils still in situ *on the skulls (C).*

of grain caryopses, a seed from the cucurbitaceae family (which includes cucumbers, melons and similar products) and a single Cornelian cherry stone (an edible wild berry) should be noted. In the second category, the botanical finds identified include the wooden boards for placing the bodies, personal objects such as rosaries, elements of furniture, furnishings of the crypt or part of the accumulated waste, branches and twigs. These latter mainly include white birch and Rosaceae, specifically the evergreen rose (or St John's rose) and dog rose, both wild shrubs known for their showy flowers and aesthetic appeal. These plants possess medicinal properties, rich in vitamin C, tannins, organic acids, pectins, carotenoids and polyphenols. Both the fruits, dried and chopped, and the petals are used for making infusions and decoctions, as well as fresh in jams and to add flavour to honey. They are recognised for their intestinal astringent, antidiarrheal, vasoprotective, anti-inflammatory, antioxidant, and antiparasitic benefits. The presence of these remains could be associated with the simple disposal of dead plants or the deposition of flowers inside the crypt, but they could also be connected to the production of beads for rosaries, made of white birch wood but also of Rosaceae (dog rose in particular) in a polyhedral and spherical shape. Finally, the wooden boards on which the bodies of the nuns were laid (approximately 123 cm long, 24 cm wide and 2 cm thick) were all crafted from silver fir. The population of this species, whether pure or in association with beech, is still concentrated in the Apennine area and the Gargano, the promontory on which Monte Sant'Angelo is located. Additionally, fragments of furniture crafted exclusively from poplar/willow were also found. These are species that typically inhabit areas along streams or bodies of water, potentially originating from the plain.

The dry fragmented insects identified among the collected material, belong to the orders Coleoptera, Lepidoptera, Hymenoptera and Diptera. In addition, other arthropods belonging to the class Arachnida were also found.

Among the identified insects, puparia of *Hydrotaea capensis* (Wiedemann 1818) (Diptera: Muscidae) accounted for over 90% of the specimens analysed. Additionally, a few puparia belonging to Calliphoridae, Sphaeroceridae and Phoridae were also noted. Coleoptera, mainly elytra and fragmented bodies, were identified as belonging to the families Ptinidae, Salpingidae, Mycetaeidae, Aderidae, Latrididae, Cryptophagidae, Dermestidae, Staphylinidae, Histeridae, Curculionidae, Carabidae and Tenebrionidae. Some of the beetles are mould-feeding, and, as already observed in other crypts, they are generally associated with the development of mould during the decomposition process. However, some specimens are wood eating, such as the

Figure 8.6. Palaeobotanical specimens (A–B) and white fir boards (C).

weevils in the family Curculionidae, revealing the presence of wood in the crypt. Among Arachnida only mites, spider exuviae and pseudoscorpions were isolated.

Discussion

The osteological sample refers to a small female community, in accordance with Donna Constantia Jordana's diary, which notes approximately 40 nuns in her community, along with others buried nearby. This is consistent with the calculated minimum number of individuals (MNI) of 22 buried bodies and approximately 30 in accumulations.

From the biological perspective, the group is composed of older women in their 50s and 60s and over, with the exception of one nun in her 30s. The average height of the group is approximately 1.5 m. Their daily activities very likely included cooking, sewing, painting and silver-working, alongside strenuous tasks such as praying in a kneeling position and lifting weights related to horticulture, common aspects of cloistered life. Pathologies identified in the sample are typical of elderly women, including osteoarthritis, fractures from falls and benign tumours. Notably, Donna Constantia also mentions an 84-year-old abbess and a nun who required a chair to attend mass due to her inability to walk, further emphasising the age-related health issues present in this community. Many of them were edentulous due to advanced age, but the evidence of dentoalveolar diseases, such as caries, abscesses and wear, suggests a varied diet that included foods with elevated levels of sugar, which may not have always been well processed.

The palaeobotanical data, in fact, refer to cherries, grape pips, nuts, almonds and chestnuts, caryopses of wheat grains, a seed of cucurbits and a seed of Cornelian cherry (an edible wild berry), which reveals a rich array of plant foods consumed by the monastic community. The study of plant remains highlights not only dietary habits but also the potential for utilising plants in rituals and daily life: in addition to the identification of the materials used to make rosaries and personal objects, the presence of sprigs of ornamental plants is suggestive, perhaps used by the community for their properties or offered as gifts to deceased sisters. Finally, the woods used to make the furniture are soft woods, suitable for carving and turning, and provide information on local wood craftsmanship.

As Donna Constantia relates in her diary, the nuns' dress was comprised of a white cotton and linen cap to cover the face and neck, a dark cloth veil worn over the cap, a wool tunic with a cord at the waist and the scapular apron that

Table 8.2 List of information contained in archive sources and those obtained from the study of the crypt

	Historical sources	*Archaeological study*	*Anthropological study*	*Specialised studies*
Monastery history	X	X		
Number of nuns	X	X	X	
Biological profile	X	X	X	X
Health status	X	X	X	X
Daily activities	X	X	X	X
Funeral	X	X	X	X
Use of lime		X		X
Clothing	X	X		X
Objects	X	X		X

covered the body in front and behind, worn over the tunic; in the colder months they also wore a cloak. She also specifies that all these garments were also worn during funerals.

At the time of death, after the funeral, the bodies of the nuns were taken down to the crypt where a ritual of corpse drainage most likely took place before they were laid on silver fir boards or on the ground. Rich rosaries with votive medals were placed in their hands or fastened to their belts, and a penitential cilice may have been positioned on their chests.

The insects collected from the bodies reveal that the whole process of decomposition occurred directly in the place where the bodies were found. This hypothesis is supported by the fact that insects belonging to all the 'waves of colonisation' of bodies in a hypogean environment were collected in significant numbers. The presence of woods in the crypts is also supported by the collection of wood-feeding specimens such as Curculionidae and Ptinidae.

As the available space on the ground was filled to capacity, with a total of just under 30 individuals present, it became necessary to relocate the bones to the corners in order to accommodate new burials. The considerable number of corpses was likely to have resulted in significant hygiene issues (including odour and the presence of bacteria) caused by the decomposition of the bodies under the nave of the church. This prompted the application of quicklime on the walls and remains. The use of lime could have influenced the state of conservation of the remains, partially damaging some of them or contributing to their desiccation, due to the general conditions of constant temperature and humidity of the crypt.

Finally, the crypt ceased to serve as a burial site and the last room became a repository for discarded objects (such as broken furniture), food (animal remains with signs of butchery) and pottery (fragments of kitchen and glazed pottery, such as cooking pots and plates).

In the early 1900s, the so-called Abbess, a skeleton with bones mounted and stacked to stand on a metal and wooden pole, identified as an adult man, was displayed during some religious festivities outside the church. This skeleton, along with various architectural elements salvaged from the church attic, was ultimately left in the first room of the crypt.

The bodies remained in the crypt untouched until the recent excavation, the official commencement of this project, which aims to study the history of these women and relocate the bones to their original place.

Conclusions

The Church of the Holy Trinity of Monte Sant'Angelo stands as a significant testament to the Poor Clare community, who established it and resided there until the late 1800s. The recent discovery of the crypt, located beneath the 17th-century church, offers invaluable insights into the lives of the nuns buried there. These recorded data serve as a unique and unprecedented documentary basis for enriching our understanding of the Poor Clare nuns' daily existence and practices, confirming and enriching the information of Donna Constantia Jordana's diary (Table 8.2).

The study is the fundamental basis for continuing the ongoing project, which aims to restore the structure in collaboration with the Municipality of Monte Sant'Angelo. The objective is to reintroduce this site to the public in a manner that reflects its profound religious and humanistic value by telling the story of the women who made the history of this monument and the town itself. The strong connection of the Monte Sant'Angelo community with this initiative is clearly demonstrated by their active involvement which encompasses attendance at open days and participation in all the events that promote the project.

Over the past two years the project has also been involved in 'Mònde - Festival del cinema dei cammini', organized by MAD-Memorie audiovisive della Daunia (director L. Toriello). Firstly, a masterclass focusing on the ethical implications of using audiovisual media to enhance the understanding of funerary contexts was created. Additionally, the history of the crypt has also inspired a video documentary 'Atto d'amore' (director L. Lo Muzio), which includes interviews with a present-day cloistered nun, and has won a contest at the Festival. Finally, a questionnaire was distributed to evaluate interest in and knowledge of this place, inviting residents to share their memories and photographic and audiovisual testimonies. The ultimate objective is to create a digital archive of these memories. The project thus demonstrates how the study of human remains can advance scientific knowledge about past communities while also fostering a dialogue with society. This dialogue reinforces collective memory, cultural identity and respect for ancestors.

Acknowledgements

We thank the Confraternity of the Holy Trinity for funding the project and part of the laboratory analyses. The authors also thanks ArchaeChemis for supporting the pXRF analysis; the team of Experimental Zooprophylactic Institute of Puglia and Basilicata (Dr D.A. Raele, Dr V. Nardelli, O. Miedico, C. Pompa, A. Accettulli, M. Iammarino), Dr G. Sarcinelli (University of Salento) and Dr D. Moretti (University of Bologna) for communicating their preliminary results, which will be further explored soon.

Notes

1 Preliminary investigation by Dr G. Sarcinelli (University of Salento).

2 Archaeobotany and Palaeoecology Laboratory of the University of Salento.

3 A total of 16 samples were also subjected to molecular research to detect the presence or absence of DNA from bacterial pathogens (*Brucella spp, Mycobacterium tuberculosis, Treponema pallidum, Rickettsiae* of the spotted fever group, *Bartonella spp.* and *Plasmodium spp.*) at the Experimental Zooprophylactic Institute of Puglia and Basilicata (Dr D.A. Raele) but the samples returned negative results. Additionally, analyses carried out at the same institution to detect the presence of heavy metals using ICP-MS are currently under further investigation (Dr V. Nardelli, O. Miedico, C. Pompa, A. Accettulli, M. Iammarino).

4 Universitat de València, Departament de Prehistòria, Arqueologia i Història Antiga. Doctorate in Geography and History of the Mediterranean from Prehistory to Modern Science, Line of Bioarchaeological Investigation (Operativa Multidisciplinary Research Unit Archaechemis). Doctoral student Ginevra Anna Panzarino, directors José Luis Jiménez, Gianni Gallello and Agustín Pastor.

5 In particular, the bones from the accumulations in Room 2 (SU 19 and SU 23) have only been catalogued due to their advanced state of degradation for safety reasons. The NMI in these Stratigraphic Units is approximately 29, based on the single bones found.

6 Preliminary study and restoration by Dr D. Moretti of the Alma Mater University of Bologna.

7 These were measured over the course of a year using a Govee Thermometer Hygrometer (model H5075).

References

Alapont, L., Gallello, G., Martinon-Torres, M., Osanna, M., Amoretti, V., Chenery, S., Ramacciotti, M., Jiménez, J. L., Morales Rubio, A., Cervera, M. L. and Pastor, A. (2023) The casts of Pompeii: post-depositional methodological insights. *PLoS ONE* 18(8), e0289378. https://doi.org/10.1371/journal.pone.0289378

Buikstra, J.E. and Ubelaker, D.H. (1994). *Standards for Data Collection from Human Skeletal Remains*. Fayetteville, Arkansas, Arkansas Archeological Survey Research Series no 44.

Cavallini, A. (2012) *L'antico complesso clariano di Monte Sant'Angelo e la Confraternita della SS. Trinità*. Monte Sant'Angelo, Edizione a cura della Confraternita della SS.ma Trinità.

Cavallini, A. (2015) *Voci e vicende delle Donne Monache 'dlu Jùnn'*. Monte Sant'Angelo, Edizione a cura della Confraternita della SS.ma Trinità.

Fornaciari, A. (2013) Scheletrizzare o mummificare: pratiche e strutture per la sepoltura secondaria nell'Italia del Sud durante l'Età moderna e contemporanea. *Medicina nei Secoli. Arte e Scienza* 25(1), 205–38.

Giordani, G., Grzywacz, A. and Vanin, S. (2019) Characterization and identification of puparia of Hydrotaea Robineau-Desvoidy, 1830 (Diptera: Muscidae) from forensic and archaeological contexts. *Journal of Medical Entomology* 56(1), 45–54.

Gutiérrez Galiacho, A., Nociarová, D., Malgosa, A., Fernández-Jalvo, Y. and Armentano, N. (2023) What does lime tell us about cadaveric remains? *Historical Biology*. Early online Publication. https://doi.org/10.1080/08912963.2023.2297911

Martin, R. and Saller, K. (1956–1959) *Lehrbuch der Anthropologie*. Stuttgart, G. Fischer.

Mariotti, V., Facchini, F. and Belcastro, M.G. (2007) The study of entheses: proposal of a standardized scoring method for twenty-three entheses of postcranial skeleton. *Collegium Antropologicum* 31(1), 291–313.

Ortner, D.J. (2003) *Identification of Pathological Conditions in Human Skeletal Remains*. 2nd ed. Washington, Academic Press.

Papa, V., Varotto, E., Del Pizzo, S., Troisi, S., Di Ciaccio, F., Staiano S.S., Lubritto, C., Vanin, S., Vaccarezza, M. and Galassi, F.M. (2021) Preliminary notes on a human eared skull. *Journal of Anatomy* 240(4), 785.

Pezzini, F. (2006) Doppie esequie e scolatura dei corpi nell'Italia meridionale d'età moderna. *Medicina nei Secoli. Arte e Scienza* 18(3), 897–924.

Porta, A. (1923–1934) *Fauna Coleopterorum Italica*. Piacenza, Stabilimento Tipografico Piacentino.

Porta, A. (1949) *Fauna Coleopterorum Italica*. Sanremo, Stabilimento Tipografico Soc. Au. G. Gandolfi.

Schotsmans, E.M.J., Denton, J., Dekeirsschieter, J., Ivaneanu, T., Leentjes, S., Janaway R.C. and Wilson, A.S. (2012) Effects of hydrated lime and quicklime on the decay of buried human remains using pig cadavers as human body analogues. *Forensic Science International* 217(1–3), 50–9. doi: 10.1016/j.forsciint.2011.09.025.

Schotsmans, E.M.J., Denton, J., Fletcher, J.N., Janaway, R.C. and Wilson, A.S. (2014a) Short-term effects of hydrated lime and quicklime on the decay of human remains using pig cadavers as human body analogues: laboratory experiments. *Forensic Science International* 238, 142, e1–142.e10. doi: 10.1016/j.forsciint.2013.12.047.

Schotsmans, E.M.J., Fletcher, J.N., Denton, J., Janaway, R.C. and Wilson, A.S. (2014b) Long-term effects of hydrated lime and quicklime on the decay of human remains using pig cadavers as human body analogues: field experiments. *Forensic Science International* 238, 141, e1–141.e13. doi: 10.1016/j.forsciint.2013.12.046.

Schotsmans, E.M.J., García-Rubio, A., Edwards, H.G.M., Munshi, T., Wilson, A.S. and Ríos, L. (2016) Analyzing and Interpreting lime burials from the Spanish Civil War (1936–1939): a case study from La Carcavilla cemetery. *Journal of Forensic Science* 62(2), 498–510. doi: 10.1111/1556-4029.13276.

Skidmore, P. (1985) The biology of the Muscidae of the world. *Series Entomologica* 29, 1–550.

Tancredi, G. (1940) *Folclore garganico*. Manfredonia, Armillotta e Marino tipografi.

Trotter, M. and Gleser, G.C. (1977) Corrigenda to 'Estimation of stature from long limb bones of American Whites and Negroes'. *American Journal of Physical Anthropology* 47, 355–6.

9

The tomb of Prioress Jerònima de Gort (1586–1601): an interdisciplinary approach to the female Hospitaller Monastery of Santa Maria d'Alguaire

Maria Soler Sala, Walter Alegría Tejedor, Izaskun Ambrosio Arcas, Sílvia Marín Ortega and Araceli Coll Sanabra

The female monastery of Santa Maria d'Alguaire, the most important convent of the Hospitaller Order in Catalonia, has been the subject of archaeological excavations since 2016. One of the most significant finds corresponds to the location of a burial with a tombstone inscribed with the name of the prioress Jerònima de Gort, indicating her date of birth and death, and bearing her family's coat of arms.

An interdisciplinary team of archaeologists, anthropologists, historians and restorers participated in the opening of the tomb. Their work made it possible to record the wooden coffin, the prioress's remains, fragments of her mortuary vestments and various objects. Besides this, the study of the archaeological evidence and written records has enabled us to establish that she was a central figure in the history of the convent, and a staunch defender of enclosure when customs relaxed. She actively participated in the internal investigation that took place during the priorate of her predecessor Jerònima de Montgay (1576–1583), accusing some of her colleagues of leading a life full of scandal, which violated the observance of the community's rules concerning isolation and spiritual life.

Our work aims to offer a comprehensive interpretation of the archaeological remains, based on an interdisciplinary approach that takes into account archaeology, physical anthropology, written documentation and restoration techniques within the framework of a true dialogue between sources. Such an approach renders this tomb a particularly compelling case study for understanding the living conditions and funerary practices of a female community within the Hospitaller Order.

Keywords: Hospitaller Order, female monasticism, anthropological study, textual analysis, restoration techniques.

Introduction

The aim of this paper is to present the archaeological research undertaken at the female Hospitaller Monastery of Santa Maria d'Alguaire (Segrià, Catalonia). The study focuses on the location, excavation and analysis of a unique tomb. According to the inscription on the tombstone, the burial is that of Jerònima de Gort, prioress of the Monastery of Alguaire between 1586 and 1601.

The tomb was opened by an interdisciplinary team composed of archaeologists, anthropologists, historians and restorers. The archaeological team documented the wooden coffin, the prioress's remains, fragments of her funerary vestments and various objects. Moreover, the analysis of the archaeological evidence and written sources has enabled us to establish that she was a key figure in the history of the convent and a staunch proponent of enclosure during a period when the observance of customs was becoming more relaxed.

The archaeological wealth of the tomb, combined with the availability of archival documents that shed light on the life of the woman buried in it, has enabled us to adopt an interdisciplinary approach. This approach incorporates the archaeological record into reflection on female spirituality during medieval and early modern times, employing the gender perspective in both the study of the funerary remains and the analysis of the space in which they were found.

The Hospitaller Commandery of Alguaire represents an intriguing subject for research for several reasons. First, it offers a unique opportunity to examine the role of women within such an intensely masculinised group as the Order of Saint John of Jerusalem. Second, the convent's archaeological remains and preserved texts provide a valuable source of information about the daily life of these communities and their influence over vast territories.

The Monastery of Santa Maria d'Alguaire

The Monastery of Santa Maria d'Alguaire represents the most significant female convent of the Order of the Hospital built in Catalonia during the Middle Ages. It is situated in the province of Lleida, in the municipality of Alguaire, on a crag known as 'Serra del Convent', which dominates the extensive Segrià plain from a commanding position.

The establishment can be considered the successor to the female monastery that was founded shortly beforehand in Cervera, some 60 km to the east. This monastery subsequently relocated to Alguaire between 1250 and 1262 (Alturo 2010, 20–1; Sans i Travé 2011, 213). The community of nuns was composed of the daughters of the highest ranks of the Catalan nobility, led by a prioress who exercised considerable authority over both the convent and the assets that constituted the Commandery of Alguaire (Bertran 1977, 25–56; Bonet 2014, 60). It is evident that the monastery's territorial domains were progressively expanded throughout the late Middle Ages. Its holdings extended beyond the municipality of Alguaire to include other settlements in the Segrià and Cervera plains (Coll 2016; Soler *et al.* 2025), effectively turning the monastery into a territorial lordship.

The monastery's structures were damaged during the Catalan Civil War in the second half of the 15th century (Lladonosa 2007, 96–106). However, the successive hostilities of the Reapers' War (1640–1652) had a more significant impact, forcing the community to seek refuge in the episcopal palace of Lleida in 1640 (Palacios 1977, 29). The monastery's proximity to the frontline and the economic difficulties it faced led the community to relocate to Barcelona in 1644. Nevertheless, in 1651, the city was struck by a plague, forcing the nuns to go back to their monastery after 11 years.

Their return was not straightforward, as the monastery had fallen into disrepair and required extensive renovation. Archaeological research has identified evidence of these repairs (Soler 2021, 609). In 1684, the economic difficulties and the poor state of the monastery prompted the community to request their definitive return to Barcelona (Lladonosa 2007, 207–8). Despite the initial opposition of the Bishopric of Lleida, the departure of the nuns was authorised in 1699. This marked the final abandonment of the monastery of Alguaire, which subsequently fell into a state of decay and collapse, eventually leading to its disappearance.

In 2016, an archaeological excavation was initiated through a collaboration agreement between the University of Barcelona and Alguaire Town Council. The objective of this collaboration was to study the historical and archaeological trajectory of the site. The research currently forms part of two projects entitled 'Prioresses, Lords, and Princes: Spaces of Power in Medieval and Early Modern Catalonia' (ARQ001SOLC-127-2022) and 'FREILAS. Women in Military and Hospitaller Orders in the Crown of Aragon (12th–16th c.)' (PID2022-136266NB-I00).

The archaeological excavations conducted to date have enabled the documentation of the structures associated with the monastic church (Fig. 9.1). The church itself was a large, imposing structure, measuring 36 by 24.60 m. It was built in the mid-13th century in the late Romanesque style, the origins of which can be traced to that of the cathedral of La Seu Vella de Lleida. This style also influenced the design of other nearby churches, including Sant Pere de Cubells, Santa Maria la Major in Tamarit de Llitera, or the Trinitarian Convent of Viganya (Adell 2002, 91–5; Conejo 2002, 233–5).

The original church (Fig. 9.1) comprised the nave with several side chapels (A4, A5, A9, A11, A12, A13, A14 and A15) and a main apse (A1) flanked by two apses, one to the north (A2) and the other to the south (A3 and A8). The conventual church had a transept, which was separated from the nave by an iron grille that delimited the nuns' choir, located in the centre of the nave. The church had three entrances: one to the south-east, used by priests officiating at the religious ceremonies; one to the south, providing access for the nuns from the cloister; and another to the west, through a monumental doorway, allowing general access to the church.

It is not the purpose of this article to provide a detailed account of the various construction phases of the building. However, it should be noted that between the end of the 15th and mid-16th century, a significant structural reform was carried out (Soler and Alegría 2024). The chronological assignment of this phase is based on the late Gothic sculptural decoration on the walls that repaired the original structure of the temple. One of the reformed chapels (A4) yielded a unique tomb that was excavated in 2021. This funerary unit (UF1) is associated with the nun who is the subject of the present study.

The tomb of Prioress Jerònima de Gort

As previously mentioned, the tomb of Jerònima de Gort, who held the position of prioress of the monastery of Santa Maria d'Alguaire, was situated in Chapel A4 (Fig. 9.1). The construction of this chapel was carried out as part of the

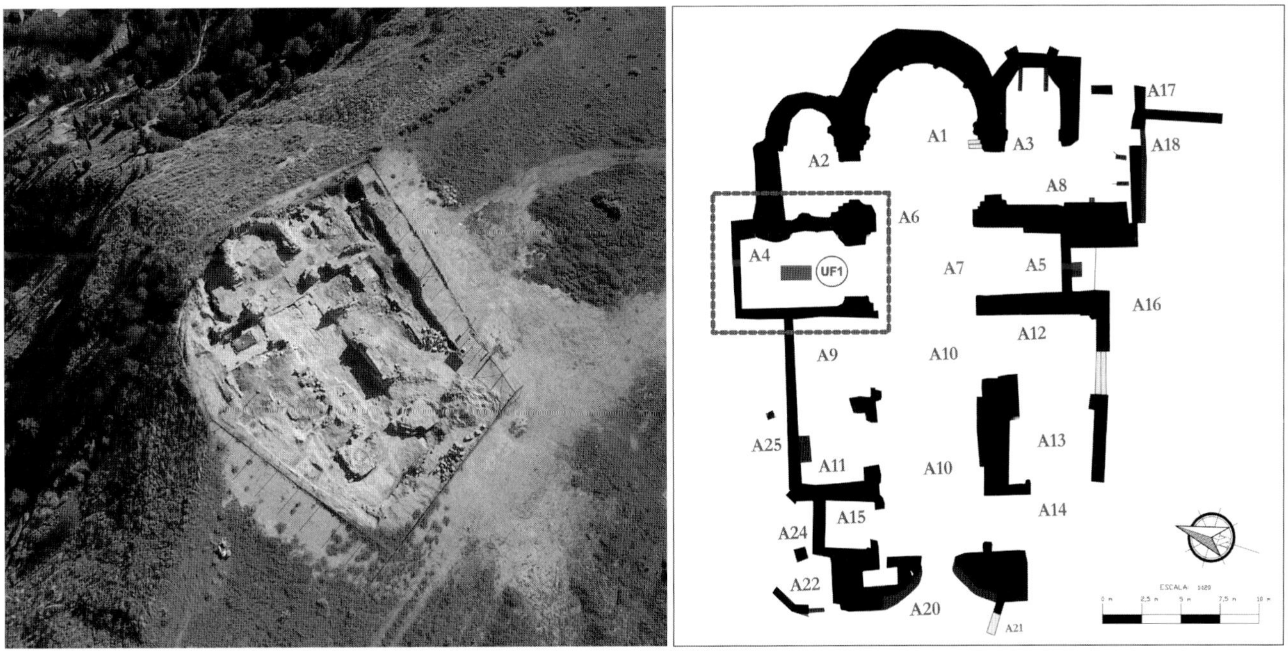

Figure 9.1. Current state of the excavation of the conventual church of Santa Maria d'Alguaire. Location of Chapel A4 and the tomb of Prioress Jerònima de Gort (UF1). Authors: Maria Soler and Walter Alegría (University of Barcelona).

extensive late Gothic reform of the church, which involved the removal of an earlier cemetery situated to the north of the archaeological area. Soil is known to have been removed from this burial area and the exhumed bones were placed as wall fillings and levelling material under the new floors of some chapels at this time (Soler *et al.* 2023). Chapel A4 provides clear evidence of this reform (Fig. 9.2). Firstly, the presence of stones with sculptural decorations, reused as ashlars, confirms the reuse of these materials. Secondly, the discovery of human remains, both inside the walls of this area and in the earth used to prepare the slab paving of the chapel, confirms the funerary origin of the construction filling materials used during the reforms.

From a chronological perspective, it can be stated that the construction of Chapel A4 must have taken place no later than the mid-16th century, while the construction of tomb UF1 occurred at the end of the same century. This conclusion is based on the funerary inscription on the slab covering the tomb (SU 1115), which indicates a burial date of 1601. To construct this burial pit, it was necessary to partially excavate the existing slab floor, exhuming the layers of earth and bones located beneath it. After the excavation of the grave pit, a brick and lime mortar wall (SU 1120) was built around its inner perimeter. The stone slab covering the tomb (SU 1115), sealed with lime and sand mortar, was also supported by this wall.

The tombstone measures 1.95 m in length and 0.80 m in width (Fig. 9.2). It bears the following inscription: 'Sepultura de dona Hyerònima de Gort, priora del monestir de Alguayre. Nasché a XXV de Nohembre [de] 1527. Morí a X de Noh[m]ebre [de] 1601', which can be translated as follows: 'Sepulchre of Dona Jerònima de Gort, who served as Prioress of the Monastery of Alguaire. Born on 25 November 1527. Died on 10 November 1601'. The lower portion of the tombstone displays the Gort family coat of arms (Biblioteca de Catalunya, ms. 161), which is depicted here with an elephant bearing a tower on a blanket. This evidence indicates that the tomb belonged to Jerònima de Gort, a noblewoman who served as prioress of the Monastery of Alguaire between 1586 and 1601. It is situated within one of the most prominent chapels of the conventual church, and immediately adjoining the nuns' choir.

Once the gravestone (SU 1115) had been fully recorded and the mortar that sealed it had been excavated, it was lifted using a system of straps and the extendable arm of a mechanical excavator. A protective layer of restoration fabric (tissue) was applied between the straps and the surface of the slab to prevent damage to the stone during the handling process. After being lifted, the slab was placed on a wooden platform and transported to a secure location for further study, restoration and preservation. Inside the funerary space, a wooden coffin was discovered (Fig. 9.2), believed to be the main burial block of the tomb (SU 1122). This coffin was accompanied by multiple skeletal remains, a considerable quantity of fabric fragments, and a range of grave goods, including rings, rosaries and medals. Additionally, more bones were identified near the main burial block, suggesting the presence of an ossuary.

The presence of a notch in one corner of the slab and a break in the brick wall (SU 1120) indicates that the tomb

Figure 9.2. Excavation process of the tomb of Prioress Jerònima de Gort (UF1). 1) General view of Chapel A4; 2) Detail of the tombstone and the excavation process of the tomb and wooden coffin; 3) Detail of the prioress's hands, with a rosary of bronze and glass beads entwined in her fingers, two rings and remnants of the shroud. Authors: Maria Soler and Walter Alegría (University of Barcelona).

was opened at least once after its initial construction. Nevertheless, it seems reasonable to conclude that the tombstone was not plundered after its final use. Following the community's abandonment in 1699, the roofs and walls of the church collapsed and crumbled. However, the tombstone and the interior of the tomb remained undamaged, in contrast to other burials in the nave, which were looted for building material. Similarly, the construction of a trench during the Spanish Civil War (1936–1939) resulted in the plundering of the apse (A2) and part of the Chapel (A4), yet the stratigraphy of the tomb remained intact until its archaeological excavation.

Osteological remains

The process of excavating the wooden coffin (considered to be the main burial block in the tomb, SU 1122) permitted the identification of several individuals. The body of a nun was the first to be documented (SU 1122.1), with the remains of her clothing, trousseau and shroud placed above the rest. In the light of the abundance of material remains (SU 1126), the meticulous way in which they were positioned on the deceased's body, and her skeletal characteristics, we conclude that they are consistent with the remains of Prioress Jerònima de Gort, who passed away in 1601 at nearly 74 years of age.

The excavation also revealed the remains of another female individual, believed to be those of a young woman (SU 1122.2), as well as another female of advanced age (SU 1125), situated in a partially secondary position and preceding the burial of the prioress. Finally, in a peripheral position relative to the main burial block, the reinterment of a fourth skeleton was observed (SU 1121). This incorporation into the tomb is likely to have occurred at a subsequent point in time.

Furthermore, within the tomb, specifically in proximity to and beneath the wooden coffin, bones that, given their secondary position, were probably within an ossuary were identified (SU 1119, 1128, 1130, 1131 and 1132). The results of the anthropological study of all the skeletal material (including the partially articulated remains of individuals and an ossuary) uncovered during the archaeological excavation of the tomb are presented below (Ambrosio 2021; 2023).

Individuals in anatomical connection

As previously stated, the fieldwork yielded the identification of multiple skeletons. The skeletal remains from the main burial (SU 1122) were exhumed en bloc, together with the mortuary clothing and grave goods, for transfer to the conservation laboratory. The individual was in a supine decubitus position within the coffin, with their arms flexed and hands placed on their abdomen, holding a rosary. The orientation of the burial was aligned on a north–south axis. Once the bones had been separated from the habit and the associated trousseau (comprising rings, rosary beads and medals), they could be studied.

The study of the skeletal remains recovered from inside the coffin revealed that some bones did not belong to the main skeleton but corresponded to two other possible individuals (supernumerary bone: left ulna). In the laboratory, the main skeleton was renumbered as SU 1122.1 (Fig. 9.3), while the second was identified as SU 1122.2. The third individual was represented only by a distal fragment of the left ulna and fragments of the coxal bone, which were discarded, as they were considered intrusive elements.

The damp and acidic environment generated inside the grave, due to the degradation of organic matter (tissue and

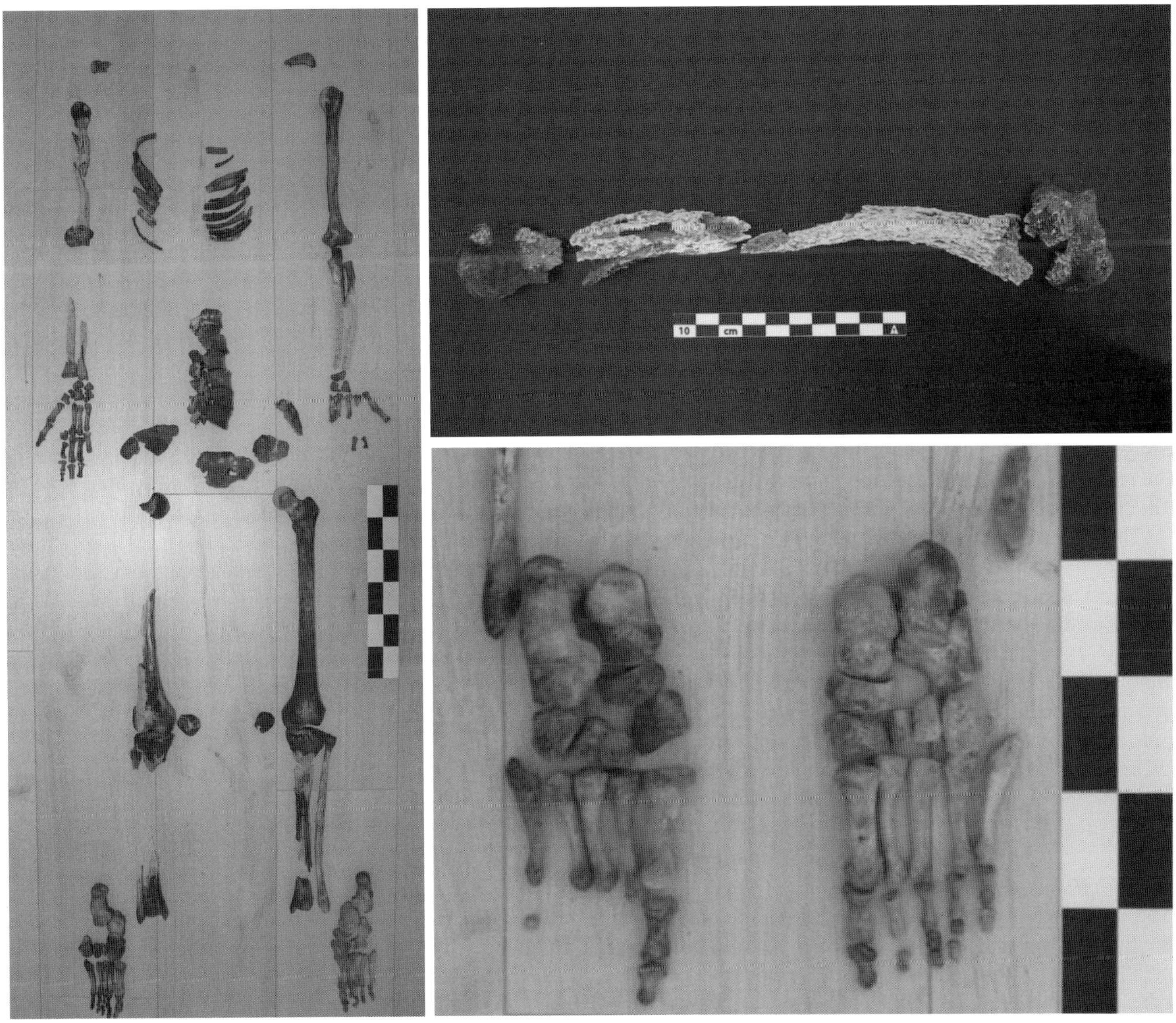

Figure 9.3. Skeletal remains of Prioress Jerònima de Gort (SU 1122.1). On the right, detail of the right humerus and the nun's feet. Author: Izaskun Ambrosio (Iltirta Arqueologia).

wood), led to the deterioration of the bones of the main skeleton (SU 1122.1). Nevertheless, the anthropological study enables us to indicate that this individual was an elderly woman (Fig. 9.3). This is based on the observation of signs of ageing, including calcification of the costal cartilages and slight arthrosis in the spinal column. These characteristics render the individual SU 1122.1 compatible with the remains of Prioress Jerònima de Gort. Regarding skeleton SU 1122.2, it has been determined that it corresponds to a younger female (17–22 years old), based on signs of bone development and maturation. Consequently, this skeleton is incompatible with the identity of Prioress Jerònima de Gort.

As previously mentioned, an additional individual (SU 1125) was recorded below the main burial coffin in a partially secondary position, preceding the burial of the prioress. The skeletal remains correspond to a woman who lived for over 60 years and display degenerative signs in the spine, including grade 2 spondylosis, as well as osteoporosis, characterised by decreased cortical bone mass and more fragile trabecular tissue with low density in the long bones. Additionally, Schmörl's nodes are evident at the dorsal level.

Furthermore, in a peripheral position, individual SU 1121 was documented. This corresponds to a woman of mature age (45–49 years old) buried in the supine position and oriented on the south–north axis. This is an intriguing case, as the skeleton was not fully preserved and was situated in the south-west corner of the grave, in an orientation opposite to that of the main burial. This could suggest that it is a reburial, very likely originating from the cemetery, like the rest of the bones in the ossuary. The fact that the bones were still articulated is compatible with the transfer of the remains a short time (3–5 years) after their original burial, as the ligaments and soft tissues would still have been present to maintain the connection of the bones. Therefore, taking into account its location, it can be interpreted that it must have been the last individual to be deposited in the grave.

Ossuary remains

The study of the bones recovered in a secondary position from the ossuary has made it possible to identify the sex and age of 20 different individuals inside the grave from the skulls and coxal bones present in the stratigraphic units located around and under the prioress's coffin (SU 1119, 1128, 1130, 1131 and 1132). Therefore, taking into account the individuals described above, the grave contains the remains of 24 individuals, the minimum number being 18 individuals (supernumerary bone: right humerus).

The presence of these skeletal remains corresponds to the reuse of the tomb as an ossuary after the burial of Prioress Jerònima de Gort (1601) and aligns with a chronology linked to the abandonment of the monastery by the community (1699). The age and sex distribution of the bones from the ossuary is not uniform: nine males, four females, one allophyseal (with characteristics of both sexes) and six indeterminate individuals were identified. In terms of age, it seems that the peak of mortality is found among individuals of adult age (8 individuals between 21 and 40 years old) and advanced age (6 of senile age) (Table 9.1).

Palaeopathological study

The palaeopathological study corroborates the findings of the biological profile outlined above, as the most prevalent pathologies are those associated with degenerative changes inherent to the age of the individuals. A total of five skeletal anomalies compatible with arthritic processes in the bone joints and five cases of possible osteoporosis have been identified, including the mature female SU 1121 and the senile individual SU 1125. Considering the ages of the individuals for whom diagnoses were made, it may be posited that the group under examination reached physical maturity and advanced age in a relatively sound state of health, as the degenerative pathologies observed are not severe. It can therefore be inferred that these individuals are likely to have led a sedentary lifestyle with minimal physical exertion.

Two cases of Schmörl's nodes and two instances of herniated discs were observed within the traumatic pathology category, probably caused by excessive strain of the spinal column. The most significant traumatic pathology observed is the perimortem incision on the right humerus diaphysis of SU 1131 (ossuary). As this is an isolated fragment of the diaphysis, and the remainder of the skeleton is not available for assessment, it is not possible to determine whether the individual sustained additional injuries. It cannot be excluded that this was a violent event.

As regards the dental remains, a high prevalence of antemortem tooth loss has been identified, which may be associated with the presence of alveolar recession, dental calculus and caries. In some cases, the recession is quite severe and may even have resulted in tooth loss. Furthermore, the total loss of dentition antemortem in some individuals or jaws studied reflects the presence of an older population. Furthermore, eight cases of dental hypoplasia have been documented, a pathology related to nutritional deficiencies (metabolic deficits), although its causes may be diverse (Oms 2016).

The associated material

The material recorded in association with the skeletal remains is composed of a variety of elements. Primarily, the material record comprises textile remains, including those of clothing, the shroud and the coffin lining. Additionally, the material record includes metallic elements associated with the textiles, both those found *in situ* and those that have been removed from their original context. These items include hooks and eyes, together with pins. In contrast, the material record associated with the grave comprises personal and devotional objects, such as rings, rosaries, a crucifix, a medal and decontextualised rosary beads.

Table 9.1. Age and sex distribution of data obtained from skulls and coxal bones from the ossuary

	Male	*Female*	*Allophyseal*	*Indeterminate*	*Total*
Infant I (1–6 years)	1				1
Juvenile–adolescent (13–20 years)			1		1
Young adult (21–30 years)	3			1	4
Adult (31–40 years)	2			2	4
Mature (41–60 years)		2		1	3
Senile (>60 years)	2	2		2	6
Adult undetermined	1				1
Total	9	4	1	6	20

The textiles have been classified according to their function, namely as clothing, shroud elements and coffin linings. The clothing associated with the skeleton of Prioress Jerònima de Gort (SU 1122.1) was found in a highly deteriorated state (Fig. 9.4), with only small fragments of hems and stitches remaining (Fig. 9.4b–i), along with seams (Fig. 9.4c–f). Additionally, the material record comprises various textile fragments, including ribbons (Fig. 9.4g, k), cords (Fig. 9.4d), stocking remnants (Fig. 9.4h, j), a fragment of a coarse garment (Fig. 9.4i), and buttons and buttonholes (Fig. 9.4c), which are associated with both the undergarment and the outer garment. The positioning of the hooks and cords indicates that the garment was designed to fasten at the front and to have long sleeves (Rodríguez and Camí 2022). The fabric of the garment's weave has been identified as thick taffeta, while the ribbons and laces have been identified as thinner taffeta. The buttons are double taffeta trimmings, and the stockings are knitted. Additionally, the remains of a folded, pinned veil were observed on the torso. This item had two small pockets inside, which contained remains of gilt paper (Fig. 9.4b, Fig. 9.5c–d). These paper remnants may have formed part of a devotional scapular or prayer card. The veil's binding is also taffeta but woven with a finer thread and an open weft that gives it a certain transparency. The fibres have been analysed by optical microscopy and SEM electron microscopy (Rodríguez and Camí 2022) (Fig. 9.5), but the results were inconclusive due to the state of preservation of the fibres. However, it can be confirmed that the veil (Fig. 9.5i), the ribbons, and the cords are of plant-based origin, and one of the few fabric samples from the clothing is wool.

With regard to the elements of the shroud with which the Prioress Jerònima de Gort was buried (SU 1122.1), there is no evidence to suggest that a sheet or linen shroud was used. Instead, the deceased appears to have been buried in her usual clothing, specifically the monastic habit, as was customary during this period in monastic contexts in the Iberian Peninsula (García Fernández 1996; Vizuete Mendoza 2014).

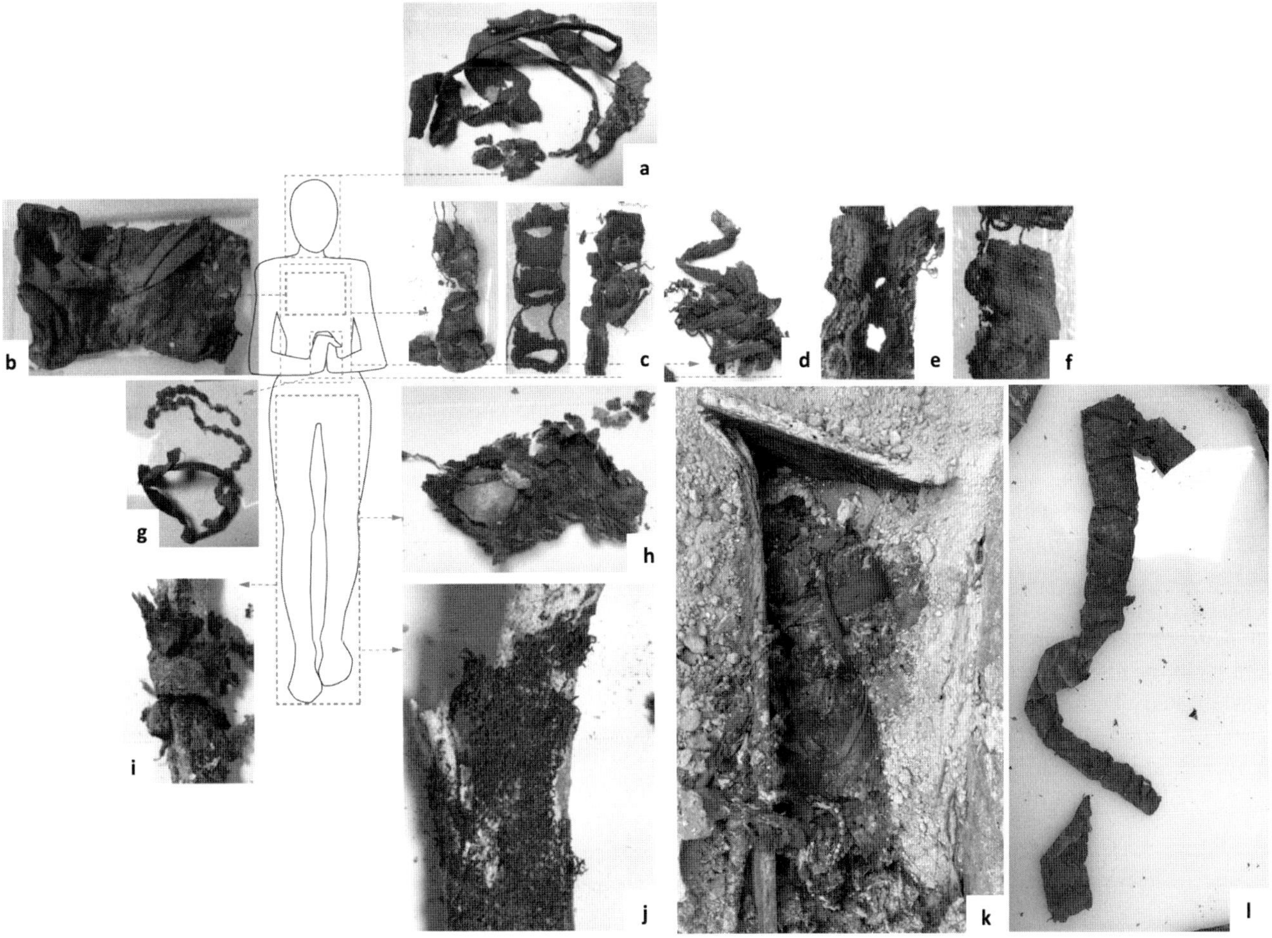

Figure 9.4. a) Ribbons of the skull shroud. b) Veil. c) Buttons and buttonholes as a fastening system. d) Buttonhole and cord fastening. e–f) Hook and eye fastening system. g) Rosary and ribbon wrapped round the Prioress's hands. h–j) Stocking remnants attached to the bones. i) Clothing fabric remains. k) General view of the upper third of the burial. l) Coffin lining border. Authors: Rosa F. Rodríguez and Eva Camí (Context Restauració).

More unusual and less well-documented, however, is the practice of using multiple ribbons to secure the legs, wrists, skull and jaw with knots or ties (Fig. 9.4a, g, i). This practice is also recorded in the tombs of other prioresses from the site, such as Helena de Remolins (1601–1604), currently under study. The best-preserved ribbons are those that held the skull and jaw (Fig. 9.4a, 9.5a). One ribbon bordered the facial oval, while another was wrapped around the forehead and extended down the neck until it was fastened to the veil by means of a pin (Rodríguez and Camí 2022). Additionally, ribbons were used to fasten the wrists and legs (Fig. 9.4g, i). The shroud bands are composed of fine taffeta and are of plant origin, as determined by optical microscopy analysis (Rodríguez and Camí 2022) (Fig. 9.5e, f).

Regarding the lining of the wooden coffin in which the prioress was buried, the findings comprise small remains of taffeta edging that lined the upper edges of the coffin planks (Fig. 9.4k, l). In this case, the identification of the fibres by microscopy has been possible, and it can be confirmed that they are linen (Rodríguez and Camí 2022) (Fig. 9.5g, h, j).

The personal and devotional objects that can be associated with the body of the prioress (SU 1122.1) include two rings, a rosary and a medal (SU 1126). The rings (Fig. 9.6a, c) are not made of metal but rather of a carved and polished mineral, probably jet, based on their appearance, colour and the parallels of the same period and religious context. Both rings are of the half-round type and are of identical dimensions, with a diameter of 16 mm and a thickness of 3 mm. The rosary is constructed from bronze filigree and flattened transparent glass beads, with a Caravaca cross at the terminal (Fig. 9.6b, d). It is worth noting that in 16th-century tombs from Hispanic monastic contexts, glass-paste rosaries and jet objects, as well as pilgrimage souvenirs such as medals, engraved prints and Caravaca crosses, were very common (Andrio *et al.*1992; García Fernández 1996).

Additionally, other objects of significant devotional interest were discovered within the tomb, associated with skeleton SU 1125, which is believed to have been buried prior to that of Prioress Jerònima de Gort (SU 1122.1). Firstly, a rosary (Fig. 9.6g) comprised of iron loops and two types of flattened beads: smaller beads of black vitreous paste and larger beads of black vitreous paste marbled in white, was located. These beads are believed to correspond to those used to mark the transition to the next mystery during prayer (Malgouyres 2017). Additionally, a bronze crucifix (Fig. 9.6e) of noteworthy dimensions (88 mm in height by 63 mm in width) was discovered beneath the skeleton SU 1125. Given its proximity to the body and the displacement of the object under its own weight, it is plausible that this crucifix may have belonged to the skeleton mentioned and may have been placed there at the time of its burial. Considering the object's prominence, it seems reasonable to suggest that this crucifix must have belonged to an important nun, possibly one of the prioresses of the community. The fact that remnants of a ribbon from the shroud have been preserved around the skull of this body (SU 1125) further supports this interpretation.

With regard to the remaining grave goods recovered in a secondary position and related to the other remains in the tomb, we found a circular bronze medal, the corrosion of which has obscured its iconography (Fig. 9.6i). The object measures 15 mm in diameter and 1 mm in thickness. Additionally, an iron rosary and wooden beads were recovered, though they have not been preserved, with only minor traces being visible adhering to the metallic corrosion (Fig. 9.6f). Finally, a set of decontextualised rosary beads, crafted from wood and spherical in shape, was discovered. The set comprised three beads measuring 5 mm in diameter and one measuring 10 mm in diameter (Fig. 9.6h).

The Prioress in the light of documents

According to information provided by written texts, Jerònima de Gort served as prioress of the Alguaire Hospitaller Women's Monastery from 1586 until her death in 1601. Until now, we had little information on her time at the priory (Miret i Sans 1899, 37–40; 1910; Lladonosa 2007, 161–5). The interdisciplinary research carried out in this study has provided new insights into this historical figure and the events that preceded her election as prioress in 1586, five years after her involvement in the visitation process – really an inspection – led by Friar Agustí d'Anglesola, who at that time was the Grand Prior (or Grand Master) of Catalonia for the religious military order of the Knights of the Hospital of Saint John of Jerusalem, also known as the Hospitallers.

The discovery of Jerònima de Gort's tombstone within the body of the monastic church of the Alguaire monastery is a significant find. Being buried or having a tombstone within the church itself are not ordinary events; only prominent individuals, or those regarded as such, enjoyed this distinction. What distinguishes Jerònima de Gort in this context, in the light of the data published by historians such as Miret i Sans (1899; 1910) and Lladonosa (2007), are the events and facts that preceded her election, as well as the historical context in which these developments took place. Fortunately, sources for this specific period – yet to be published – provide us with complementary information on these little-known or previously undisclosed events.

The conflicts that occurred during the disputed priorate of Jerònima de Montgai (1576–1583) as well as the brief priorate of her successor Anna Desvalls (1583–1585) were described as 'scandals' caused by the presence – in the cloister of the monastery, and contrary to cloister rules – of some men, some relatives and some alleged bandits fleeing justice. But it seems that underlying reasons for the 1581 (and 1584 as well) visitation concern a general issue; the existence of opposing parties at the heart of the

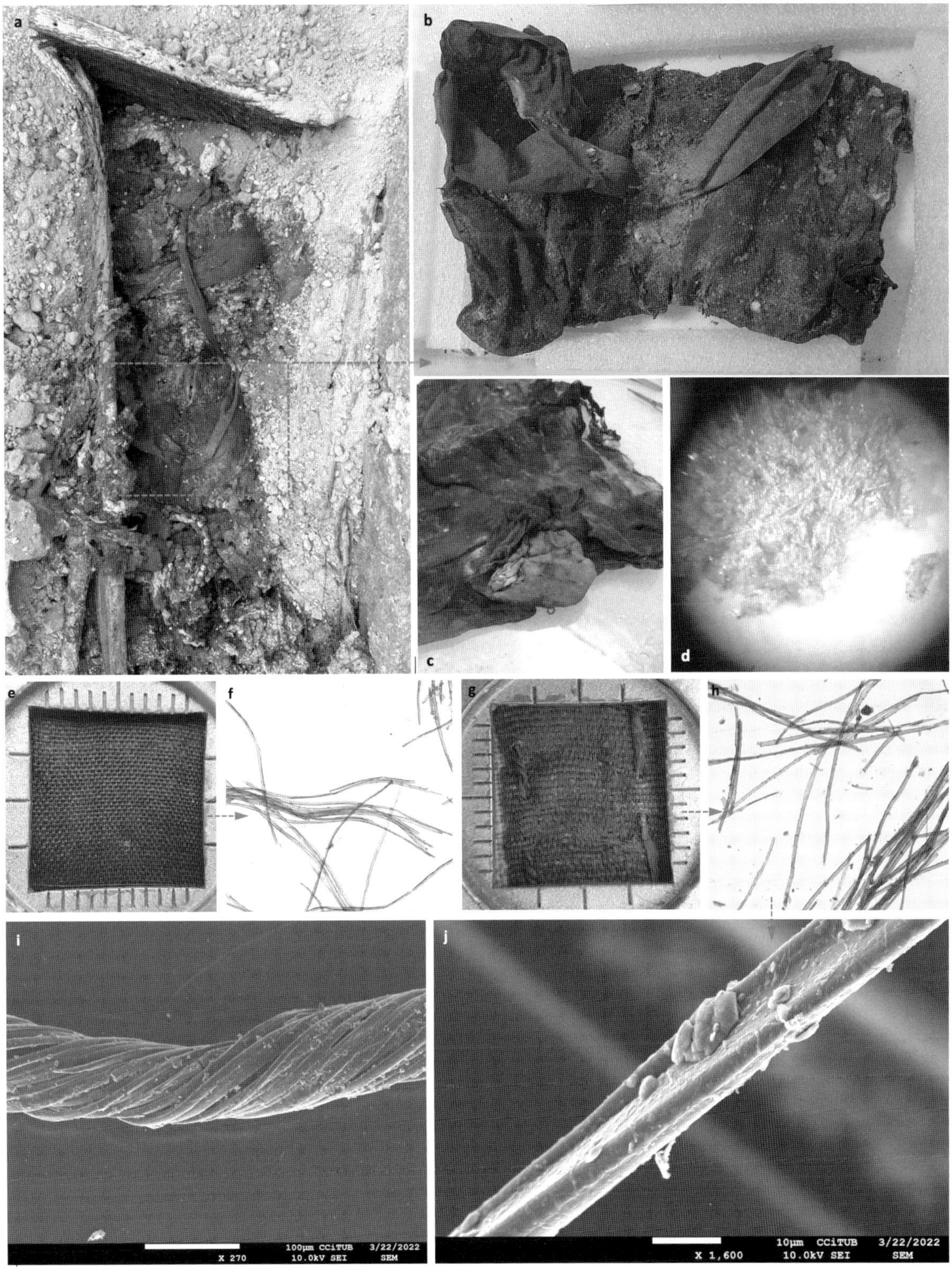

Figure 9.5. a) General image of the upper third of the remains before excavation, with the position of the veil highlighted. b) The veil in its original position after extraction. c) One of the inner pockets of the veil. d) A binocular magnifying glass image at 50x magnification of a piece of gold paper found inside one of the pockets, corresponding to a devotional scapular or prayer card. e–f) Image of the ribbon holding the skull using a Leica DM 1000 optical microscope (50x). g–h) Image of a coffin lining border using a Leica DM 1000 optical microscope (50x). i–j) SEM images of a fibre from the veil (270x) and a fibre from the sewing thread of the coffin edge (1600x). Authors: Aleix Barberà (Resguard), Silvia Marin (ESCRBCC), Rosa F. Rodríguez and Eva Camí (Context Restauració).

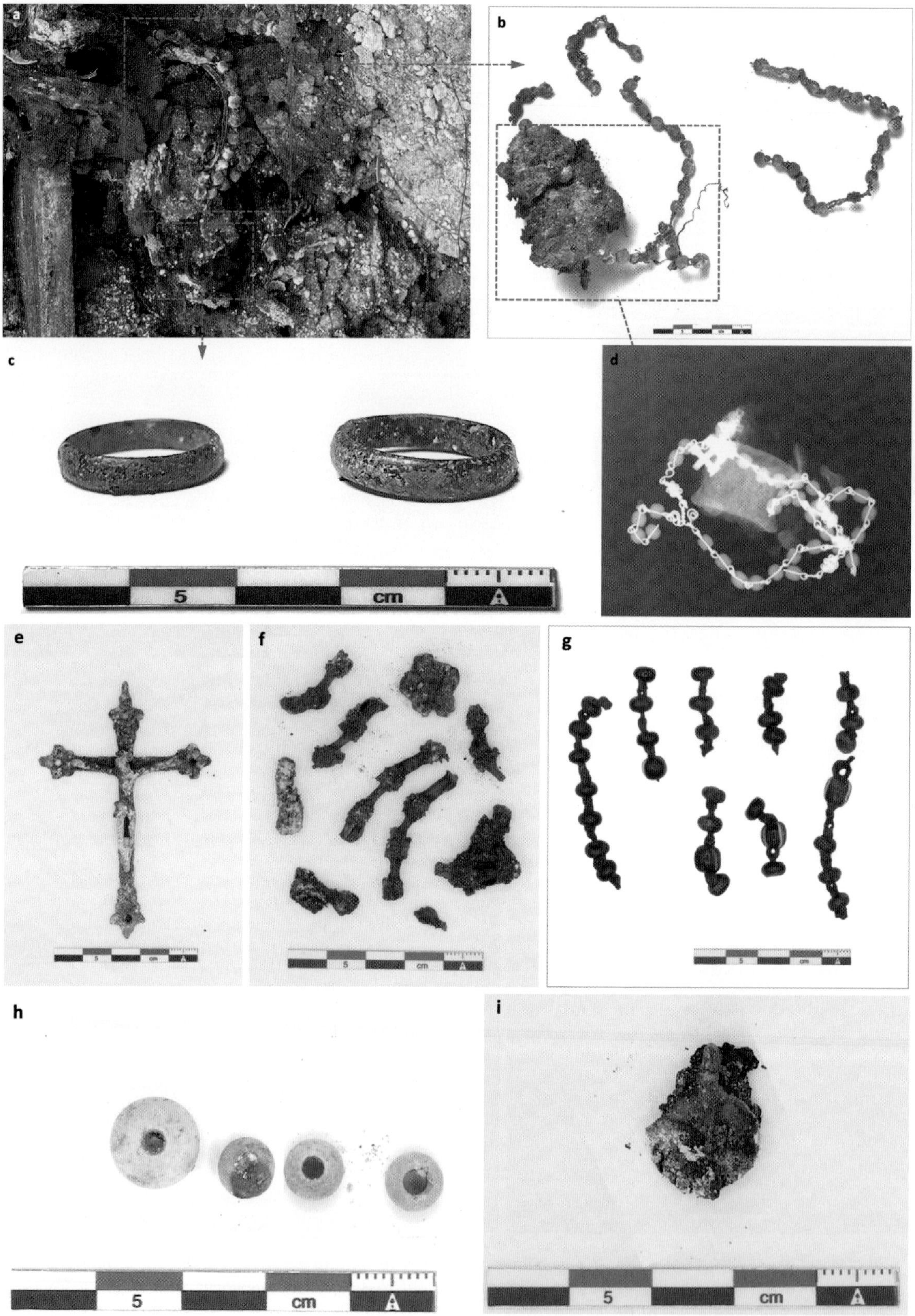

Figure 9.6. a) The rosary and rings in situ *around the hands and wrists of the prioress. b) The rosary after extraction, fused together with bone remains and concretions. c) Detailed image of the rings. d) X-ray showing a Caravaca cross under the fused part of the rosary. e) Bronze crucifix. f) Fragments of an iron and wood rosary. g) Fragments of an iron and vitreous paste rosary. h) Decontextualised rosary beads. i) Bronze medal. Authors: Silvia Marín, Esther Gual, Sonia Mironiouk and Marta Gómez Sanabria (ESCRBCC).*

community itself regarding the strict application of cloister rules. Furthermore, some of the violent incidents that took place in this period were caused by gangs of a political and social character that were dichotomously used by knights and gentlemen of one party or the other. And the same occurred with the Alguaire nuns connected by kinship with one or the other of those opposing parties (Lladonosa 2007, 163–5).

We can confirm the existence of nuns known as *extra vaggantes* – who spent long periods of time outside the Convent – while still exercising the rights associated with their status as Hospitaller sisters, and formally delegating their vote when Convent chapter meetings were held. All the same, they failed to observe the cloister regulations (Coll 2016, 171–2). This appears to be an expression of an attitude of resistance on the part of some Alguaire nuns towards the strict application of the rules of cloistered life.

Almost from the time of its foundation by the Marquesa de la Guàrdia, the *commendatrix*, the freedom enjoyed by Hospitaller nuns in the Alguaire monastery was progressively reduced by the Order's supreme authorities through the promulgation of instructions and regulations to direct the convent, both internally and externally, an event that usually occurred after every visitation by the authorities of the Order. The enactment of these ordinances tended to be the consequence of a visitation by a friar, ordered by those authorities of the Order who habitually examined any conflict or scandal in the community, as happened in Jerónima de Montgay's priorate (1576–1583). More specifically, unauthorised men and other people – including bandits, others involved in a robbery in the priory itself, as well as builders during rest hours – were said to be present in the cloister at untimely hours. The examination was undertaken in the visitation process of 1581 undertaken by the Grand Prior of Catalunya, Friar Agustí d'Anglesola, with the assistance of other friar-priests, friar-knights and a notary to certify the process. Among the leading participants in this visitation, the aforementioned Grand Prior and his delegates, on the one hand, were in favour of a strict application of monastic rules in female Hospitaller communities, whereas on the other hand, the prioress, Jerònima de Montgay, resisted such a move.

Jerònima de Gort also acted as a leading participant in the same visitation – five years before being elected prioress – expressing her support for the application of cloister rules in the Alguaire convent. Consequently, she answered the Grand Prior of Catalonia's delegates as follows:

> 'Domina Hieronyma Gort, monialis interrogatus ut supra, dixit *que Aldonça Mediona (religiosa del convent) ha tingut moltes voltes a don Joseph Marles en la casa donant-li entrada i aço ha durat set o vuyt mesos y que Ana Castro tenia en sa casa bandolers y entre altres hun que's deia Barberet ... la senora priora en los priorat molta gent dolenta y entre los altres ... un home que's diu Massaguer lo qual es troba en un robo que's feu en lo priorat del present monestir.*
>
> Item interrogada dixit, *que la senora priora no ha corregit lo que era menester, que ho hagués fet no foren vingudes en los scandols que son vingudes. Item interrogata dixit, que fins al dia de la Assensió proppassada may se ha servada la clausura en lo present monestir y eren vingudes a termens que alguna senora anava ma per ma ab homens per lo present monestir y de aso may la Sa. Priora no feia cars y aquestos abusos ab importunació [meua] se son remediats de la Assensió fins ara*'
>
> Arxiu de la Corona d'Aragó, ORM, Gran Priorat. vol. 653, f. 18, 23 and 32 (year 1581).

Other answers follow – *medio iuramento* (under oath) – given by Jerònima de Gort in the course of her interrogation, demonstrating her commitment to the observance of strict monastic rules and placing herself alongside her own inquirer, the delegate of the aforementioned Grand Prior of Catalonia, Agustí d'Anglesola. This fact also explains the succession of scandalous events involving both sides. It would appear that the election of Jerònima de Gort as prioress in 1586 was a further step in the gradual implementation of the conclusions and decrees of the Council of Trent. Her determined attitude throughout this process of inquiry and, ultimately, her subsequent election make her the main figure in a specific period in the history of the Alguaire convent. In this respect, the discovery and archaeological study of her tomb is a significant find that helps us to place, in a very precise way, the historical evolution of cloister observance in the monastery of Santa Maria d'Alguaire, and more broadly, in women's nunneries of the Order of Saint John of Jerusalem.

Conclusions

The archaeological, anthropological and documentary study carried out at the monastery of Santa Maria d'Alguaire has significantly enhanced our understanding of the historical figure of Jerònima de Gort. A woman of noble origin, she served as prioress of the monastery of Santa Maria d'Alguaire from 1586 until her death on 25 November 1601, a few days before her 74th birthday. Her tomb was located in Chapel A4, which was renovated in the mid-16th century and situated in proximity to the nuns' choir and the transept, which are of significant symbolic value. This discovery is exceptional not only because of the wealth of material remains recorded within the tomb, but also for the new insights it provides into the life and death of women in such a highly masculinised religious and military order as the Order of the Hospital.

The interpretation of the diverse, rich and complex set of archaeological remains identified inside the tomb has undoubtedly constituted a significant challenge for the research team responsible for the site. It has required interdisciplinary collaboration between archaeologists, anthropologists, documentalists, historians and restorers, with the aim of developing a more nuanced interpretation than that which might be inferred from the inscription on the tombstone. Once again, the commitment to interdisciplinary research and the dialogue between sources has proved essential for the archaeological understanding of this complex sepulchre. This approach has allowed us to articulate a global explanation of the diversity of processes that converged in this tomb before and after the burial of the prioress.

The archaeological stratigraphy and biological profile of the bones, along with the ostentatious trousseau with which she was buried (personal objects: rings; devotional objects: rosary, medal, devotional scapular or prayer card; veil carefully folded over her chest), and the delicate shrouding process to which her body was subjected, allow us to conclude that the elderly woman documented inside the wooden coffin (SU 1122.1) is that of Prioress Jerònima de Gort. It is reasonable to claim that the tomb had previously been used for the burial of a young woman and another elderly woman (possibly also a prioress, given the notable crucifix associated with her and the fabric remains recorded on her forehead). Nevertheless, it seems plausible that, following the interment of Jerònima de Gort in 1601, the tomb was sealed for a period until it was reopened for use as an ossuary.

It is a well-established fact that the reuse of funerary structures was a common practice in medieval and early modern times, driven by both spatial optimisation and symbolic significance. Similarly, in Alguaire, as in numerous other female monasteries of medieval origin, such as Sant Daniel in Girona or Santa Maria de Pedralbes in Barcelona, the space adjacent to the transept served as a ritual and symbolic area designated for the burial of individuals of significance to the community. On some occasions, these tombs were later subject to repeated re-inhumations. This phenomenon has been observed in the tomb of Prioress Jerònima de Gort, as well as in other tombs situated within the nave and transept of the Alguaire convent church. In the same way that some noble families maintained family sarcophagi within conventual churches, it is possible that the reuse of tombs in Alguaire was related to the interment of women belonging to the same family connected to the monastery. It is hoped that the DNA study of the bone fragments of the skeletons associated with these tombs will shed light on the existence, or absence, of kinship ties between them.

The application of archaeological, anthropological and object-based analysis has enabled us to ascertain the manner in which Prioress Jerònima de Gort was shrouded and buried, as well as the circumstances of her final years. These findings indicate the presence of slight arthrosis in her spine, consistent with the physically undemanding existence of a nun who had spent almost 74 years in a conventual environment. Furthermore, the archival documents provide a deeper comprehension of the historical significance of this woman as a defender of the observance of cloister rules during a period when customs tended to become relaxed. Her active participation in the internal investigation process initiated during the tenure of her predecessor, in which she accused several of her colleagues of leading a 'scandalous' lifestyle, was a key factor in her election as prioress in 1586. This laid the foundation for a 15-year tenure as superior, during which she exercised considerable control over the community of nuns at the monastery of Alguaire.

In conclusion, the interdisciplinary study presented in this article has offered valuable and unique insights into the burial practices and funerary rituals of female communities within the Hospitaller Order, an area of study that has received little attention to date. These findings are particularly significant within the unique context of the years following the Council of Trent (1545–1563) and the progressive implementation of strict enclosure in European monastic settings, phenomena that had a profound impact both on the monastery of Santa Maria d'Alguaire and on the female communities of its time.

Acknowledgements

This chapter is one of the principal outcomes of the projects: 'FREILAS. Women in Military and Hospitaller Orders in the Crown of Aragon (12th–16th c.)' (Proyectos de Generación del Conocimiento 2022, Ministerio de Ciencia e Innovación – PID2022-136266NB-I00) and 'Prioresses, Lords and Princes: Spaces of Power in Medieval and Early Modern Catalonia' (Projectes Quadriennals d'Arqueologia i Paleontologia 2022, Generalitat de Catalunya – ARQ001SOLC-127-2022).

References

Adell, J.A. (2002) La incidencia de les seus de Lleida i Tarragona. In *L'art gòtic a Catalunya I. Catedrals, monestirs i altres edificis religiosos,* 91–98. Barcelona, Enciclopèdia Catalana.

Alturo i Perucho, J. (2010) *Diplomatari d'Alguaire i el seu monestir duple de l'orde de Sant Joan de Jerusalem (1245–1300).* Barcelona, Fundació Noguera.

Ambrosio, I. (2021) *Informe antropològic: restes esquelètiques exhumades a la sepultura de la Priora Hyeronima de Gort.* Unpublished archaeological report, Iltirta Arqueologia.

Ambrosio, I. (2023) *Informe antropològic: restes esquelètiques associades a l'hàbit de la priora Hyeronima de Gort. Monestir de Santa Maria d'Alguaire.* Unpublished archaeological report, Iltirta Arqueologia.

Andrio Gonzalo, J., Loyola Perea, E., Martínez Flórez, J. and Moreda Blanco, J. (1992) *El conjunto arqueológico del monasterio de San Juan de la Hoz de Cillaperlata (Burgos).* Miranda de Ebro, Junta de Castilla y León.

Bertran, P. (1977) Les ordinacions del convent d'Alguaire. *Cuadernos de Historia económica de Cataluña* 17, 25–56.

Bonet Donato, M. (2014) Identidad de las hospitalarias en la Corona de Aragón (siglos XII–XIII). *Memoria y Civilización* 7, 43–87.

Coll Sanabra, A. (2016) *Diplomatari del Monestir Santjoanista d'Alguaire (1301–1350): estudi i edició*. PhD thesis, Universitat Autònoma de Barcelona.

Conejo da Pena, A. (2002) El convent trinitari de Viganya. In *L'art gòtic a Catalunya I. Catedrals, monestirs i altres edificis religiosos*, 233–5. Barcelona, Enciclopèdia Catalana.

García Fernández, M. (1996) *Los castellanos y la muerte. Religiosidad y comportamientos colectivos en el antiguo régimen*. Valladolid, Junta de Castilla y León, Consejería de Educación y Cultura.

Lladonosa i Pujol, J. (2007) *Història de la Vila d'Alguaire i el seu Monestir Santjoanista*. Lleida, Gràfiques Larrosa.

Malgouyres, P. (2017) *Au fil des perles, la prière comptée. Chapelets et couronnes de prières dans l'Occident chrétien*. París, Somogy.

Miret i Sans, J. (1899) *Noticia histórica del monestir d'Alguaire de la orde sagrada y militar del Hospital de Sant Joan de Jerusalem*. Barcelona, Tip. L'Avenç.

Miret i Sans, J. (1910, reed. 2006) *Les cases de templers y hospitalers a Catalunya*. Lleida, Pagès.

Oms, J.I. (2016) *Testimonio de los dientes. Antropología dentaria de la transición neolítica hasta el presente*. Barcelona, Bellaterra arqueología.

Palacios Sánchez, J.M. (1977) *La sagrada, soberana e ínclita orden militar de San Juan de Jerusalén (Orden de Malta) y sus monasterios de religiosas en España*. Logroño, Ochoa.

Rodríguez Rodríguez, R.F. and Camí Fernández, E. (2022) Informe de restauració de les restes tèxtils de l'enterrament de la Priora del Monestir d'Alguaire Jerònima de Gort. Unpublished restoration report, Barcelona, Context Restauració.

Sans i Travé, J.M. (2011) The military orders in Catalonia. *Catalan Historical Review* 4, 53–82.

Soler, M. (2021) Arqueología de los espacios monásticos medievales: el monasterio femenino de Alguaire (El Segrià). In *Actas del VI Congreso de Arqueología Medieval (España-Portugal)*, 607–13. Ciudad Real, Asociación Española de Arqueología Medieval.

Soler, M. and Alegría, W. (2024) Arqueologia medieval i paisatge espiritual: el monestir femení hospitaler de Santa Maria d'Alguaire (segles XIII–XVII). *Tribuna d'Arqueologia*, 2021–2024.

Soler, M., Alegría, W. and Sancho, M. (2023) *Memòria d'Excavació Arqueològica. Castell i Monestir Santjoanista d'Alguaire (Abril 2021)*. Unpublished archaeological report, University of Barcelona.

Soler, M., Lluch, R. and Mallofré, B. (2025) Del recinto conventual a la gestión del patrimonio. Estrategias de análisis espacial en el estudio del monasterio femenino hospitalario de Santa Maria de Alguaire (el Segrià). *SVMMA. Revista de Cultures Medievals* 25.

Vizuete Mendoza, J.C. (2014) Morir en comunidad. Usos, costumbres y rituales en torno a la muerte en monasterios femeninos hispánicos. In F.J. Campos (coord.) *El mundo de los difuntos: culto, cofradías y tradiciones*, vol. 2, 643–58. San Lorenzo del Escorial, Instituto Escurialense de investigaciones históricas y artísticas.

10

'The greatest evil is physical pain': an exploration of suffering from a medieval Austin Friary in Cambridge, UK using an 'Avatar Model'

Benjamin Neil

This study investigates the osteobiographical and osteological analysis of 11 individuals from the Cambridge Austin Friary, spanning AD 1302–1538. The research contextualises skeletal remains within broader historical, ethical and sociocultural frameworks, emphasising the complex interplay between empirical data and interpretative reconstruction. The proposed 'Avatar Model' allows for imaginative reconstructions of lived experiences while mitigating the risk of speculative narratives. Integrating osteological evidence with historical and clinical knowledge, the study provides insights into pain, suffering and behaviour, enriching the understanding of past human experiences in religious and urban medieval contexts. This study also addresses ethical concerns regarding the treatment and interpretation of human remains, highlighting the need for a careful balance between scientific inquiry and respect for past populations.

Keywords: Osteobiography, avatar model, osteology, monasticism, pain, suffering.

Introduction

St Augustine's call to 'love the suffering' presents a profound challenge when viewed through the lens of osteology. The ethical dilemma arises from attempting to reconcile fixed empirical data with the experience of a temporally distant individual. Diagnosing trauma or pathology in the archaeological record is fraught with complexity, as emotions like suffering are transient and intangible, making it ethically problematic to attribute emotional states or personhood to human remains. In modern contexts, personal identity is seen as unique and sacred, and efforts to humanise past individuals risk reducing them to mere academic objects or, at worst, tools for political agendas.

This ethical complexity extends into the broader discussion of postmortem rights within the context of archaeological skeletal remains in the UK. Legal frameworks, such as the Burial Act and the Human Tissue Act, and professional policies from bodies like Historic England, BABAO and APABE, provide guidelines for the respectful treatment of human remains. However, their application is often less straightforward in practice, especially in balancing the pursuit of knowledge with the preservation of dignity and cultural sensitivity. The interpretation of these remains, including the narratives assigned to them, can be contentious, raising the question of how modern ethical and legal standards apply to individuals from centuries or millennia ago. While some argue for universal respect for all human remains, others contend that the historical and scientific value of these remains may warrant different standards of treatment (Moon 2020). Researchers must therefore navigate the fine line between scientific objectivity and the imposition of contemporary values on past populations, reflecting broader debates about the rights of the dead, the responsibilities of the living, and the role of archaeology in interpreting human history.

To address these issues, I propose the concept of an 'Avatar Model'. The term 'Avatar' comes from the Sanskrit word 'avatāra' meaning 'descent'. Within Hinduism, it means a manifestation of a deity in bodily form on earth, such as a divine teacher. Outside Hinduism, it technically means 'an incarnation, embodiment, or manifestation of a person or idea'. In this context, these manifestations allow researchers to explore humanity within a constructed model of the historical environment. This leverages specialists' often unacknowledged emotional connections to remains as an asset. By contextualising osteological findings within broader frameworks, researchers can reflect on several possible lived experiences. By combining ethical considerations, trace-based reasoning using clinical and palaeopathological analysis, multi-scalar analysis using historical context, and crucially imaginative exploration, the 'Avatar Model' allows for a more nuanced osteobiographical narrative.

The Cambridge Friary context

The Cambridge Austin Friars, part of the Augustinian order, were established in the late 13th century and quickly became a significant presence in the town of Cambridge. The earliest documented reference to the friary dates to October 1289, indicating the provision of royal support for 20 friars. By 1297, the community had grown to 36 members, and by 1326, it expanded to 70 friars. The friary continued to develop its infrastructure throughout the 14th and 15th centuries, adding a church, cloister, and possibly a second cloister, reflecting its increasing influence in the region.

The Cambridge friary was not only a religious institution but also an important educational centre. In the early 14th century, it hosted chapters of the Augustinian national province, attracting friars from across Continental Europe to study and teach. Theology was the primary focus of their studies, and many friars became integral to Cambridge's intellectual community. Notably, some members of the friary emerged as influential Evangelical and early Protestant figures in the early 16th century, indicating the friary's role in the broader religious transformations of the period.

The friary was deeply integrated into both the town of Cambridge and the University. It benefited from property donations and hosted various university events, underscoring its central role in the community. The friary church was actively involved in local religious and social activities, as evidenced by wills and other documents. These records highlight the friars' engagement in preaching, hearing confessions, and collecting alms, indicating their significant influence on the spiritual life of Cambridge.

Despite its success, the Cambridge friary faced challenges, particularly during the Black Death in the 14th century, which had a devastating impact on its community. It must also be acknowledged that whilst the Austin friars played a vital role in providing spiritual guidance and care to marginalised communities, their behaviour was known to spark social tensions. Their engagement with morally complex situations damaged their reputation, fuelling anti-fraternal sentiment and, in some cases, incited violent retaliation from the lay population (Geltner 2010; 2012a; 2012b). Nevertheless, the friary remained a key institution, housing a diverse population of clerics and laypersons who maintained close ties with the broader Cambridge population.

The friary's dissolution in 1538 marked the end of its active presence in Cambridge. The master and only three friars signed the surrender document, bringing an era to a close. However, some friars received dispensations to continue serving as secular priests, and the friary's legacy endured through its contributions to education, religious practice and community life in Cambridge (Cessford and Neil 2022).

Health in archaeology

Understanding health in archaeological populations is complex, often due to incomplete skeletal remains, limiting the visibility of diseases affecting only bones and teeth. Conditions impacting soft tissues often remain undetected without additional historical or biomolecular evidence. To address these limitations, researchers incorporate contextual factors such as occupation, diet and population density (Roberts 2009). Bio/osteoarchaeology excels in revealing insights into ancient pain and suffering by integrating skeletal data with cultural and environmental contexts, providing a nuanced view of human adaptation and survival. Leveraging modern medical knowledge enhances the ability to infer past experiences of pain, offering an empathetic understanding of ancient lives. Despite inherent biases in skeletal collections, osteology remains invaluable in illuminating human resilience, particularly through the study of survivors, demonstrating how individuals and communities navigated and overcame hardships (Martin and Harrod 2016).

Pain

Pain has been perceived differently across historical periods, shaped by prevailing beliefs and knowledge systems. In primitive societies, pain was seen as external and magical, linked to supernatural causes. Ancient Egyptians combined empirical knowledge with religious mysticism, viewing pain as influenced by spirits. Greek medicine, from Hippocrates to Galen, conceptualised pain through natural laws, balancing humours and anatomical studies. In the Renaissance, advancements in dissection and anatomy, particularly by Vesalius and Leonardo da Vinci, challenged earlier dogmas. The Early Modern Era, influenced by Newtonian ideas, continued evolving pain theories, contributing to modern physiological and psychological understandings (Todd 1993).

Today, we understand pain as a sensory and emotional experience triggered by nociceptors, specialised receptors that detect harmful stimuli whether it be chemical, thermal or mechanical. This nociceptive process involves four stages: transduction, transmission, modulation and perception. Transduction converts harmful stimuli into electrical signals, which are transmitted to the spinal cord and brain, where modulation can either amplify or diminish these signals, and perception involves the brain's interpretation of the signals as pain (Beaulieu 2008).

In acute pain, such as an arm, rib or leg fracture (see F.191, F.332 and F.311 below), inflammatory mediators such as prostaglandins and bradykinin are released, sensitising nociceptors and causing sharp, intense pain. Persistent inflammation continues to activate nociceptors, leading to sustained, aching pain. These pain signals travel through pathways like the spinothalamic tract to reach brain regions that process pain perception and its emotional and cognitive aspects. If acute pain is not effectively managed, it can lead to chronic pain conditions (Beaulieu 2008).

Chronic pain, usually involving repetitive biomechanical disorder and degeneration (see F.237, F.302, F.230, F.311, F.232, F.146 and F.343), involves more complex mechanisms, including peripheral sensitisation – where ongoing inflammation keeps nociceptors in a heightened state – and central sensitisation, which makes spinal cord neurons hyperexcitable. This can cause normal stimuli to be perceived as painful (allodynia) and painful stimuli to become more intense (hyperalgesia) (LeBovits 2008). Chronic pain is associated with structural and functional changes in the nervous system, such as altered gene expression and nerve sprouting. Genetic factors also influence pain sensitivity, with variations in genes related to voltage-gated sodium channels affecting pain thresholds (Aouizerat and Miaskowski 2008).

Effective pain management requires a comprehensive approach that addresses physical, emotional and cognitive dimensions. In clinical settings, this incorporates strategies like cognitive-behavioural therapy, mindfulness and physical therapy (Schaible 2008). In lieu of these modern protocols, there lie parallels in how pain might have been controlled within an Austin monastic framework, as outlined below.

Suffering

Suffering is an experience that extends beyond physical pain to include emotional, social and spiritual dimensions. Unlike pain, which can be objectively measured, suffering is shaped by an individual's experiences and internal states. In religious traditions, particularly monastic asceticism, suffering is not merely endured but embraced as a path to spiritual growth and purification. Practices such as self-flagellation are employed to achieve mastery over the body and mind, increasing pain tolerance and, paradoxically, reducing the perception of pain over time. The biochemical release of endorphins during self-inflicted pain can make this experience more bearable, sometimes even transcendent (Easterling 2015).

A significant outcome of this disciplined suffering is the cultivation of empathy. By voluntarily enduring pain, ascetics develop a deeper understanding of suffering, fostering greater compassion and ethical behaviour. This empathy strengthens social bonds and contributes to moral development within both religious communities and broader society. However, empathy can also be manipulated, as control over pain may be used to dominate or influence others. (Easterling 2015; Sheridan and Gregoricka 2020)

In monastic communities, ascetic practices carried ethical, social and religious significance, extending beyond personal penance. Late medieval Christian ascetics, who practised extreme forms of self-wounding, were seen as both martyrs and symbols of communal atonement. Their suffering highlighted the fragility of the human body and raised ethical concerns about extreme devotion. The distinction between personal and communal responsibility often blurred, with ascetic suffering serving both as private devotion and public demonstration of piety (Sheridan and Gregoricka 2020).

Saint Augustine's views on asceticism further illuminate the relationship between the body, soul and suffering. For Augustine, bodily abstinence and the mortification of the flesh were essential for spiritual purification and alignment with the divine. Ascetic discipline, for him, was not only personal but also performative, intended to inspire the Christian community. His emphasis on visible acts of piety reinforced the belief that true holiness required rigorous bodily discipline (Nightingale 2011).

The experience of suffering within asceticism provides profound insights into the nature of empathy and transformation. When suffering is imbued with spiritual meaning, it reveals its potential to foster personal growth and reinforce communal responsibility.

Behaviour

The behaviour of Austin friars during the medieval period often resulted in social tensions within the communities they served (Geltner 2010; 2012a; 2012b) despite the honourable intentions of their work. Their close interaction with marginalised groups exposed them to morally complex situations, occasionally leading to scandalous behaviour, such as engaging with sex workers. These transgressions, under public scrutiny, fuelled mistrust and criticism of the mendicant orders, contributing to anti-fraternal sentiment (Geltner 2012a).

Additionally, friars faced hostility and violence, particularly in 15th-century Britain, where accusations of economic exploitation, hypocrisy and corruption led to resentment. Many laypeople believed the friars exploited them through tithes and donations, further exacerbating tensions and leading to violent confrontations. Moreover, the friars' involvement in the Inquisition and their political entanglements further exacerbated public anger. As agents of papal and royal policies, friars were often seen as enforcers of unpopular decisions, which made them targets of public ire. The severity of the violence they faced varied, from verbal abuse and public humiliation to physical assaults and, in extreme cases, murder. Urban areas, where friars were more visible and active, witnessed more frequent and severe acts of violence, including mob attacks on friaries and public beatings. This widespread hostility highlights the fraught relationship between the friars and the communities they served (Geltner 2012b).

The friars' perceived hypocrisy, particularly in relation to their moral failings and economic exploitation, played a significant role in inciting violence against them. Reports of immoral behaviour, such as sexual misconduct and the sale of fake relics, undermined their spiritual authority and led to a loss of public trust. This erosion of their moral standing was particularly damaging given their role as religious leaders, and it contributed to a cycle of resentment and retaliation that further strained their relationship with the lay population (Geltner 2012a; 2012b).

The behaviour of the Austin friars, shaped by the difficult circumstances of their mission, had far-reaching implications for both their personal integrity and their public reputation. Their close contact with vice and suffering likely had a psychological impact that influenced their actions, leading to behaviours that were widely criticised and sometimes violently opposed. Understanding these dynamics requires a nuanced perspective that considers the friars' challenging roles, the psychological effects of their work, and the complex social and economic contexts in which they operated. The result was a strained relationship with the communities they served, marked by mistrust, hostility and, ultimately, calls for reform.

Osteobiographical techniques and the 'Avatar Model'

In osteology, methodologies like trace-based reasoning and multi-scalar analysis are integral to constructing osteobiographies, which reconstruct the lives of individuals from their skeletal remains. These methods enable researchers to infer past life conditions by analysing the physical traces on bones and contextualising them within broader cultural and temporal frameworks. Integrating imagination enhances these reconstructions, offering a means to balance scientific rigour with the creation of ethically reasoned and humanised narratives.

Trace-based reasoning involves interpreting skeletal evidence to deduce aspects of an individual's life, such as health, occupation and social status. Physical markers on bones, like wear patterns or isotopic signatures, reveal insights into daily activities, dietary habits and geographic mobility (Currie and Killin 2019). Multi-scalar analysis complements this by linking individual findings to broader population-level data, situating the individual's life within the wider social and environmental context. This dual approach ensures that osteobiographies reflect both personal and societal dimensions, creating a more holistic understanding of past lives (Hosek and Robb 2023).

Imagination plays a crucial role in bridging gaps left by incomplete archaeological records. It allows researchers to construct more engaging and relatable narratives by envisioning the daily activities, social interactions and personal experiences of individuals. This process goes beyond creative interpretation; it is a necessary tool for contextualising data that cannot be fully recovered through empirical methods alone. By incorporating imagination, osteobiographies become more accessible to both academic and public audiences, enriching the understanding of historical human experiences (Robb 2019).

However, the use of imagination in osteobiographies carries risks, particularly the potential for speculative reconstructions that lack empirical support. This can lead to narratives that blur the line between fact and hypothesis, potentially misrepresenting the past. Additionally, imaginative reconstructions are vulnerable to contemporary biases, where modern perspectives inadvertently influence the interpretation of historical data. This is problematic in a field that prioritises objectivity and scientific accuracy.

To address these challenges, I propose an 'Avatar Model' that incorporates and modulates all three techniques with ethics. This promising solution creates a simulated framework in which avatars, based on osteobiographical data, interact within a reconstructed archaeological environment. The environment reflects the cultural and temporal context inferred from historical and osteological evidence, providing a dynamic, yet controlled space to explore various life scenarios. The Avatar Model allows for systematic testing of different interpretations of the data, highlighting the range of possible lived experiences within the historical context.

The benefits of the Avatar Model are twofold. First, it acknowledges the inherent uncertainty in osteological reconstructions by generating multiple possible narratives, each rooted in empirical data. Second, by embedding avatars in a carefully designed environment, it mitigates the biases introduced by modern perspectives, ensuring that reconstructions remain anchored in the known conditions of the past. Additionally, the model distinguishes between empirical data and imaginative interpretation, enhancing the credibility of the resulting narratives.

Nonetheless, the Avatar Model is not without its own challenges. Illustrating all possible narratives in full would be prohibitive, so the reader has no choice but to rely on the researcher's judgement. The solution here is two-fold. To avoid an over-reliance on the reconstructed environment, any use of the Avatar Model must be accompanied by a rigorous and transparent rationale ensuring that empirical evidence remains central to the osteobiographical narrative. By providing synopses of different narratives and the reasons why they were not fully explored, the reader has the option to consider different perspectives or formulate their own based on the data. Here, three additional synopses are outlined, below.

The combination of trace-based reasoning, multi-scalar analysis and imagination into the Avatar Model provides a powerful framework for constructing osteobiographies. These methods enable a detailed and empathetic reconstruction of past lives that purely empirical approaches might not achieve. The Avatar Model, while not without challenges, represents an alternative approach to balancing scientific rigor with the ethical and meaningful humanisation of ancient lives. This integrated methodology ultimately aims to recover not just the facts of historical existence but the human experiences behind them.

The differentiated osteological data

The following is a selected summary of the osteological data which can be found in full, with associated methodologies in the supplementary materials of Cessford and Neil (2022).

Feature 237, Skeleton 1516 James the Teacher

Burial position: SW–NE, extended, supine. The lower limbs were straight, drawn at the ankles.
Grave goods: None
Age: Young 'Old Middle Adult'
Osteological sex: Male
Stature (cm): 167.56 ± 3.27
Percentage of skeleton present: *c.* 35%
A sub-circular area of bone resorption exposing the underlying trabecular structure superior to the fovea capitis is indicative of a condition disrupting the blood supply to the femoral head. This disruption can be caused by various risk factors, including trauma and secondary conditions associated with hypercoagulation such as alcoholism, hematologic and metabolic diseases (*e.g.* gout), as well as other factors like smoking (Malizos *et al.* 2007). Given the absence of evidence for fracture or gout in this context, alcohol abuse emerges as a plausible contributing factor. The initial symptom of this condition is typically hip pain, which can progress to a dull ache or throbbing pain localised in the groin or buttock area.

Feature 302, Skeleton 1735 Clement the Horseman

Burial position: SW–NE, extended, supine, the skull was in neutral position. The upper limbs were slightly abducted at the shoulder and flexed at the elbow to bring the left hand around the right side of the thorax and the right hand over the right chest. The lower limbs were straight, drawn at the ankles.
Grave goods: Buckle: early 14th century
Age: Old 'Old Middle Adult'
Osteological sex: (Male)
Stature (cm): 175.41 ± 3.27
Percentage of skeleton present: *c.* 98%
Large enthesophytes over the odontoid process of the axis vertebra may be linked to the alar ligament, crucial for stabilising the craniocervical junction (CVJ) and limiting axial rotation. Variations in alar ligament attachment to the odontoid process include fibres orienting horizontally, upward or downward, expanding on typical anatomical descriptions. Enthesophyte formation may indicate Crowned Dens Syndrome (CDS), where the alar ligament calcifies or ossifies, potentially causing occipital pain and neck stiffness. Anteriorly oriented fibres could reduce the ligament's biomechanical efficiency in restricting lateral and sagittal flexion, contributing to symptoms (Sardi *et al.* 2017). Enthesophyte expressions presented along the medial edge of the femoral linea aspera and medially directed enthesophytes over the lesser trochanter. The former corresponds to the insertions of adductor magnus and adductor longus muscles, which flex, adduct and medially rotate the thighs at the hips. The latter corresponds to the activity of the iliopsoas muscle, which flexes the trunk and thigh at the hip and externally rotates the thigh at the hip. These expressions may relate to the habitual isometric contraction of muscles during activities such as horseback riding (Djukic *et al.* 2018), which is supported by historical accounts of friars using them as a means of transport (Webb 1921; Fizzard 2007; Miller 2012). Periarticular osteophytes around the right radial tuberosity, talus, calcaneus and navicular bones may relate to the overuse of these joints.

Feature 332, Skeleton 1880 Andrew the Beggar

Burial position: SW–NE, extended, supine, the cranium was absent. The left arm was flexed at the elbow to bring the hand forearm across the lower thorax. The right arm was missing. Both lower limbs were in anatomical position at the femoroacetabular and tibiofemoral joints. Although drawn at the ankles, the lower limbs were rotationally displaced inferior to mid femorae fractures; thus, the right side was presented posteromedially, and the left was presented posterolaterally.
Grave goods: Buckle: 14th century
Age: Young Adult
Osteological sex: (Male)
Stature (cm): 160.00 ± 4.05
Percentage of skeleton present: *c.* 79%
Individual F.332 exhibits an extraordinary array of injuries, including neck fractures, bilateral leg fractures, and a healed collarbone fracture. Notably, the femoral fractures present complex patterns. The right leg shows a small butterfly segment on the posterolateral side with posteromedial comminution and an anterior spiral fracture. The left leg has a small, incomplete butterfly fracture posteromedially, with comminution posterolaterally and a transverse anterior fracture.

These patterns suggest a high-impact trauma from behind and to the right, a type of injury typically seen in young males (Court-Brown *et al.* 2015, 2151).

Bilateral femoral fractures are life-threatening, particularly with potential vascular injury, and can lead to significant blood loss and haemorrhagic shock (Court-Brown *et al.* 2015, 2204; Lane *et al.* 2015; Sharma and Sharma 2023). The absence of osteogenic remodelling in F.332 indicates death occurred within weeks post-injury due to complications from shock or inadequate immobilisation and realignment of the fractures (Tortora and Derrickson 2014, 183).

The trauma to the cervicothoracic junction also lacks healing evidence, suggesting severe hyperflexion or hyperextension, possibly whiplash, as the injury mechanism (Court-Brown *et al.* 2015, 1672; Dash 2019). Additionally, a healed type I fracture on the lateral end of the left clavicle shows minimal displacement due to its stable position lateral to the coracoclavicular ligaments (Holder *et al.* 2017).

Feature 260, Skeleton 1863 Richard the Novice

Burial position: SW–NE, extended, supine. The skull was in forward and lateral flexion towards the left shoulder, resting against the wall of the grave cut. The upper limbs were straight against the body to bring the hands under the respective sides of the ischium. The lower limbs were straight.

Grave goods: Buckle pin fragments

Age: Young Adult

Osteological sex: (Male)

Stature (cm): 171.17 ± 4.05

Percentage of skeleton present: *c.* 88%

Individual F.260, among seven friars, exhibited pathology characterised by periosteal new bone and sclerosis localised to the parietal tubers and occipital planum, without extension below the temporal lines. This condition, comprising 18% of the assemblage, is hypothesised to relate to tonsuring, a practice involving the shaving of the head's crown, which marked an individual's transition from lay life to ecclesiastical obedience (Cullum 2004; Miller 2012). The severity in F.260 included sclerotic bone and venous sulci involvement. With no historical evidence directly linking tonsure to scalp infection, possible causes include severe folliculitis from bacterial infection due to unclean blades (Karthikeyan 2009) or ringworm (Tinea Capitis) caused by dermatophytes such as Trichophyton tonsurans (Piggott and Friedlander 2012). This condition might also be misdiagnosed skeletal tuberculosis, as seen in individual F.343.

Additionally, F.260 exhibited lower back degeneration, specifically spondylolysis and spondylolisthesis, common in individuals up to 20 years old (Ohmori *et al.* 1995). The intervertebral disc degeneration (IDD) between L4 and L5 aligns with young adult prevalence (Takatalo *et al.* 2009; Zobel *et al.* 2012), potentially influenced by genetic factors and BMI in males (Kelempisioti *et al.* 2011; Takatalo *et al.* 2013). Observations of psoas muscle attachments suggest biomechanical compensation for lumbosacral instability, not an anatomical variant like 'Cupid's bow' (Bogduk *et al.* 1992).

Feature 311, Skeleton 1782 Robert the Scholar

Burial position: SW–NE, extended, supine, the skull was in neutral position. The upper limbs were flexed at the elbows to bring the hands around the opposite sides of the abdomen. The lower limbs were straight, drawn at the ankles.

Grave goods: Buckle: medieval

Age: mature Adult

Osteological sex: Male

Stature (cm): 176.84 ± 3.27

Percentage of skeleton present: *c.* 98%

Individual F.311 exhibited three rib fractures on the right side of their thorax, with two healed but misaligned fractures between the fifth and seventh ribs and an ununited fracture on the tenth rib. Rib fractures commonly result from blunt trauma, falls, osteoporosis, or severe coughing (Roberge *et al.* 1984; Anavian *et al.* 2009). Fractures of ribs 4–9 are often associated with heart and lung injuries, while ribs 5–12 fractures risk abdominal organ injury (Skubic *et al.* 2018). Most rib fractures heal without long-term issues due to the well-vascularised environment (Anavian *et al.* 2009); however, the displacement and non-union in this case suggest complex injuries, potentially reducing thorax volume and stability (Edwards and Hunt 2018). Rib fractures can cause significant pain and quality of life disruptions, and pneumonia risk increases if pain impedes full lung inflation (Brasel *et al.* 2006; Marasco *et al.* 2015; Bugaev *et al.* 2016; Musonza and Todd 2018). Osteoarthritic changes in several joints imply a lifestyle with habitual loading, increasing fracture risk.

Diffuse Idiopathic Skeletal Hyperostosis (DISH) involves ossification of the anterior longitudinal ligament and vertebral entheses, presenting as enthesophytes and syndesmophytes (Mader *et al.* 2017). Though the aetiology is unknown, DISH is variably associated with osteoarthritis (OA) and considered non-inflammatory. Molecularly, new bone growth from osteophytes, enthesophytes and syndesmophytes involves mesenchymal cell and chondrocyte stimulation, with leptin playing a significant role (Cava 2017; Pillai and Littlejohn 2014). These formations co-exist with metabolic/endocrine disorders like diabetes, increased BMI, and cardiovascular disease, suggesting DISH as a systemic condition influenced by genetic, metabolic and inflammatory factors (Katz *et al.* 2010; Kortyna 2017). Symptoms include back pain, stiffness and impaired thoracic spine motion affecting daily activities (Katzman *et al.* 2017).

Feature 232, Skeleton 1482 John the Scholar

Burial position SW–NE, extended, supine, the skull was in neutral position. The upper limbs were slightly abducted at the shoulder and flexed at the elbow to bring the hands right over left over the sternum. The lower limbs were straight, drawn at the ankles.

Grave goods: None

Age: Young 'Old Middle Adult'

Osteological sex: Male

Stature (cm): 167.32 ± 3.27

Percentage of skeleton present: *c.* 99%

The sacrum exhibits sexual dimorphism, with males displaying a greater degree of continuous curvature, including a slight forward projection of the coccyx (Saukko and Knight 2016). Individual F.232 presented an abnormal sacrococcygeal angle approaching 90°, potentially due to idiopathic causes or a healed flexion fracture. Such fractures at the sacrococcygeal junction result from external trauma, such as falls or direct blows (Maigne *et al.* 2020). Although the coccyx is small and vestigial, it is significant as an attachment site for the pelvic diaphragm (Tortora and Derrickson 2014). Opinions vary on its redundancy (Lee *et al.* 2016), but abnormal sacrococcygeal angles are linked to coccydynia, where pelvic floor strain and soft tissue irritation cause pain, especially when sitting or transitioning from sitting to standing (Nathan *et al.* 2010; Foye *et al.* 2017). Additionally, anterior displacement of a fractured coccyx can lead to cerebrospinal fluid leakage, peroneal numbness, and bowel and bladder dysfunction (Cawley *et al.* 2016).

Feature 146, Skeleton 1164 Amice the Benefactor

Burial position: SW–NE, extended, supine, the skull was in neutral position. The upper limbs were straight against the body and pronated to bring the hand next to the hips. The lower limbs were straight, drawn at the ankles.

Grave goods: Coin and jetton: *c.* 1500+

Age: Mature Adult

Osteological sex: Female

Stature (cm): 157.84 ± 3.72

Percentage of skeleton present: *c.* 98%

Individual F.146, the sole adult female in this assemblage, exhibited distinct musculoskeletal adaptations indicative of habitual resistance to downward forces, evidenced by rugosity and enthesophyte formations on the humeri, forearms and right fingers. These adaptations, involving the deltoid and interosseous muscles, suggest regular engagement in heavy lifting or similar activities requiring upper limb stability (Panchal-Kildare and Malone 2013; Matthias and Wright 2016).

Additionally, F.146 displayed notable bilateral bone formation on the tibiae and fibulae, particularly distoanteriorly, suggesting a possible infection exacerbated by prolonged standing. The inflammation's characteristics align with osteitis and osteomyelitis, which are inflammatory conditions often resulting from bacterial infections such as staphylococcus aureus (Ciampolini and Harding 2000; Masters *et al.* 2019). Osteitis typically results from soft tissue trauma, whereas osteomyelitis involves a primary infection of the bone marrow, potentially due to vascular infection or open fractures (Tiemann and Hofmann 2009).

The bilateral nature of the infection suggests a systemic origin, possibly linked to chronic leg ulcers caused by venous or arterial insufficiency (Agale 2013; Valencia *et al.* 2001). Other differential diagnoses include osteopetrosis, a rare metabolic bone disease characterised by bone overgrowth and brittleness (Arumugam *et al.* 2015), and treponemal diseases like syphilis, although the latter is unsupported by cranial evidence in this case. Furthermore, chronic bone infections could precede squamous-cell carcinoma, albeit at a low incidence (Bauer *et al.* 2007; Tavares *et al.* 2011).

Feature 343, Skeleton 1918 Eustace the Orphan

Grave dimension: Burial position SW–NE, extended, supine, the skull was in neutral position. The upper limbs were flexed at the elbow to bring right hand over the right chest and left hand over the sternum. The lower limbs were straight, drawn at the ankles.

Grave goods: Buckle: medieval

Age: Adolescent

Osteological sex: Male

Stature (cm): 178.03 ± 3.27

Percentage of skeleton present: *c.* 95%

The most common/classic representation of skeletal tuberculosis is the cavitation and hypervascularisation of the vertebral bodies – particularly in the thoracic (as seen in these individuals) then less frequently in the lumbar and cervical region (Weaver and Lifeso 1984; Spekker *et al.* 2020). Systemic (also known as extrapulmonary/haematologically disseminated) manifestations of tuberculosis are generally not known to occur in the absence of pulmonary involvement (Morris *et al.* 2002). When they do, Mycobacterium tuberculosis usually affects the metaphysis of tubular bones due to their vascularised nature (Edeiken *et al.* 1963).

Enamel hypoplasia in individuals results from physiological stress during early tooth crown development (Nelson 2018), with a higher prevalence in men than women (Larsen 2015, 20). The exact cause is uncertain; historically, it has been linked to weaning and associated nutritional deficiencies (Lanphear 1990; Blakey *et al.* 1994). However, epidemiological studies reveal that factors such as fever, disease, maternal health and environmental influences also play significant roles (Katzenberg *et al.* 1996). Consequently, weaning is just one of several factors creating a suboptimal growth environment leading to enamel defects (Larsen 2015, 51).

Feature 191, Skeleton 1460 Geoffrey the Busy Friar

Grave dimension: Burial position SW–NE, extended, supine, the skull was in neutral position. The right arm was slightly abducted. Both upper limbs were flexed at the elbow to bring the left hand around the right side of the abdomen and the right hand over the left pelvis. The lower limbs were straight, drawn at the ankles.

Grave goods: Buckle: late 14th–15th century

Age: Mature Adult

Osteological sex: Male

Stature (cm): 178.74 ± 3.27

Percentage of skeleton present: *c.* 99%

Individual F.191 exhibited a healing isolated fracture of the left ulna, commonly known as a 'night stick' fracture, typically caused by a low-energy, direct blow as the arm is raised to shield the body (Mackay *et al.* 2000; Cai *et al.* 2013). These injuries, often referred to as 'parry' fractures in palaeopathological contexts, imply interpersonal or extramural violence (Lovell 1997). It is crucial to distinguish these from other forearm injuries such as those resulting from falls, which present different fracture patterns. For example, Monteggia fractures involve proximal oblique ulna fractures and radial head dislocation, while Galeazzi fractures entail proximal radial fractures and dislocation of the distal ulna from the carpus. Colles and Smith's fractures occur at the distal radius metaphysis. The characteristics of F.191's fracture align with a 'parry' fracture: the ipsilateral radius was unaffected, the fracture line was less than 45°, located below the midshaft, and exhibited minor misalignment (Judd 2008). Additionally, the low-energy impact is a significant factor, as higher energy fractures of the ulna diaphysis often result in non-union (Szabo and Skinner 2009).

The Osteobiographies

Model 1 (synopsis)

Brother Richard grapples with chronic pain and the recent death of Brother Andrew. Seeking guidance, Richard turns to Brother Robert, an elder scholar who has long suffered from unhealed rib fractures resulting from a horse kick whilst tending the stables. Through Robert's teachings, Richard learns to accept physical and emotional pain as part of life, drawing lessons from the friars around him. Brother Eustace, an orphan suffering from illness, finds joy in his fragility, while Brother John endures pain and embarrassing continence issues through dedicated scholarship. Brother Geoffrey, who had a passion for crafting and selling Christian iconography bears the scars of past discontent from the outside community, continues his duties with pride. Richard also reflects on Amice, a laywoman who endures her leg ulcers with faith, offering her suffering as devotion. As Richard embraces these examples, he learns that pain, though heavy, can be carried with grace, transforming suffering into a path toward spiritual strength and peace.

Reasons for not choosing this narrative:

Focusing on Richard's narrative to illustrate the osteobiography of the Cambridge Friary raises ethical concerns due to its bias toward the friars' perspective. The narrative centres on Richard's personal suffering, portraying it as spiritual devotion while minimising the experiences of laypeople affected by the friars, such as Amice, and those harmed by unethical behaviour. This selective focus risks romanticising suffering without critically addressing the friars' power dynamics or the broader social implications of their actions. Additionally, by framing suffering primarily through a religious lens, the narrative may gloss over the ethical complexities and systemic issues that contributed to both the friars' and the community's struggles.

Model 2 (synopsis)

Brother Clement, a seasoned friar, silently deals with the emotional pain of watching young Brother Eustace, a frail orphan, succumb to declining health. Eustace, born to a woman Clement had a deep connection with in an outlying village, embodies grace in suffering – an ability Clement both admires and struggles to understand. While Eustace's godmother, Amice, insists that her alms go toward his care despite her own afflictions, Clement is torn by his unspoken bond with the boy and the promise he made to his late mother. Seeking solace, Clement confides in Brothers Robert and James, who offer wisdom on accepting suffering through their own physical misfortune. Clement uses asceticism and horseback riding to manage his own physical and emotional pain, but ultimately realises that, like Eustace, he must learn to embrace his suffering. Through this journey, Clement comes to terms with his emotional burden, finding peace in knowing that Eustace's suffering is not his to control.

Reasons for not choosing this narrative:

Clement's narrative, while engaging, takes significant imaginative liberties that compromise its suitability for illustrating the osteobiography of the Cambridge Friary. The speculative relationships with Eustace's mother and Amice, along with the dramatised internal conflict, shift the focus from historical accuracy to personal drama. This risks creating a romanticised and fictionalised portrayal rather than a fact-based reconstruction grounded in osteological evidence.

The overemphasis on Clement's individual emotional journey detracts from the broader collective experiences of the friars and their socio-religious context. Moreover, the narrative's speculative elements lack verifiable data, undermining the integrity of the osteobiographical method, which prioritises evidence over creative storytelling. By focusing on Clement's emotional landscape, the narrative narrows its scope, failing to represent the friary community, thus making it unsuitable for accurately reflecting the group's lived realities.

Model 3 (synopsis)

Amice, a longstanding benefactor of the friary, struggles with internal conflict as she witnesses the contradictions in the friars' behaviour, particularly concerning Brother Geoffrey, who was assaulted by an aggrieved father after inappropriate conduct with his son. Though devoted to the friary for spiritual guidance and solace amidst her physical pain, Amice finds herself increasingly unable to justify the friars' actions to her friends in town. Through conversations with Brother Robert, she seeks to reconcile her faith in the friary's mission with the flawed behaviour of its members, all while ensuring that her alms provide the means for the friars to serve the sick and frail in the nearby hospital. Despite her own ailments, Amice remains resolute, determined to hold the friars accountable while maintaining her devotion to their spiritual ideals.

Reasons for not choosing this narrative:

Amice's narrative, while offering a valuable perspective as the only female layperson, presents challenges for illustrating the friary's osteobiography. As a benefactor, her privileged relationship with the friars may skew the narrative toward overemphasising the influence of wealth and power, while underrepresenting the experiences of less privileged townspeople who interacted with the friary. Her singular viewpoint as a woman does not adequately reflect the diverse gender dynamics of the community. Additionally, her personal conflict regarding the friars' behaviour, particularly her struggle to reconcile their actions with her spiritual devotion, narrows the narrative's focus on individual moral dilemmas, detracting from a broader understanding of the friary's collective identity and social role. The narrative risks becoming imbalanced by centring too much on Amice's unique position, while failing to fully capture the varied social and gender interactions that shaped the friary's relationship with the wider community.

Chosen osteobiography

In the bustling medieval town of Cambridge, within the sacred walls of the Austin Friary, lived a community bound by faith, service and a common life of austerity. Among these friars, each carried their own burdens – burdens not just of the spirit, but of the flesh. In their devotions, they sought to rise above their earthly suffering, yet their bodies reminded them daily of the greatest evil: physical pain.

James, an old middle-aged friar, woke to the sound of the bell signalling matins. The dull ache in his right hip was always worse in the morning, a pain he had grown accustomed to over the years. His affliction, avascular necrosis, was possibly due to his regular consumption of the communion wine. Yet, it was this very pain that had turned his heart to God, a penance he could neither escape nor reject. 'Pain is the way of purification,' he would often tell himself, though in the quiet of dawn, when the cold crept through his bones, he sometimes wondered if his suffering was not a test, but a punishment.

Each morning, despite his discomfort, James knelt before the high altar, offering prayers for strength and guidance. Pain was his constant companion, but it had not broken his spirit. As a Eucharistic minister, teacher and healer, he found solace in his work and numbness in the wine, though the demands of his body never lessened. It was said among the friars that 'the greatest evil is not sin, but the suffering of the flesh,' and in James's own trials, he felt the truth of those words keenly.

After matins, James would often cross paths with Clement on his way to the library. Clement, too, bore the weight of his own suffering, though he was less inclined to speak of it. The stiffness in his neck, elbow and foot, the remnants of years of horse riding and travel, seemed to do little to deter his passion for preaching in the surrounding villages. Yet, Clement's proud bearing and wealthy upbringing had always set him apart. James admired his zeal but could not help but question whether Clement's charm and confidence were at odds with the friary's mendicant ethos, where humility and poverty were the cornerstones of their faith. Clement's pain, though hidden behind his strong demeanour, was another testament to the friary's unspoken understanding that suffering shaped the soul.

As James walked through the cloister, he met Andrew, a young friar whose bright spirit belied the hardship he had already endured. James had once taught him logic, and though Andrew had proven an eager pupil, his true dedication now lay in the more demanding duty of begging for the friary's sustenance. James admired Andrew's resilience, where a past shoulder injury did not deter him from his study or community duties. Andrew's suffering, like that of others, revealed how the friars' behaviour was shaped by pain: it hardened some, while breaking others, and Andrew passed this test with dignity.

Beside Andrew stood Richard, another young friar plagued by his own afflictions. Chronic lower back pain and recurrent sores from his tonsured scalp left Richard in constant discomfort, but like Andrew, he soldiered on with the unspoken belief that suffering was God's will. Richard's devotion to communal duties, despite his pain, mirrored the friary's belief that the body's torment was secondary to the soul's salvation, yet James could not ignore the toll it took on the young man.

In the library, the silent figures of Robert and John were hunched over their manuscripts. Robert, reclusive since a brawl that left him with lingering pain and a persistent cough, found refuge in the study of sacred texts, though his physical limitations hampered his ability to serve the community. John, similarly burdened, bore the scars of a fall that left him with a painful, healed coccyx injury. Their scholarly pursuits provided purpose, but as James watched them, he could not help but reflect on the friary's shared conviction: physical suffering was not just a burden, but a means of grace.

Yet there were days when James doubted. As he passed through the herb garden, he saw Amice, the friary's elderly benefactor, hobbling with her swollen, ulcerated legs. Despite the ointments James prepared for her, her condition worsened, and her body became a testament to the relentless march of time. Amice's pain was as much a part of her life as her devotion to the friary, but unlike the friars, her suffering was not chosen; it was inflicted by the cruel indifference of age. James admired her resilience but could not help question why the evil of pain was endured by the innocent and the devout alike.

Later, James prepared a herbal remedy for Eustace, the young orphaned friar whose persistent cough and fatigue worried him deeply. Eustace's malnourished past now manifested in troubling symptoms that James feared were signs of tuberculosis. The boy's frail frame seemed to embody the very heart of the friary's ethos – that the weakest among them bore the heaviest cross, yet did so without complaint. As he sat by Eustace's bedside, James marvelled at the boy's quiet acceptance of his condition. In Eustace's gentle suffering, James saw a reflection of his own battle with pain – a battle waged in silence, with only God as a witness.

And then there was Geoffrey, the most outwardly vigorous of the friars, though his strength had been tested by a violent attack during a confession. His arm, still weak from the fracture, had never fully healed. But Geoffrey, like the others, soldiered on, his faith his armour.

James, as he moved through the friary, saw the signs of physical decay in each of his brothers – the quiet groans, the stiff limbs, the hollow coughs. It was inescapable. And while they believed their suffering would bring them closer to God, there were moments, in the quiet of night or the stillness of prayer, when James could not help but wonder if the greatest test of their faith was not in their devotion, but in their ability to endure the unrelenting torment of their own bodies.

A sudden commotion shattered the stillness. The sound of hurried, laboured footsteps echoed through the cloister, followed by the breathless arrival of young Richard, his face pained and as pale as a shroud.

'Brother James! Brother Andrew… he has been struck by a cart on the road near the stables!' Richard gasped, his voice quivering with fear and confusion. Without a moment's hesitation, James and several other friars rushed to the scene, their hearts heavy with dread.

They found Andrew lying motionless on the cobblestone road, the wooden cart, now halted, looming ominously beside him. A small crowd of townsfolk had gathered, their murmurs of concern mixing with whispers of something darker. Brother Clement, who had been passing by on horseback, knelt beside Andrew, his face a mask of sorrow as he examined the young friar.

'He's gone,' Clement said softly, his words falling like stones into a deep well. The friars stood in stunned silence; their breaths held in a collective prayer for Andrew's soul.

The positioning of Andrew's body, the speed of the cart, and the unusual time it travelled down that quiet road – it all felt wrong. James's mind turned over the possibilities as he knelt before the altar that evening, his prayers mingling with a growing sense of unease.

That night, after a solemn Mass for Andrew's soul, James sought out Clement and Geoffrey. Together, they sat in the library, where the flickering light of the candles cast long shadows on the stone walls.

'I fear Andrew's death was no mere accident,' James began, his voice low. 'The cart that struck him belongs to Master Piers, who is known to avoid this path. Something does not sit right in my heart.'

Clement, ever perceptive, nodded. 'I saw a man standing near the stables as I arrived. He disappeared quickly when the crowd gathered.'

Geoffrey added, 'And there was a smell of spirits about Andrew – not enough to suggest he had been drinking, but perhaps someone else had. And this –' Geoffrey held up a small, broken piece of wood, stained with something dark. 'It could be part of a cudgel. I fear someone struck Andrew before the cart hit him.'

James noted Geoffrey's sour tone and the fear behind his eyes, perhaps colouring his perspective. The friars sat in heavy silence, the weight of their suspicions pressing down on them. Who would want to harm Andrew, a friar dedicated to service and alms? Their investigation led them into the town, where whispers of discontent had grown louder over the years. The friary, once seen as a pillar of faith and charity, had become a target for those who saw the friars as hypocrites living off the sweat of the poor.

It was Amice, the elderly benefactor of the friary, who provided the first clue that shifted their investigation. She was carrying a bucket of water in each hand, as she did three times a day, to the infirmary to help the sick. James had a freshly prepared salve for her legs. She spoke in a low voice.

'Brother James,' she began, her tone grave, 'I have heard whispers in the market. There is talk of Andrew being seen inappropriately close with a young girl – a girl of his own age, but more troubling, the daughter of Master Piers, the merchant.'

James paused, his hands stilling over the ointment jar. 'Are you certain of this, Amice?'

'I did not see it with my own eyes, but the townsfolk talk. They say Master Piers was furious when he learned of it. A girl of such standing… you know how these things can unravel, Brother.'

The weight of her words settled heavily in James's mind. If Andrew had been seen in such a compromising position, and if the girl's father had caught wind of it, it could explain the deliberate nature of the cart's path. Master Piers, a man of influence and wealth, would not tolerate such behaviour, especially from a lowly friar.

With this new information, James returned to Clement and Geoffrey. Together, they pieced together a motive that went beyond mere anger. This was personal – a father's wrath against a friar who had crossed a line.

As night fell, James, Clement and Geoffrey ventured to the outskirts of the town. There, they sought out the meeting place of the 'Fellows of the Free', a group of townsmen who had taken up the cause of ridding Cambridge of perceived oppressors. But this was no ordinary group of dissenters – they were emboldened by a sense of righteousness that teetered on the edge of violence.

When they arrived, they found a small gathering of men, and among them, to James's dismay, was Master Piers himself. The merchant's face was etched with bitterness, his eyes darkened by a hatred that James now understood.

'Master Piers,' James called out, stepping forward with the calm authority that had guided him through many trials. 'We know of your anger, of what you believe Brother Andrew did. But this is not the way of justice.'

Piers turned sharply; his expression twisted in a sneer. 'You speak of justice, friar? That boy – he defiled my daughter's honour. And you expect me to let it pass?'

James held his gaze, steady and unyielding. 'I expect you to remember that vengeance is not ours to claim. Andrew's blood is on your hands now, and though you believe you acted in righteousness, you have condemned your own soul to darkness.'

For a moment, it seemed as though violence would erupt. The men behind Piers gripped their weapons tightly, but the merchant hesitated. Slowly, his hand fell to his side, the fight draining from him as James's words sank in.

'Go,' Piers spat, his voice hollow. 'But this is not over, friar.'

The friars left in silence; their hearts heavy with the knowledge of the dangers that still lurked. But they had uncovered the truth, and in doing so, they had confronted the darkness that threatened to consume their community.

Andrew was laid to rest with all the rites of the Church, his death a sombre reminder of the perils that came not only from the outside world but from within their own community. The friars, though saddened, found solace in their faith, knowing that they had faced the storm and uncovered the truth.

As the friary returned to its quiet rhythm, James reflected on the lessons learned from Andrew's tragic death. In a world filled with anger and pain, it was their faith that would guide them through, even as shadows gathered around them.

Conclusion

The Avatar Model allows for a nuanced approach to osteological analysis, by balancing scientific rigour with the ethical humanisation of historical lives. Its strength lies in offering multiple narrative possibilities grounded in empirical data, thereby acknowledging the inherent uncertainties of reconstructing the past. This model capitalises on the emotional connections that researchers often have with skeletal remains, transforming these connections into an asset rather than a liability. In the case of the Cambridge Austin Friary, the model's application is further enhanced by the good preservation of the remains and the structured, communal lifestyle of the Augustinian friars. This allows for more reliable reconstructions of lived experiences; an advantage not always present in other assemblages. However, it remains essential to maintain a clear distinction between empirical data and imaginative interpretation, ensuring that the resulting narratives remain credible and anchored in historical reality. By integrating trace-based reasoning, multi-scalar analysis, and imagination within the Avatar Model, Osteologists can enrich our understanding of the past while respecting the complexities and dignity of ancient human lives. Ultimately, this approach has the potential to contribute to a deeper, more empathetic engagement with history, transforming skeletal remains from mere objects of study into thoughtful representations of possible lived experience.

References

Agale, S.V. (2013) Chronic leg ulcers: epidemiology, aetiopathogenesis, and management. *Ulcers* 2013, 1–9.

Anavian, J., Wijdicks, C.A., Schroder, L.K., Vang, S. and Cole, P.A. (2009) Surgery for scapula process fractures. *Acta Orthopaedica* 80(3), 344–50.

Aouizerat, B.E. and Miaskowski, C.A. (2008) Genetics of chronic pain: crucial concepts in genetics and research tools to understand the molecular biology of pain and analgesia. In H.-G. Schaible (ed.) *Clinical Pain Management: Chronic Pain*, 48–64. London, Hodder Arnold.

Arumugam, E., Harinathbabu, M., Thillaigovindan, R., Prabhu, G. and Arumugam, D.E. (2015) Marble bone disease: a rare bone disorder. *Cureus* 7(10), e339.

Bauer, T., David, T., Rimareix, F. and Lortat-Jacob, A. (2007) Ulcère de Marjolin sur ostéite chronique: diagnostic et résultats du traitement: 7 cas. *Revue de Chirurgie Orthopédique et Réparatrice de l'Appareil Moteur* 93(1), 63–71.

Beaulieu, P. (2008) Applied physiology of nociception. In A. Rice (ed.) *Clinical Pain Management: Acute Pain*, 4–19. London, Hodder Arnold.

Blakey, M.L., Leslie, T.E. and Reidy, J.P. (1994) Frequency and chronological distribution of dental enamel hypoplasia in enslaved African Americans: A test of the weaning hypothesis. *American Journal of Physical Anthropology* 95(4), 371–83.

Bogduk, N., Pearcy, M. and Hadfield, G. (1992) Anatomy and biomechanics of psoas major. *Clinical Biomechanics* 7(2), 109–19.

Brasel, K.J., Guse, C.E., Layde, P. and Weigelt, J.A. (2006) Rib fractures; relationship with pneumonia and mortality. *Critical Care Medicine* 34(6), 1642–6.

Bugaev, N., Breeze, J.L., Alhazmi, M., Anbari, H.S., Arabian, S.S., Holewinski, S. and Rabinovici, R. (2016) Magnitude of rib fracture displacement predicts opioid requirements. *Journal of Trauma and Acute Care Surgery* 81(4), 699–704.

Cai, X.-Z., Yan, S.-G. and Giddins, G. (2013) A systematic review of the non-operative treatment of nightstick fractures of the ulna. *The Bone & Joint Journal* 95-B(7), 952–9.

Cava, A.L. (2017) Leptin in inflammation and autoimmunity. *Cytokine* 98, 51–8.

Cawley, D.T., Power, F. and Murphy, M. (2016) Cerebrospinal fluid leak after coccyx fracture. *The Spine Journal* 16(11), e735–6.

Cessford, C. and Neil, B. (2022) The people of the Cambridge Austin friars. *Archaeological Journal* 179(2), 383–444.

Ciampolini, J. and Harding, K.G. (2000) Pathophysiology of chronic bacterial osteomyelitis. Why do antibiotics fail so often? *Postgraduate Medical Journal* 76(898), 479.

Court-Brown, C.M., Court-Brown, J., Heckman, D., McQueen, M.M., Ricci, W.M., Tornetta, P. and McKee, M.D. (2015) *Rockwood and Green's Fractures in Adults*. Philadelphia, Wolters Kluwer Health.

Cullum, P., (2004) Boy/man into clerk/priest: the making of the late medieval clergy. In N. McDonald and W. Ormrod (eds) *Rites of Passage. Cultures of Transition in the Fourteenth Century*, 51–65. York, York Medieval Press.

Currie, A. and Killin, A. (2019) From things to thinking: cognitive archaeology. *Mind & Language* 34(2), 263–79.

Dash, C., (2019) Cervicothoracic junction fractures management: an overview of literature. *Indian Journal of Neurotrauma* 16(01), 27–32.

Djukic, K., Miladinovic-Radmilovic, N., Draskovic, M. and Djuric, M. (2018) Morphological appearance of muscle attachment sites on lower limbs: horse riders versus agricultural population. *International Journal of Osteoarchaeology* 28(6), 656–68.

Easterling, J.S., (2015) Ascetic blood: ethics, suffering and community in late-medieval culture. In L. Tracy and K. DeVries (eds) *Wounds and Wound Repair in Medieval Culture*, 369–88. Leiden, Brill.

Edeiken, J., DePalma, A., Moskowitz, H. and Smythe, V. (1963) 'Cystic' tuberculosis of bone. *Clinical Orthopaedics and Related Research* 28(NA;), 163–8.

Edwards, J.G. and Hunt, W.J. (2018) Non-union. In M. de Moya and J. Mayberry (eds) *Rib Fracture Management: A Practical Manual*, 123–34. Switzerland, Springer Nature.

Fizzard, A.D. (2007) Shoes, Boots, Leggings, and cloaks: the Augustinian canons and dress in later medieval England. *Journal of British Studies* 46(2), 245–62.

Foye, P.M., Shah, J.M. and Sinha, D. (2017) Coccyx fracture and dislocation. In S.B. Kahn and R.Y. Xu (eds) *Musculoskeletal Sports and Spine Disorders. A Comprehensive Guide*, 461–3. Switzerland, Springer.

Geldof, M.R. (2015) 'And to describe the shapes of the dead': making sense of the archaeology of armed violence. In L. Tracy and K. DeVries (eds) *Wounds and Wound Repair in Medieval Culture*, 57–80. Leiden, Brill.

Geltner, G. (2010) Mendicants as victims: scale, scope and the idiom of violence. *Journal of Medieval History* 36(2), 126–41.

Geltner, G. (2012a) Violence: friars under fire. In G. Geltner (ed.) *The Making of Medieval Antifraternalism: Polemic, Violence, Deviance, and Remembrance*, 46–77. Oxford, Oxford University Press.

Geltner, G. (2012b) Deviance: brethren behaving badly. In G. Geltner (ed.) *The Making of Medieval Antifraternalism: Polemic, Violence, Deviance, and Remembrance*, 81–102. Oxford, Oxford University Press.

Holder, J., Kolla, S. and Lehto, S. (2017) Clavicle fractures: Allman and Neer classification. *Journal of Advances in Radiology and Medical Imaging* 2(1), 102.

Hosek, L. and Robb, J. (2023) Osteobiography: the bioarchaeology of individuals, reference module in social sciences. In E. Nikita and T. Rehren (eds) *Encyclopedia of Archaeology*, 917–25. New York, Academic Press.

Judd, M.A. (2008) The Parry problem. *Journal of Archaeological Science* 35(6), 1658–66.

Karthikeyan, K. (2009) Tonsuring: myths and facts. *International Journal of Trichology* 1, 33–4.

Katz, J.D., Agrawal, S. and Velasquez, M. (2010) Getting to the heart of the matter; osteoarthritis takes its place as part of the metabolic syndrome. *Current Opinion in Rheumatology* 22(5), 512–19.

Katzenberg, M.A., Herring, D.A. and Saunders, S.R. (1996) Weaning and infant mortality: evaluating the skeletal evidence. *American Journal of Physical Anthropology* 101(S23), 177–99.

Katzman, W.B., Huang, M., Kritz-Silverstein, D., Barrett-Connor, E. and Kado, D.M. (2017) Diffuse Idiopathic Skeletal Hyperostosis (DISH) and impaired physical function: the Rancho Bernardo Study. *Journal of the American Geriatrics Society* 65(7), 1476–81.

Kelempisioti, A., Eskola, P.J., Okuloff, A., Karjalainen, U., Takatalo, J., Daavittila, I., Niinimäki, J., Sequeiros, R.B., Tervonen, O., Solovieva, S., Kao, P.Y., Song, Y.-Q., Cheung, K.M., Chan, D., Ala-Kokko, L., Järvelin, M.-R., Karppinen, J. and Männikkö, M. (2011) Genetic susceptibility of intervertebral disc degeneration among young Finnish adults. *BMC Medical Genetics* 12(1), 153.

Kortyna, R. (2017) Diffuse Idiopathic Skeletal Hyperostosis: a review. *JBJS Journal of Orthopaedics for Physician Assistants* 5(4), e27-1–6.

Lane, M.K., Nahm, N.J. and Vallier, H.A. (2015) Morbidity and mortality of bilateral femur fractures. *Orthopedics* 38(7), e588-92.

Lanphear, K.M. (1990) Frequency and distribution of enamel hypoplasias in a historic skeletal sample. *American Journal of Physical Anthropology* 81(1), 35–43.

Larsen, C.S. (2015) *Bioarchaeology. Interpreting Behaviour from the Human Skeleton.* Cambridge, Cambridge University Press.

LeBovits, A.H. (2008) The psychological assessment of pain in patients with chronic pain. In H.-G. Schaible (ed.) *Clinical Pain Management: Chronic Pain*, 123–31. London, Hodder Arnold.

Lee, J., Gil, Y., Shin, K., Kim, J., Joo, S., Koh, K. and Song, W. (2016) An anatomical and morphometric study of the coccyx using three-dimensional reconstruction. *The Anatomical Record* 299(3), 307–12.

Lovell, N.C. (1997) Trauma analysis in paleopathology. *American Journal of Physical Anthropology* 104(S25), 139–70.

Mackay, D., Wood, L. and Rangan, A. (2000) The treatment of isolated ulnar fractures in adults: a systematic review. *Injury* 31(8), 565–70.

Mader, R., Verlaan, J. J., Eshed, I., Jacome, B.-A., Puttini, P.S., Atzeni, F., Buskila, D., Reinshtein, E., Novofastovski, I., Fawaz, A., Kurt, de V. and Baraliakos, X. (2017) Diffuse Idiopathic Skeletal Hyperostosis (DISH): where we are now and where to go next. *RMD Open* 3(1), e000472.

Maigne, J.-Y., Doursounian, L. and Jacquot, F. (2020) Classification of fractures of the coccyx from a series of 104 patients. *European Spine Journal* 29(10), 2534–42.

Malizos, K.N., Karantanas, A.H., Varitimidis, S.E., Dailiana, Z.H., Bargiotas, K. and Maris, T. (2007) Osteonecrosis of the femoral head: etiology, imaging and treatment. *European Journal of Radiology* 63(1), 16–28.

Marasco, S., Lee, G., Summerhayes, R., Fitzgerald, M. and Bailey, M. (2015) Quality of life after major trauma with multiple rib fractures. *Injury* 46(1), 61–5.

Martin, D.L. and Harrod, R.P. (2016) The bioarchaeology of pain and suffering: human adaptation and survival during troubled times. In *Archaeology of the Human Experience* (Archaeological Papers of the American Anthropological Association), 161–74. The American Anthropological Association.

Masters, E.A., Trombetta, R.P., Bentley, K.L. de M., Boyce, B.F., Gill, A.L., Gill, S.R., Nishitani, K., Ishikawa, M., Morita, Y., Ito, H., Bello-Irizarry, S.N., Ninomiya, M., Brodell, J.D., Lee, C.C., Hao, S.P., Oh, I., Xie, C., Awad, H.A., Daiss, J.L., Owen, J.R., Kates, S.L., Schwarz, E.M. and Muthukrishnan, G. (2019) Evolving concepts in bone infection: redefining 'biofilm', 'acute vs. chronic osteomyelitis', 'the immune proteome' and 'local antibiotic therapy'. *Bone Research* 7(1), 20.

Matthias, R. and Wright, T. (2016) Interosseous membrane of the forearm. *Journal of Wrist Surgery* 5(3), 188–93.

Miller, A.G. (2012) To 'frock' a cleric: the gendered implications of mutilating ecclesiastical vestments in medieval England. *Gender & History* 24(2), 271–91.

Moon, C. (2020) What remains? Human rights after death. In K. Squires, D. Errickson and N. Márquez-Grant (eds) *Ethical Approaches to Human Remains. A Global Challenge in Bioarchaeology and Forensic Anthropology*, 39–58. Switzerland, Springer Nature.

Morris, B., Varma, R., Garg, A., Awasthi, M. and Maheshwari, M. (2002) Multifocal Musculoskeletal Tuberculosis in Children: Appearances on Computed Tomography, *Skeletal Radiology* 31, 1–8.

Musonza, T. and Todd, S.R. (2018) Medical management of rib fractures. In M. de Moya and J. Mayberry (eds) *Rib Fracture Management: A Practical Manual*, 55–67. Switzerland, Springer Nature.

Nathan, S.T., Fisher, B.E. and Roberts, C.S. (2010) Coccydynia. *Bone & Joint Journal* 92-B(12), 1622–7.

Nelson, J.S. (2018) An examination of the differential susceptibility pattern of the dentition to linear enamel hypoplasia. *COMPASS* 2(1), 54–69.

Nightingale, A., (2011) *Once Out of Nature. Augustine on Time and the Body*. Chicago, University of Chicago Press.

Ohmori, K., Ishida, Y., Takatsu, T., Inoue, H. and Suzuki, K. (1995) Vertebral slip in lumbar spondylolysis and spondylolisthesis. Long-term follow-up of 22 adult patients. *The Journal of Bone and Joint Surgery*. British volume 77(5), 771–3.

Panchal-Kildare, S. and Malone, K. (2013) Skeletal anatomy of the hand. *Hand Clinics* 29(4), 459–71.

Piggott, C.D.S. and Friedlander, S.F. (2012) Dermatophytes and other superficial fungi. In S.S. Long, L.K. Pickering and C.G. Prober (eds) *Principles and Practice of Pediatric Infectious Diseases*, 1246–50. Amsterdam, Elsevier.

Pillai, S. and Littlejohn, C. (2014) Metabolic factors in Diffuse Idiopathic Skeletal Hyperostosis – a review of clinical data. *The Open Rheumatology Journal* (8), 116–28.

Robb, J. (2019) Beyond individual lives: using comparative osteobiography to trace social patterns in classical Italy. *Bioarchaeology International* 3(1), 58–77.

Robb, J., Inskip, S., Cessford, C., Dittmar, J., Kivisild, T., Mitchell, P., Mulder, B., O'Connell, T., Price, M., Rose, A. and Scheib, C. (2019) Osteobiography: the history of the body as real bottom-line history. *Bioarchaeology International* 3(1), 16–31.

Roberge, R.J., Morgenstern, M.J. and Osborn, H. (1984) Cough fracture of the ribs. *The American Journal of Emergency Medicine* 2(6), 513–17.

Roberts, C. (2009) Health and welfare in medieval England: the human skeletal remains contextualized. In *Reflections: 50 Years of Medieval Archaeology, 1957–2007*, 307–325. Leeds, Maney.

Sardi, J.P., Iwanaga, J., Schmidt, C., Rustagi, T., Chapman, J.R., Oskouian, R.J. and Tubbs, R.S. (2017) Anatomy of the alar ligament: Part II: variations of its attachment onto the dens. *World Neurosurgery* 107, 1007–11.

Saukko, P. and Knight, B. (2016) *Knight's Forensic Pathology*. Florida, CRC Press.

Schaible, H.-G. (2008) Applied physiology: persistent musculoskeletal pain. In A. Rice (ed.) *Clinical Pain Management: Chronic Pain*, 24–36. London, Hodder Arnold.

Sharma, R. and Sharma, S. (2023) Physiology, blood volume. *StatPearls [Internet]*. Available from: https://www.ncbi.nlm.nih.gov/books/NBK526077/

Sheridan, S.G. and Gregoricka, L.A. (2020) *Purposeful Pain, The Bioarchaeology of Intentional Suffering* (Bioarchaeology and Social Theory). Switzerland, Springer Nature.

Skubic, J.J., Okafo, B.U. and Nehra, D. (2018) Pathophysiology of rib fracture and lung contusion. In M. de Moya and J. Mayberry (eds) *Rib Fracture Management: A Practical Manual*, 35–44. Switzerland, Springer Nature.

Spekker, O., Hunt, D.R., Paja, L., Molnár, E., Pálfi, G, and Schultz, M. (2020) Tracking down the White Plague: The skeletal evidence of tuberculous meningitis in the Robert J. Terry Anatomical Skeletal Collection. *PLoS ONE* 15(3), e0230418.

Szabo, R.M. and Skinner, M. (2009) Isolated ulnar shaft fractures: retrospective study of 46 cases. *Acta Orthopaedica Scandinavica* 61(4), 350–2.

Takatalo, J., Karppinen, J., Niinimäki, J., Taimela, S., Näyhä, S., Järvelin, M.-R., Kyllönen, E. and Tervonen, O. (2009) Prevalence of degenerative imaging findings in lumbar magnetic resonance imaging among young adults. *Spine* 34(16), 1716–21.

Takatalo, J., Karppinen, J., Taimela, S., Niinimäki, J., Laitinen, J., Sequeiros, R.B., Paananen, M., Remes, J., Näyhä, S., Tammelin, T., Korpelainen, R. and Tervonen, O. (2013) Body mass index is associated with lumbar disc degeneration in young Finnish males: subsample of Northern Finland birth cohort study 1986. *BMC Musculoskeletal Disorders* 14(1), 87.

Tavares, E., Martinho, G., Dores, J.A., Vera-Cruz, F. and Ferreira, L. (2011) Úlcera de Marjolin associada a ulceração e osteomielite crônicas. *Anais Brasileiros de Dermatologia* 86(2), 366–9.

Tiemann, A.H. and Hofmann, G.O. (2009) Principles of the therapy of bone infections in adult extremities. *Strategies in Trauma and Limb Reconstruction* 4(2), 57–64.

Todd, E.M. (1993) Pain: historical perspectives. In C. Warfield (ed.) *Principles and Practice of Pain Management*, 1–9. London, McGraw-Hill.

Tortora, G.J. and Derrickson, B. (2014) *Principles of Anatomy and Physiology*, 14th edition. Chichester, Wiley.

Valencia, I.C., Falabella, A., Kirsner, R.S. and Eaglstein, W.H. (2001) Chronic venous insufficiency and venous leg ulceration. *Journal of the American Academy of Dermatology* 44(3), 401–24.

Weaver, P. and Lifeso, R.M. (1984) The radiological diagnosis of tuberculosis of the adult spine. *Skeletal Radiology* 12(3), 178–86.

Webb, E.A. (1921) *The Records of St. Bartholomew's Priory and St. Bartholomew the Great, West Smithfield.* London, Humphrey Milford, Oxford University Press.

Zobel, B.B., Vadalà, G., Vescovo, R.D., Battisti, S., Martina, F.M., Stellato, L., Leoncini, E., Borthakur, A. and Denaro, V. (2012) T1ρ; Magnetic resonance imaging quantification of early lumbar intervertebral disc degeneration in healthy young adults. *Spine* 37(14), 1224–30.

11

The ossuary of the Teutonic monastery of San Leonardo di Siponto in Manfredonia (Puglia, southern Italy): anthropological and palaeopathological data for the reconstruction of the monastic community

Ginevra Panzarino, Elena Varotto, Francesco Maria Galassi, Donatella Pian and Elena Dellù

This study presents the results of the archaeological and anthropological study of the human remains recovered from the ossuary of the Teutonic Abbey of San Leonardo di Siponto in Manfredonia (Puglia, Southern Italy). The abbey, located along the pilgrim road that led to the sanctuary of Saint Michael the Archangel in Monte Sant'Angelo and to the port of Siponto to embark for the Holy Land, was an important religious and medical centre during the Middle Ages. From 1260 onwards, it was held by the Teutonic Knights of St Mary's hospital in Jerusalem. As part of an ongoing recovery and display project in the area, an archaeological excavation was undertaken in different parts of the complex, especially around the church, renowned for its Romanesque portal and the frescoes depicting the emblems of the religious chivalric order. In addition to the external cemetery intended for common burials (including two cases caused by the Black Death), within the church, a rectangular structure was discovered in the left-hand aisle. It consisted of an underground chamber of bell-shaped section, to accommodate the disarticulated skeletal remains of 12 individuals. These remains were randomly arranged, a situation that may also have been a result of episodes of water rising and infiltrating through the rock. The bones belong to adult males, potentially clerics from the Abbey, and were originally accompanied by clothing and devotional items. The bodies were affected by several pathologies, offering new insight into the lifestyle and health of the monastic community, for which our knowledge has until now mainly been restricted to historical, religious and architectural information, derived from the available documentary sources.

Keywords: Teutonic Order, ossuary, male monasticism, bioarchaeology, multidisciplinary.

Introduction

The Abbey of San Leonardo di Siponto, also known as San Leonardo in Lama Volara, in Manfredonia (Puglia, Southern Italy, Fig. 11.1), was an important religious and medical centre during the Middle Ages. The abbey comprises three main buildings: the church of San Leonardo, the hospital and the convent. Strategically, it is located along the pilgrim road (Via Francigena) that led to the sanctuary of Saint Michael of Monte Sant'Angelo and to the port of Siponto, from which the Holy Land could be reached. The abbey has a long, rich history (D'Ardes 1999; 2000; 2006), as documented by various sources (Camobreco 1913; Ventura 1978; Mazzoleni 1991; Pensato 2000; Houben and Pascazio 2010).

Founded in the early 12th century, the abbey was initially entrusted to the Augustinians (Andenna 2006). At the end of the same century, the abbey experienced its most prosperous period during the Norman era, marked by the completion of the church and the construction of the northern portal, a remarkable example of Romanesque art (Fig. 11.2) (Houben 2002).

As from 1260, the abbey was entrusted to the *Ordo Fratrum Domus Hospitalis Sanctae Mariae Teutonicorum*

Figure 11.1. Location of the monastery of San Leonardo di Siponto (Manfredonia, Southern Italy).

Figure 11.2. Church of San Leonardo and details of the Romanesque portal (right) and Teutonic frescoes (left).

in Jerusalem (Toomaspoeg 2002; Houben 2006; D'Ardes 2021). Puglia, along with Sicily, became an important province, or bailiwick, of the Teutonic Order, and the commandery of San Leonardo with its possessions (houses, lands, animals, etc.) represented the most important religious and economic centre in northern Puglia. During this period, the church was renovated, including a north–south orientation and the painting of the frescoes featuring the Order's emblems (Fig. 11.2). The abbey was fortified with walls and a tower, and additional structures such as the cells and the chapel of San Leonardo (14th–15th century) were built. The north side of the church was closed to create a communal cemetery for men, women and children (between the late 13th and the early 14th centuries) where two individual graves contained numerous coins dating back to the 14th century. These finds, hidden in clothes and in a bag tied to the waist (Sarcinelli *et al.* 2019) suggest they were victims of the Black Death (1347–1353), as confirmed by genetic analysis of DNA extracted from teeth (Raele *et al.* 2021). The hospital was also rebuilt, providing refuge for pilgrims, the poor and the sick, as well as merchants and travellers (Carbone 2019). The presence of the Teutonic Order in San Leonardo is attested not only by architectural evidence but also by other artifacts, such as tombstones, inscriptions in Gothic script and pottery sherds from the nearby Torre Alemanna kiln, which bear the distinctive emblem of the 'Tau' or T-shaped cross. This is a symbol linked to the Order that represents a worldview centred on Jerusalem and was used by lay brothers to distinguish themselves from knights, who bore a full cross on their shields (D'Ardes 2021, 298 with bibliography).

After the death of the last representative of the Teutonic Order in 1480, the abbey passed into the possession of various bishops and cardinals (Sarnelli 1680), including figures such as Caetani (1561–1642), Barberini (1643–1690) and Acquaviva d'Aragona (1704–1763). These leaders undertook various restoration projects, particularly following the earthquakes that affected the region. In 1781, the abbey became state property and in 1809 the hospital was officially abolished.

After years of abandonment, the religious community of the 'Ricostruttori nella Preghiera' settled in the monastery and took care of worship and the maintenance of the church and the hospital while the monastery fell into ruin. In the first decade of the 21st century, urgent structural interventions were carried out and from 2014 to 2015 a significant recovery project was initiated leading to the discovery of new information about the abbey's buildings, their construction phases and the individuals interred here.

Area studied

The church of San Leonardo is situated on the north-eastern side of the Abbey. It is a rectangular space, oriented east–west, terminating in apses. It is characterised by three entrances: the main entrance to the west, a second to the north related to the Romanesque portal, and a third to the south, which gives direct access to the internal courtyard of the monastic buildings. The church has a two-nave plan, with domes aligned over the main nave and lateral barrel vaults supported by cruciform pillars. Only the first bay of the right-hand nave remains, covered by a barrel vault. The remaining two bays are closed off by a wall and contain a two-storey space (built after 1261), which is covered by cross vaults.

Immediately below the modern floor of the church, there is a limestone surface with stone blocks marked with hundreds of hand- and footprints. This is the sign of the passage of pilgrims through the site. After the removal of the old floor, various archaeological remains were identified, including traces of the historic wall frescoes and several Baroque altars. These features were identified using documentary and archival sources: the altar of Saint Celestine the Martyr, dating from 1634 to 1693, the altar of the Caetani family of 1634, the altar of Saint Hippolytus prior to 1693, the altar of the 'locati abruzzesi' after 1693, the main altar in Baroque style, described in 1693 and 1809, and another altar in the first bay of the right-hand nave.

In the left-hand nave, a square entrance was discovered associated with two blocks on the floor arranged at right-angles to each other. This entrance was constructed from blocks and bricks bound with soil and mortar (Fig. 11.3). In the vicinity, blocks, fragments and reused sculptural elements of a Baroque stone escutcheon can be seen. This modification includes the original enclosing wall, which is partially preserved on the northern side and composed of a row of slabs.

Beneath this roof, there is a rectangular underground chamber, oriented east–west (approx. 2.10 by 1.23 m). It is made of calcarenite blocks,[1] of bell-shaped section, built directly upon the rock on which the church stands. The walls of the ossuary display evidence of intentional perforations, which may have been used during the structure's initial construction. These openings could have served a functional purpose, facilitating descent into the chamber or accommodating wooden biers or stretchers for the bodies.

A variety of disarticulated skeletal remains were recovered from the area and spread across approximately three different contexts (Fig. 11.4). These remains were separated by thin layers of soil due to the infiltration of water through the rock, which caused the bones to be mixed.

- The upper context (38) contains the majority of skulls and long bones. These remains are arranged in a random configuration and do not occupy the entire surface area of the ossuary. Instead, they are stacked in the corners and display a slope towards the NE, which is attributed to the underlying rocky bank.
- In the middle context (40), the bones are concentrated on the western side of the ossuary, directly above the

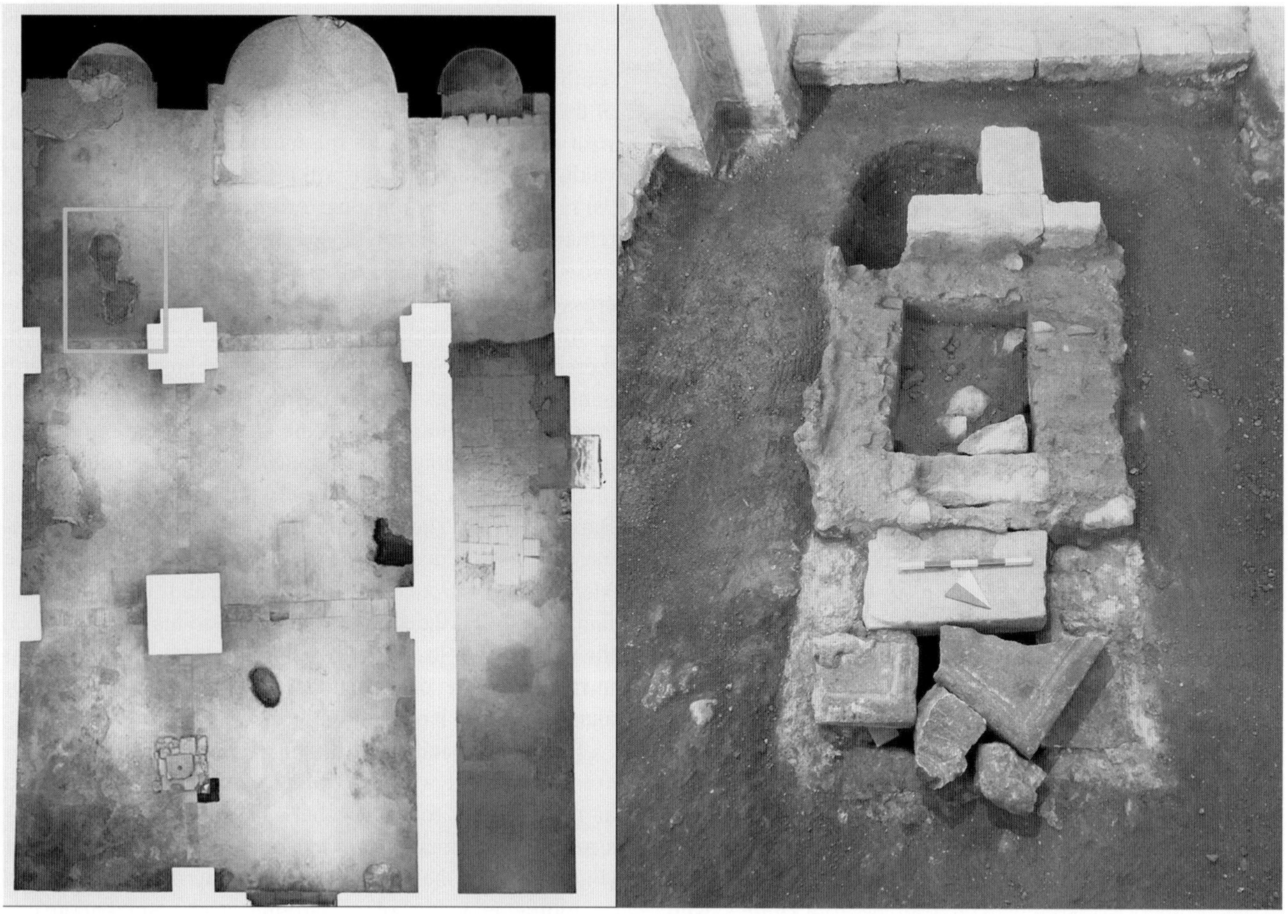

Figure 11.3. Location of the ossuary (left) and covering with fragments of escutcheons (right).

Figure 11.4. Deposition levels of disarticulated bones.

rock base on which the ossuary is located. This area of concentration extends from the base of the ossuary in the southern corner. In this context, long bones and numerous skulls are intermingled with ceramic fragments (predominantly cooking pots and glazed vessels, particularly white-glazed) and animal bones (exhibiting butchery marks). This context also yielded a number of items that are likely to be related to devotional and personal objects such as an iron cruciform element, a rosary made of glass paste beads and bronze hooks (surrounded by numerous phalanges), a bronze ring (situated around a partially articulated hand and still *in situ* on the corresponding phalanx), and a hemispherical button on the back with four eyelets.

- In the lowest context (42), a considerable number of small bones from the hands and feet are concentrated, with some bones inserted into an opening on the eastern side of the ossuary. The bones are mixed with fragments of coarse ware pottery and lenses of carbon residue. Additional metal artifacts include two circular neck links with a circular eye (potentially devotional medals), a folded hinged element and a small iron piece, probably a cloak buckle, recovered in proximity to a skull. This context continues underneath an opening in the eastern wall, where the excavation was interrupted due to substantial water infiltration from the subsoil.

Material and methodology

The bones are in an excellent state of preservation due to the micro-climatic conditions of the ossuary, which are characterised by constant humidity and rising water, as well as the presence of sandy soil. Furthermore, in many cases they display green discolouration resulting from the presence of the metallic objects interred within the ossuary. The bones are complete and measurable (Martin and

Saller 1956–1959) thus enabling the assessment of joint and pathological compatibility and enabling the reconstruction of several individuals. The remains were excavated stratigraphically, numbered and spatially positioned in order to reconstruct the movement of the individual segments. Each bone was studied to determine sex and age (Buikstra and Ubelaker 1994), stature (Trotter and Gleser 1977), pathological conditions (Ortner 2003) and biomechanical stress, including syndesmosis lesions and enthesopathies (Mariotti *et al.* 2007), as well as new articular surfaces, degenerative joint diseases and dental pathologies.

Results

The skeletal remains (MNI 13) refer to:

- 12 adult-senile male individuals (Table 11.1)
- A few bones of 1 non-adult individual.

The average height, calculated from complete osteological segments, is estimated to range between 1.65 and 1.77 m.

Based on anthropological, metric, articular and patholog ical data, the bones can be attributed to individuals.[2] Among the specimens, a particularly complete individual stands out, exhibiting ossification of the cartilages and fusion of the vertebral and costal joints (Fig. 11.5). The remains in question are located in the most superficial contexts (38 and 40), suggesting that he was one of the last individuals buried in the ossuary.

In general, the individuals display robust bones with pronounced entheses on the shoulder girdle and on long limb bones, accompanied by evidence of tendon and ligament tears. They exhibited significant osteoarthritis in the joints, spine, rib cage, elbows and ankles, along with several facets. Some evidence of trauma and numerous pathologies such as infections, periostitis and osteomyelitis were recognised, particularly in the elbows, tibiae and fibulae (Fig. 11.6). Dental pathology is also prevalent, with common diseases such as tartar, severe wear and antemortem tooth loss, sometimes resulting in complete edentulism.

The non-adult individual is estimated to be less than 13 years old and is represented by only 3 vertebrae, 1 coxal bone, 1 tibial plateau, 2 metatarsals and a phalanx. Given its age, it does not yet have fully developed secondary sexual characteristics, although a typically male form of the ischial notch can be observed.

Table 11.1. Minimal number of adult individuals (MNI) in the ossuary of the Teutonic Abbey of San Leonardo di Siponto in Manfredonia

Bone	*n*	*Left*	*Right*
Skull	11		
Mandible	8		
Manubrium and sternum	6		
Sacrum	1		
Clavicle		6	8
Scapula		3	5
Humerus		11	10
Radius		7	11
Ulna		10	8
Coxal bone		7	4
Femur		11	9
Patella		6	5
Tibia		12	12
Fibula		11	11
Metacarpal 1		1	4
Metacarpal 2		5	4
Metacarpal 3		7	5
Metacarpal 4		7	6
Metacarpal 5		6	5
Astragalus/Tallus		9	10
Calcaneus		11	7
Scaphoid		7	7
Cuneiform 1		6	6
Cuneiform 2		2	3
Cuneiform 3		5	2
Cuboid		5	6
Metatarsal 1		8	10
Metatarsal 2		7	8
Metatarsal 3		6	12
Metatarsal 4		10	12
Metatarsal 5		8	11
Total		12	

Discussion

The structure is situated directly in front of the Gothic arch connecting the church and the chapel of San Leonardo built by the Teutonic Order and was aligned perfectly with the altar, reflecting the new orientation of the church (D'Ardes 2021, 293).

Furthermore, an analysis of the tomb walls reveals that the same blocks used in the wall structure of the adjacent chapel of San Leonardo were employed (D'Ardes 2021, 293). Therefore, on the basis of the historical sources, construction technique, stratigraphical and material data, the ossuary can be dated to a period between the late 13th century or the first half of the 14th century, and 1693, when the building was closed and marked on the floor by some limestone blocks (D'Ardes 2021, 292–293).

The ossuary comprises a funerary chamber constructed directly on the bedrock, in which numerous disarticulated bones with various objects related to clothing or for devotional purposes were recovered. The human remains belong to 12 adults, elderly men, probably identifiable as members

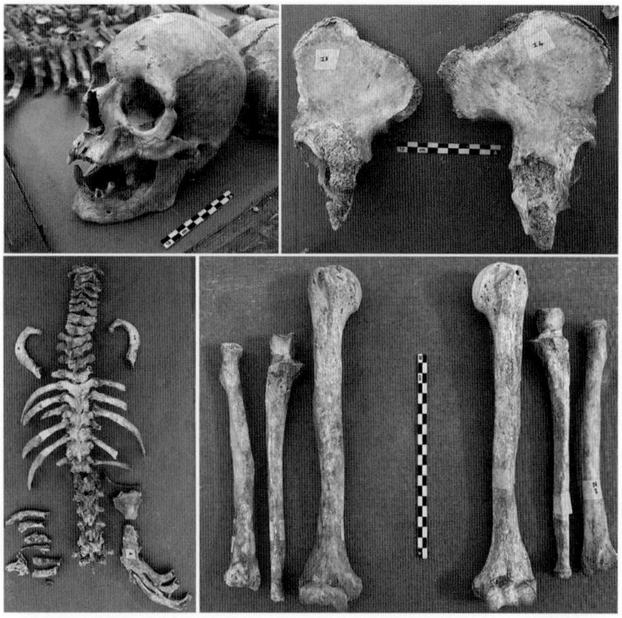

Figure 11.5. Individual reconstructed on the basis of age and pathological aspects.

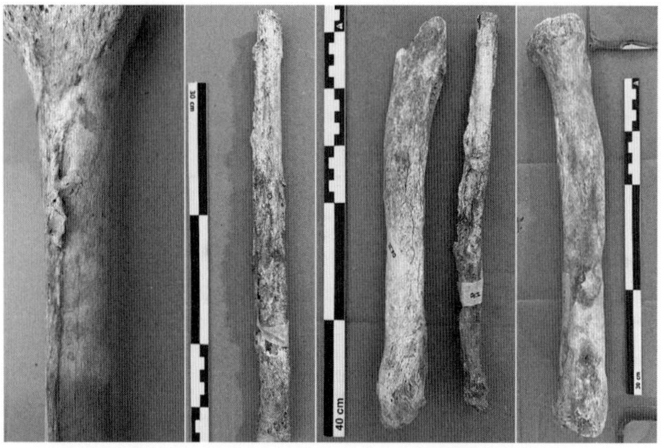

Figure 11.6. Long bone diseases.

of the Teutonic Order of the Abbey. A significant proportion of the pathologies observed are related to advanced age; however, other factors, arising from the physical demands of everyday monastic activities, both religious (such as prayer) and work-related (*e.g.* tending to the convent garden) might also have contributed to these conditions. It is also important to consider they may also have died from diseases such as malaria or other infections due to the environmental conditions of Siponto. This area was already marshy in Roman times and remained a disease-ridden swamp until the 1900s. Additionally, most of the hospital's patients in the summer and autumn months were affected by malaria (Pellegrino 2004). In fact, two cases of victims of the plague of 1350 have been documented in the cemetery outside the church (Raele *et al.* 2021).

The non-adult individual is largely incomplete and may have entered the ossuary randomly, in the same way as the animal bones and the pottery fragments. It is possible that the remains were introduced involuntarily or due to post-deposition contamination.

Once the ossuary was no longer used, it was sealed and only the northern row of its original roof and the limestone slabs resting on it remained. Next to these, a square ring was later constructed on the western side. Additionally, when the upper floor surface was laid down, the eastern part was sealed using what could have been reused stone from an escutcheon, and it was filled with soil and other fragments from the Baroque period. The ossuary was totally hidden by the installation of the clay floor and the pavement, where two blocks set at right angles to each other indicated its presence. Finally, the previous floor was completely concealed and subsequently revealed during the recent restoration work in the abbey.

Conclusions

An underground structure was investigated in the northern nave of the church of San Leonardo di Siponto, where numerous disarticulated bones and some personal and devotional objects were discovered. This archaeological and anthropological study confirmed the initial hypothesis that it was the ossuary intended for the members of the Teutonic Order. The presence of bones from only male individuals, affected by pathologies related to age and activities typical of daily monastic life, was documented. Thus, this investigation offers new insights into the lifestyle and health of this monastic community.

Acknowledgements

Many thanks to Dr Anna Maria Tunzi of the Soprintendenza Archeologica della Puglia, who directed the archaeological excavations. The authors also would like to express their gratitude to the architect Dr Antonello d'Ardes for his kind assistance and valuable recommendations.

Notes

1 This building material, known as 'tufo', is typical of the region. It consists of limestone formed from sediments precipitated due to the action of water, often containing traces of plants or fossilised shells.

2 The number of bone groups is higher than the MNI of individuals.

References

Andenna, C. (2006) *Da domus Dei a spelunca latronum.* Fortuna e declino di una canonica regolare nel mezzogiorno. In H. Houben (ed.) *San Leonardo di Siponto. In Cella Monastica, Canonica, Domus Theutonicorum. Atti del Convegno*

Internazionale, Manfredonia, 18–19 marzo, 73–90. Galatina, Congedo Editore.

Buikstra, J.E. and Ubelaker, D.H. (1994) *Standards for Data Collection from Human Skeletal Remains*. Fayetteville, Arkansas, Arkansas Archeological Survey Research Series, 44.

Camobreco, F. (ed.) (1913) *Regesto di S. Leonardo di Siponto*. (Regesta Chartarum Italiae). Rome, Istituto storico italiano-Loescher.

Carbone, A. (2019) Hospital assistance in Apulia: Institutions, archives and sources (14th–17th centuries). *RiMe, Rivista dell'Istituto di Storia dell'Europa Mediterranea* 4, 39–62. https://doi.org/10.7410/1371

D'Ardes, A. (1999) Note intorno alle vicende architettoniche del complesso abbaziale di San Leonardo in 'Lama Volara' presso Siponto. In A. Ventura (ed.) *Il Cabreo di San Leonardo 1634–1799*, 39–58. Manfredonia 1999, E.S.I. editrice.

D'Ardes, A. (2000) Interventi edilizi dei Cavalieri Teutonici nell'abbazia di S. Leonardo. *Siponto e Manfredonia nella Daunia*. In *Atti del V Convegno di studi. Manfredonia, 9–10 aprile 1999*, 100–21. Manfredonia, Andrea Pacilli Editore.

D'Ardes, A. (2006) L'antico Ospedale di San Leonardo in Lama Volara tra fondazione, riedificazione e abbandono. In H. Houben (ed.) *San Leonardo di Siponto. In Cella Monastica, Canonica, Domus Theutonicorum. Atti del Convegno Internazionale, Manfredonia, 18–19 marzo*, 269–99. Galatina, Congedo Editore.

D'Ardes, A. (2021) Gli insediamenti dell'Ordine Teutonico in Puglia nell'area sipontina: esiti dei restauri e delle campagne di scavo. In A. Baudin, S. Merli and M. Santanicchia (eds) *Gli Ordini di Terrasanta. Questioni aperte. Nuove acquisizioni (secoli XII–XVI). Atti del Convegno internazionale di studi. Perugia, 14–15 novembre 2019*, 279–314. Perugia, Fabrizio Fabbri editore.

Houben, H. (2002) «Iuxta stratam peregrinorum». La canonica di San Leonardo di Siponto (1127–1260). *Rivista di Storia della Chiesa in Italia* 56, 323–45.

Houben, H. (ed.) (2006) *San Leonardo di Siponto. In Cella Monastica, Canonica, Domus Theutonicorum. Atti del Convegno Internazionale, Manfredonia, 18–19 marzo*. Galatina, Congedo Editore.

Houben, H. and Pascazio, V. (2010) *L'inventario dell'archivio di San Leonardo di Siponto (ms. Brindisi, bibl. De Leo B 61). Una fonte per la storia dell'ordine teutonico in Puglia*. Galatina, Congedo Editore.

Martin, R. and Saller, K. (1956–1959) *Lehrbuch der Anthropologie*. Stuttgart, G. Fischer.

Mariotti, V., Facchini, F. and Belcastro, M.G. (2007) The study of entheses: proposal of a standardized scoring method for twenty-three entheses of postcranial skeleton. *Collegium Antropologicum* 31(1), 291–313.

Mazzoleni, J. (1991) *Le carte del monastero di S. Leonardo della matina in Siponto (1090–1771). (Codice Diplomatico Pugliese* 31). Bari, Società di Storia patria per la Puglia.

Ortner, D.J. (2003) *Identification of Pathological Conditions in Human Skeletal Remains*. 2nd ed. Washington, Academic Press.

Pellegrino, L. (2004) *L'Abbazia di San Leonardo di Siponto nel XIX Secolo*. Manfredonia, Nuovo centro di documentazione storica.

Pensato, G. (2000) *Il Cabreo di San Leonardo di Siponto 1634–1799*. Foggia, Edizioni Scientifiche Italiane.

Raele, D.A., Panzarino, G., Sarcinelli, G., Cafiero, M.A., Maria Tunzi, A. and Dellù, E. (2021) Genetic evidence of the Black Death in the Abbey of San Leonardo (Apulia Region, Italy): tracing the cause of death in two individuals buried with coins. *Pathogens* 10, 1354. doi.org/10.3390/pathogens10111354

Sarcinelli, G., Tunzi, A.M. and Panzarino, G. (2019) Coins in a medieval grave in Siponto (Apulia): intentional or unintentional deposition? *Journal of Archaeological Numismatics* 9, 283–92.

Sarnelli, P. (1680) *Cronologia de'vescovi ed arcivescovi sipontini*. Manfredonia 1680 (facsimile edition).

Toomaspoeg, K. (2002) Le ravitaillement de la Terre sainte. L'exemple des possessions des ordres militaires dans le royaume de Sicile au XIIIe siècle. In *L'expansion occidentale (XIe – XVe siècles). Formes et conséquences XXXIIIe Congrès de la S.H.M.E.S. Actes des congrès de la Société des historiens médiévistes de l'enseignement supérieur public. Madrid, 23–26 mai 2002*, 143–58. Paris, Éditions de la Sorbonne.

Trotter, M. and Gleser, G.C. (1977) Corrigenda to 'Estimation of stature from long limb bones of American Whites and Negroes'. *American Journal of Physical Anthropology* 47, 355–6.

Ventura, A. (1978) *Il patrimonio dell'Abbazia di San Leonardo di Siponto. Illustrazione e trascrizione del manoscritto di una 'visita pastorale' di fine secolo XVII conservato nella Biblioteca di Foggia*. Foggia, Amministrazione provinciale di Capitanata.

12

The everyday life of a female religious community from a multidisciplinary approach: the convent of Santa Clara (Pontevedra, Spain)

María Martín-Seijo, Carlos Fernández-Rodríguez, Eduardo González Gómez de Agüero, Rafael M. Rodríguez Martínez and Israel Picón Platas

The archaeological survey at the convent of Santa Clara (Pontevedra, Spain), carried out in 2022, provided an exceptional opportunity to gain insight into the daily life of a cloistered female religious community. Archaeological evidence of different episodes in the convent from the 14th century onwards was recorded and systematic sampling was carried out to recover bioarcheological remains. The archaeological data, together with archaeobotanical and zooarchaeological analyses, have provided an unprecedented amount of information on the various activities associated with life and death in an enclosed convent. Charcoal was recovered from the backfilling of the tombs and the orchard, providing information on the domestic debris associated with the firewood burnt in the convent, mainly from Fabaceae shrubs and oak, as previously attested in other religious buildings in the city. Analysis of the faunal, shellfish and fish remains provides valuable information on other aspects of the nuns' daily life. In addition, aspects related to their death and the material culture associated with funerary practices were revealed in this multidisciplinary study. The preservation of mineral-replaced wood attached to iron nails has also made it possible to identify the raw material used for making the coffins, and the recovery of a textile pseudomorph has even offered the opportunity to analyse a shroud.

Keywords: Archaeobotany, zooarchaeology, textile analysis, archaeology, cloistered life.

Introduction

According to documentary sources, the convent of Santa Clara (Pontevedra, Spain) was founded in 1310. Two of its buildings are related to this initial period, when construction of the church began. Subsequently, during 15th and 16th centuries, major building activity took place in the convent, while the apse of the church and the three wings of the cloister were completed. In the 17th century, structural modifications and new construction work took place, including the erection of the wall that encloses and delimits the grounds of the convent. During the 18th century the different spaces in the convent were reorganised at the same time as construction of a new building, more monumental than the previous ones, was started, although in fact it was never completed. The 19th century was a turbulent period in the life of the convent with nuns who entered or left the community during episodes such as the Peninsular War (1808–1814) and the ecclesiastical confiscations or disentailment of Mendizábal (Desamortización de Mendizábal) (1835–1844). By the end of the 19th century further construction activity was undertaken, involving the reorganisation of the inner areas of the convent, and in the 20th century a new cemetery was built in the eastern area of the convent. The convent was closed in 2017 and the buildings acquired by Pontevedra city council in 2021 (Fig. 12.1).

An archaeological excavation took place in 2022, supervised by two of the authors (R.M.R.M and I.P.P.). The archaeological work was carried out inside the convent buildings – including the church of Santa Clara, several buildings and the cloister – and in the grounds – including the vegetable garden, the apple orchard, the wood and the vineyard – surrounded by the wall. The archaeological work involved the excavation of 45 test pits and 2 open areas, covering a surface area of 780 m^2. The study of the cloistered convent of Santa Clara has provided an opportunity to address different aspects related to the life and death of the nuns and their relationship with the city and its surroundings. During the archaeological excavation, a systematic sampling strategy was implemented using a combination of grab and soil samples, which provided different types of bioarchaeological evidence, including plant and animal remains. The majority of the samples were taken from the fills of negative features associated with agricultural and arboricultural activity (Fig. 12.2), deposits where concentrations of debris were identified (Fig. 12.4) and burials (Fig. 12.3), as well as from building structures – pits, postholes. For this study we made a selection of grab and soil samples from 14 stratigraphic units.The aim of this research was to provide archaeological data on the various activities carried out in the enclosed convent of Santa Clara, integrating different types of evidence from a multidisciplinary approach. The specific objectives of this research were: 1) to produce information on the management and consumption of plant and faunal resources within the convent; 2) to obtain data on wood and textile crafts associated with funerary contexts; and 3) to integrate this information with the data from other medieval and post-medieval archaeological contexts in the urban area of Pontevedra.

Figure 12.1. From top to bottom, and left to right: 1) Location of Pontevedra in the north-west of the Iberian Peninsula, 2) Location of the Convent of Santa Clara in the city of Pontevedra, and 3) Aerial view of the Convent of Santa Clara.

Material and methods

Charcoal analysis

For this study, we selected four soil samples from Sector 6, from which 13 litres of sediment were processed by flotation, for analysis (Table 12.1). The charcoal remains recovered were part of the infill deposits of pits located outside the convent buildings, as part of secondary or tertiary refuse (Fuller *et al.* 2014). The archaeological materials recovered from these deposits spanned the period between the 16th and the 19th centuries. All the charcoal fragments of more than 2 mm in size were analysed. Charcoal analysis followed the standard procedure (Cartwright 2015). The charcoal pieces were manually broken, and the three sections of wood were observed under a ZEISS Axiolab 5 reflected light microscope with dark and bright fields. The anatomical features were compared with a reference collection and with atlases and specialised literature (Schweingruber 1990; Gale and Cutler 2000; Hather 2000). In tandem with the taxonomic identification, the dendrological and taphonomic attributes that broaden the information provided by the analysis of the charred wood remains were also registered: plant part, maturity of wood, alterations related to the combustion

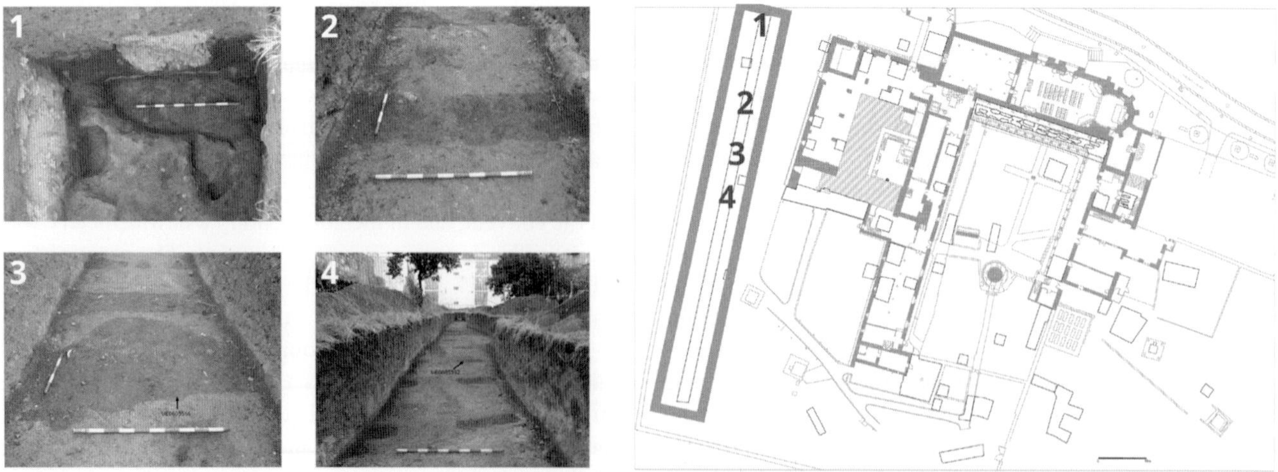

Figure 12.2. Images and locations of the pits and ditches excavated in the vegetable garden area (Sector 6-Test pit 5) (Photos by Paula Méndez and map drawn by María Méndez): 1) SU520, 2) SU552, 3) SU566 and 4) SU592.

process, and features related to biodeterioration, among others (Marguerie and Hunot 2007; Schweingruber *et al.* 2008).

Mineral-replaced wood analysis

In the graves in Sector 3-Test pit 1 and Sector 4-Test pit 2 (Fig. 12.3), nine iron nails associated with the coffins produced corroded remains that enabled wooden pseudomorphs to be preserved (Tables 12.1 and 12.2). Two radiocarbon dates obtained from two graves in Sector 3-Test pit 1 provided dates of between the mid-13th century and the end of the 14th century cal. AD for the burials in this area of the cloister. The graves recorded in Sector 4-Test pit 2 were dated to the late 18th or early 19th century. They were taxonomically identified observing the three sections of wood, by using a ZEISS Axiolab 5 reflected light microscope and a ZEISS EVO SEM equipped with EDX. The conditions of EDX analysis were 20 kV of accelerating voltage and 8.5 mm of working distance. The systematic use of EDX provides information about the mineralisation processes involved in this kind of wood preservation.

Fibre and textile analysis

A pseudomorph of a shroud was found inside one of the graves in Sector 3-Test pit 1 (SU187) (Fig. 12.3). Its fibres were observed under a ZEISS EVO SEM. To achieve identification, the surface and medulla of the fibres were observed (Cybulska *et al.* 2008; Bender Jørgensen and Grömer 2012; Gleba and Mannering 2012) following the method described by Rast-Eicher (2016). These features were compared with atlases (Hausman 1920; Rast-Eicher 2016).

The technical analysis of the textile fragment was performed with the naked eye and with the help of a stereoscopic microscope. It consisted of different stages (Bender Jørgensen and Grömer 2012); the first one, was the description of the condition of the textile (type and state of preservation, dimensions and colour). Secondly, it was followed by the description of the type and condition of the fibres, as well as evidence of fibre preparation; the number of threads per cm in warp and weft; if it was not possible to identify warp and weft, the terms system 1 and 2 were used. The description of the yarns includes twist direction, diameter of the yarns of both systems, and whether it is single or plied yarn.

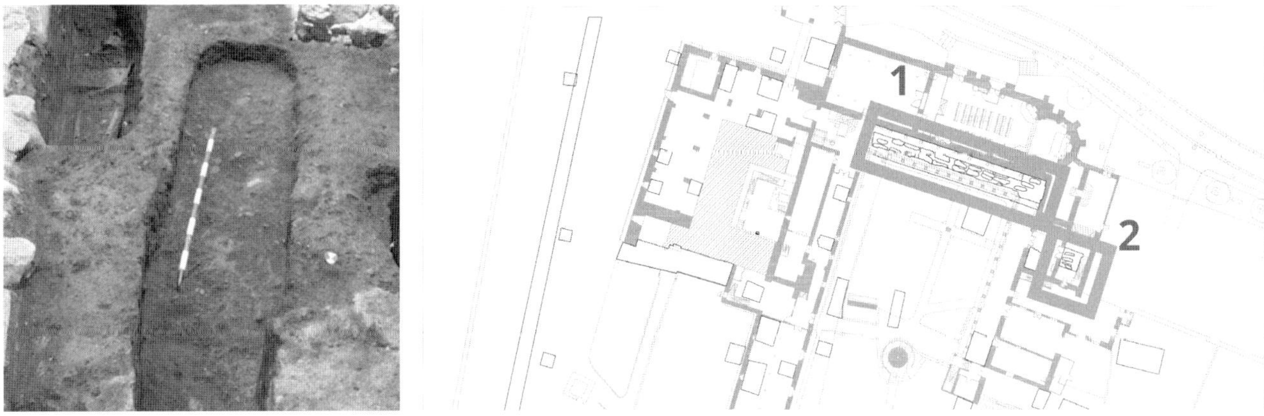

Figure 12.3. Left, detail of SU110 (Photo by Paula Méndez), infill of a tomb where mineral replaced wood and a textile pseudomorph were recovered. Right, location of sectors: 1) Sector 3-Test pit 1 and 2) Sector 4-Test pit 2 (Map drawn by María Méndez).

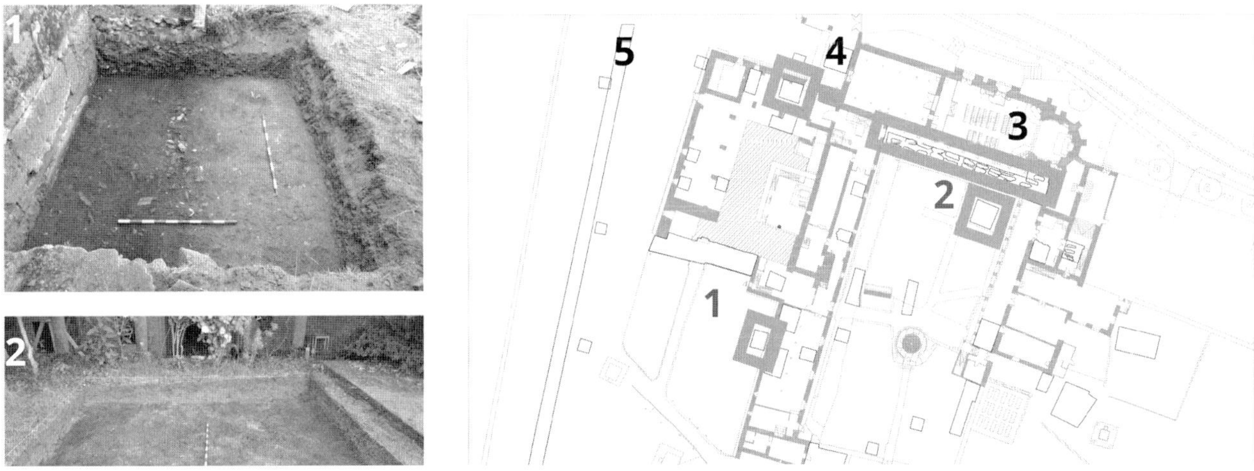

Figure 12.4. Images and locations of the test pits of provenance of the zooarchaeological remains (Photos by Paula Méndez, Map drawn by María Méndez): 1) Sector 6-Test pit 7, 2) Sector 7-Test pit 9, 3) Sector 3-Test pit 1, 4) Sector 1-Test pit 1, and 5) Sector 6-Test pit 5.

Zooarchaeology

Few faunal remains were found during the archaeological work. The assemblage is composed of 162 remains from different sectors, pits and stratigraphic units, but it is limited in view of the area affected by archaeological works (Table 12.3). The main assemblage of faunal remains was recovered from Sector 6-Test pit 7, with 82 faunal remains from SU703 and 38 from SU709. This area is outside the convent buildings, next to the wall of the structure identified as the old monastic refectory, adjacent to the orchard. Another four samples were also analysed, although they contained a smaller number of remains: one sample from Sector 1, inside the monastic building (Test pit 1, SU102: 9 faunal remains); another from the cloister Sector 3 (Test pit 1, SU105: 29 remains); another from Sector 6 (Test pit 6, SU520: 1), and the last from Sector 7, outside the building (Test pit 9, SU901: 3). These deposits have been dated by means of the archaeological materials to between the 16th and the 19th century. The remains analysed largely consist of mammal bones and teeth (142 remains, 21 unidentified). Five bird bones (one unidentified, from SU709), 3 fish vertebrae and 15 remains of marine malacofauna (10 of molluscs and 5 of a single species of cirripede crustacean) have also been examined.

For taxonomic identification the comparative reference collection of the Universidad de León, as well as osteological atlases (*e.g.* Schmid 1972; Cohen and Serjeanston 1996; Barone 1999; Trigo *et al.* 2018), and specific publications focusing on the anatomical differentiation of caprines (Boessneck 1969; Zeder and Lapham 2010; Zeder and Pilaar 2010) were used. In those cases in which it has not been possible to assign the remains of the caprine subfamily (Caprinae) specifically to sheep (*Ovis aries*) or goat (*Capra hircus*), it has been assumed that they belong to the generic group of caprines.

Regarding quantification, the standard methods to estimate the number of remains (NR), minimum number of individuals (MNI) and minimum number of elements (MNE), as defined by different scholars, have been used (Klein and Cruz-Uribe 1984; Lyman 2008), considering aspects such as laterality and age. Biometric data have followed the proposals of Von den Driesch (1976) for mammals, of Morales and Roselund (1979) for fish, and Gutiérrez Zugasti (2009) for molluscs.

The age of death of mammals has been established according to the stages of epiphyseal fusion and the appearance and wear of the dentition, defined for domestic animals by, among others, Habermehl (1975) and Silver (1969). When it has not been possible to apply these criteria, the state of development of the bone texture of the cortical surface has also been considered (Ruscillo 2014). In relation to the calculation of height at the withers, we have used the indexes established for sheep by Teichert (1975) and for goats by Schramm (1967).

The evaluation of the state of preservation of the faunal remains, considering different taphonomic processes from their death to the moment of their recovery during archaeological excavation, adopts the proposal of Fernández-Jalvo and Andrews (2016). To define butchery marks, Crader's (1990) proposal was followed.

Table 12.1. Charcoal and wood analysis results

Charcoal/Sector 6	*SU520*	*SU552*	*SU566*	*SU592*
Volume (L.)	*2*	*5*	*3*	*3*
Taxa/Sample code	*MU54*	*MU049*	*MU048*	*MU050*
Fabaceae	11	25	46	15
Quercus sp. deciduous	9	25	13	3
Erica sp.	5			
Quercus/Castanea				3
Juniperus sp.				1
Olea europaea				3
Vitis vinifera	1			
Castanea sativa		1		
Prunus sp.		2		
Rosaceae/Maloideae				1
Salix/Populus		1		
Total	26	54	59	26
Mineral replaced wood/ Sector	*3*	*4*	*4*	*4*
Taxa/SU	*SU110*	*SU207*	*SU210*	*SU216*
Castanea sativa	4	1	2	2
Total	4	1	2	2

Results

Firewood

A hundred and sixty-five charcoal fragments from the garden area were analysed. A total of 11 taxa have been identified: shrubby species of the Fabaceae family (Fabaceae), oak (deciduous *Quercus*), heather (*Erica* sp.), oak/chestnut (*Quercus/Castanea*), juniper (*Juniperus* sp.), olive tree (*Olea europaea*), grapevine (*Vitis vinifera*), chestnut (*Castanea sativa*), plums-cherries (*Prunus* sp.), Rosaceae/Maloideae and willow/poplar (*Salix/Populus*) (Table 12.1).

Strong and moderate curvatures prevail in the annual rings observed, and sporadically weak curvatures have been identified in some charcoal fragments of deciduous *Quercus*. This is consistent with the identification of Fabaceae and *Quercus* twigs as part of the samples. The firewood burnt was predominantly of small calibre.

Some alterations related to plant growth could be identified as being present. In two fragments (*Salix/Populus*, *Prunus* sp.) found in SU552, the presence of CODIT (Compartmentalisation of Decay in Trees) was identified (Shigo and Marx 1977; Shigo 1979; 1984a; 1984b). This is evidence of the burning of decayed wood obtained from living trees or shrubs. Depending on the genus, the presence of CODIT might be related to the pruning of twigs, for improving fruit growth in *Prunus* and for producing long, straight, flexible sprouts in *Salix/Populus*. The presence of alterations in the growth rhythm, with the suppression of the spring growth in two charcoal fragments of *Quercus/Castanea* found in SU592, was also identified.

Woodcrafts

Mineralised wood remains of the coffins were occasionally preserved attached to the iron nails. In all cases, the taxon identified was chestnut wood (*Castanea sativa*), with weak curvature in the annual ring, and tyloses filling earlywood vessels, suggesting that they were probably obtained from mature trunks or large branches. Despite the anatomical features having been preserved, some of them were modified during the mineralisation process, leading to the presence of iron oxides. The EDX results for the samples made it possible to determine that the replacement of organic matter by iron products is almost complete (Table 12.2).

Textiles

The presence of textiles inside the tombs was identified by indirect and direct evidence. Small metal pins were located in several graves. In tomb SU187, the pseudomorph of a textile was exceptionally preserved. The contact of the textile with fat and other organic substances produced by decomposition of the deceased's body favoured the creation of a mould of the fibres and the textiles, preserving even the folds of the shroud. The morphology of the fibres corresponds with bast fibres, probably linen (*Linum usitatissimum*). The yarns were Z-spun in system 1 and system 2, and the type of weaving was plain weave (tabby) with single yarn, a fine fabric with 20 threads per cm; no edges or selvedges were preserved.

Faunal remains

Overall, the faunal remains recovered may be associated with materials identifiable as food waste, Group 1 as defined by Gautier (1987). It is only worth mentioning the presence of 26 pig remains (*Sus domesticus*) in SU105 (Sector 1-Pit 3), 25 of which come from the same specimen, less than six months old, represented by different elements of the skull, some occasional pieces of the axial skeleton, and many more abundant ones of the limbs, without signs of any type of anthropic processing. This type of information leads us to consider that the evidence reflects the partially preserved remains of an infant pig that may have died as a result of some unexpected cause or because its meat was considered unsuitable for consumption (an illness or a sudden death, for example), which was buried without any type of recognisable exploitation having previously taken place. Accordingly, this assemblage would be assigned to Gautier's Group 3 (1987).

It should also be noted that the five goose barnacle (*Pollicipes pollicipes*) plates were traced in a sample (Sector 1, SU 102) in which construction material was recorded; their state of preservation and the limited interest that has traditionally been given to this product as a food source until very recently leads one to suggest that their presence is indicative of them possibly having been included among the materials used for the preparation of the mortars recorded, either as a contribution among the sands used or by their intentional selection as calcareous elements. If so, these pieces should be interpreted as contemporary intrusives, Gautier's Group 4 (1987), with no meaningful food significance.

The analysis has revealed the predominance of mammals, exclusively represented by the usual domestic taxa (Tables 12.3 and 12.4), among which cattle (*Bos taurus*) clearly predominate, while caprines (*Ovis aries/Capra hircus*) and swine (*Sus domesticus*) contribute in an apparently similar way, taking into account what is indicated in relation to the whole of SU105. The assemblage is completed with a more limited presence of bird remains (also domestic ones as indicated by the remains identified), and marine fish and molluscs.

The taphonomic study has recorded a high presence of butchery marks, among which, in addition to incisions and cuts, it is worth mentioning the sections (slashes) made by means of metal cutting objects (axes, machetes or similar), which are especially recorded on cattle remains. On caprine

Table 12.2. EDX analysis results for the mineralised wood of Castanea sativa from SU210

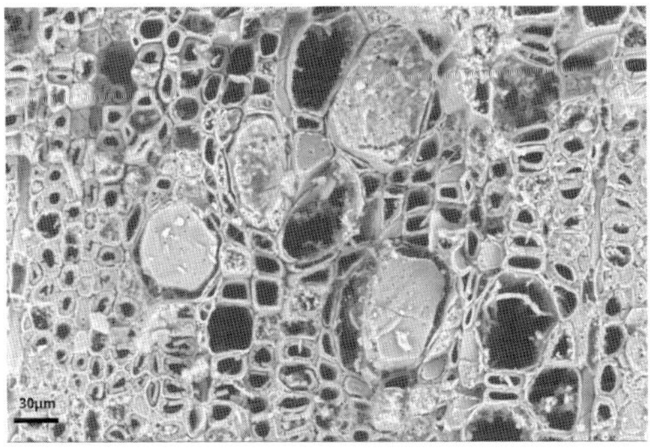

Result/weight %	*Spectrum 12*	*Spectrum 13*
Fe	84.89	86.06
C	11.28	8.53
O	3.41	4.72
Al	0.29	0.42
Si	0.07	0.10
S	0.07	0.18
Total	100.00	100.00

Table 12.3. Faunal remains: distribution by taxa. The number of remains (NR) and the minimum number of individuals (MNI) are indicated

Sector/Test pit	*1 / 1*		*1 / 3*		*6 / 6*		*6 / 7*		*6 / 7*		*7 / 9*	
Stratigraphic unit	*SU102*		*SU105*		*SU520*		*SU703*		*SU709*		*SU901*	
	NR	*MNI*	*NR*	*MNI*	*NR*	*MNI*	*NR*	*MNI*	*NR*	*MNI*	*NR*	*MNI*
Mammals												
Bos taurus					1	1	43	4	21	2	2	2
Ovis aries/Capra hircus	1	1	1	1			5	2	2	1	1	1
Capra hircus			1	1					1	1		
Sus domesticus			26	2			4	1	8	2		
Birds												
Gallus gallus domesticus	2	1					2	1				
Fishes												
Merluccius merluccius	1	1					1	1				
Conger conger			1	1								
Molluscs												
Ostrea edulis							3	3	2	1		
Pecten maximus							5	2				
Pollicipes pollicipes	5	3										
Identified	9		29		1		64		34		3	
Indeterminate					-		18		4			
Total	9		29		1		82		38		3	

and pig remains, butchery marks, much more sporadic, are identified with incisions. Also of interest is the absence of carnivore marks in the material analysed.

Regarding the age of mammals, the bovine specimens represented clearly seem to correspond to adults (over 60 months) or to sub-adults close to this age; the two dental pieces analysed do not allow for greater precision either. There are only two remains that come from younger individuals, a proximal tibia fragment (SU709) from a bovine under 36 months of age and, even more clearly, an atlas (SU901) in which the different parts that compose it had not yet fused, and which must have come from a young calf.

The small amount of material assigned to caprines reflects a wide variety of ages. Two deciduous premolars assigned to *Capra hircus* (a DP3 from SU105 and a DP4 from SU709) show reduced wear, which points to their correspondence with infantile specimens, around 6 months of age; the caprine represented by a first upper molar (M1) from SU901 must have belonged to a slightly more advanced stage (but possibly not exceeding 12 months). On the other hand, two metatarsals found in SU703 indicate the sacrifice of specimens that must have been over 24 months of age, although there is no other evidence that might allow us to be more precise.

The porcine remains, excluding the aforementioned remains from SU105, representative of a female less than 6 months old, reflect that animals were slaughtered at ages that can approximately be established as having been between 18 and 30 months, without any evidence suggesting the presence of specimens that had reached 36 months; only a distal humerus fragment (SU703) might have come from a slightly younger animal, but, whatever the case, one of at least one year of age. A canine (SU703) allows us to identify the presence of a male individual.

The four poultry remains, distributed between SU102 and SU703, come from domestic fowl (*Gallus gallus domesticus*). No butchery marks were recorded, but the evidence is consistent with their consumption. A tarsometatarsus (SU102) enables us to recognise the presence of a female.

The marine resources are represented by fish remains as well as mollusc shells and other crustacean remains (Table 12.3). The fish recorded (three precaudal vertebrae)

Table 12.4. Mammals and birds: distribution of anatomical parts represented by species (the NR is indicated; in shoulder and pelvic girdles, long bones, metapodials and phalanges, -/- refers to NR/MNE; in Total, -/- refers to NR/MNI)

	Bos taurus				*Ovis/Capra*					*Capra hircus*		*Sus domesticus*			*Gallus g.*	
SU	*520*	*703*	*709*	*901*	*102*	*105*	*703*	*709*	*901*	*105*	*709*	*105*	*703*	*709*	*102*	*703*
Horn	1															
Neurocranium		1										1				
Maxillary teeth									1							
Mandibular condyle												2				
Mandible		1										3		3		
Mandibular teeth			2							1	1	3	1	1		
Atlas				1								1				
Thoracic vertebra												1				
Lumbar vertebra		13														
Caudal vertebra		1														
Rib			2					1								
Scapula			1/1									1/1				
Humerus		2/2	2/2									4/3	2/2			
Radius		1/1	1/1				2/1					4/2	1/1	1/1		
Ulna			1/1									1/1		1/1		
Pyramidal			1													
Metacarpus		5/4										1/1				
Innominate		1/1	2/1													
Femur		9/5	5/2		1/1							4/2		1/1		1/1
Tibia		1/1	1/1					1/1						1/1	1/1	1/1
Talus							1									
Cubonavicular			1													
Metatarsus		5/4	1/1	1/1		1/1	2/2								1/1	
Distal metapodial		2														
First phalanx		1/1														
Third phalanx			1/1													
Total	1/1	43/4	21/2	2/1	1/1	1/1	5/2	2/1	1/1	1/1	1/1	26/2	4/1	8/2	2/1	2/1

correspond to two of the most abundant species in the north-west of the Iberian Peninsula since the Middle Ages, at least according to written sources: namely hake (*Merluccius merluccius*) in SU102 and SU703, and conger eel (*Conger conger*) in SU105. Despite the limited sample, the small size of the hake vertebrae is striking.

The consumption of shellfish is at least confirmed by the oysters (*Ostrea edulis*), of which five shells were found (SU703 and SU709), representative of at least four individuals. The presence of scallops (*Pecten maximus*), exclusively in SU703, may also reflect their consumption, but another purpose cannot be ruled out considering their high symbolic value, especially in religious contexts. The absence of intentional modifications (such as perforations that are indicative of a possible use as an element of adornment) and the presence of both flat (right) and convex (left) shells may suggest, in principle, that they are consumption waste, an explanation that would also be supported by their origin in a context in which a significant series of remains consistent with one or more deposits of food waste accumulated.

Finally, five barnacle remains (*Pollicipes pollicipes*) were also recovered, all from SU102. Barnacle is an edible cirripedean crustacean that was not very important during the historical period until very recently. It is necessary to evaluate the context of origin to try to rule out other possible reasons that may account for the origin of these remains, the use of which as an element in the production of mortars cannot be excluded.

Discussion

Everyday life in the convent of Santa Clara

The analysis of charcoal remains has provided an opportunity to gain a clearer idea of the firewood management strategies applied in this enclosed community of nuns. Small calibre wood was preferred for burning. This preference might be related to the legal protection of woodlands and trees as from the Late Middle Ages (Carlé 1976). The presence of shrubs such as Fabaceae and *Erica* sp. – the former as the predominant taxon in all the samples – suggests that this kind of firewood probably arrived from outside the convent walls. These plants are likely to have grown in the city's surroundings and were transported to the convent with other kinds of supplies. This wood from shrubby species was most probably mixed with other resources available inside the convent, such as trees and shrubs growing in the orchard, the garden and the wood, which were probably regularly pruned (Rosaceae/Maloideae, *Castanea sativa, Olea europaea, Vitis vinifera*, *Prunus* sp. and *Salix/Populus*). The significance of fruit-bearing trees and shrubs during medieval and post-medieval times has been attested at other sites (Martín-Seijo and Vázquez Collazo 2020).

The faunal assemblage detected, although not very numerous, allows us to recognise the presence of a significant diversity of food resources. However, there is a clear predominance of mammals and, more specifically, cattle. The presence of a significant percentage of remains of this species with butchery marks, and specifically those that show evidence of cutting processes (sectioned bones), suggests that the carcasses of these animals were prepared by specialised personnel. In addition, and although the assemblage studied is limited in size, if we evaluate the record provided by the cattle, the apparent predominance of certain sections of the skeleton (lumbar vertebrae, metapodials, femurs...) and the limited presence (such as cranial elements) or even absence of others (especially in some areas of the axial skeleton), could be the result of certain parts of the cattle being selected prior to their arrival in the conventual kitchen, reflecting a previous treatment that would have involved cutting up the bones and associated meat and a well-defined marketing process.

The smaller volume of caprine and pig remains does not allow for a proposal such as the one put forward for cattle to be made, nor is it easy for poultry. Among the material assigned to the former, only the presence of goat (*Capra hircus*) has been specifically identified, with slaughter processes that affect different age groups, from infant specimens (without having reached the age of one year) to others that clearly exceed two years of age. The consumption of pork, on the other hand, clearly seems to centre on individuals of two years of age, with ages ranging between 18 and 30 months being recorded

Whereas for cattle it is possible to propose a possible commercial acquisition limited to the meat necessary for consumption, it is not easy to propose the same for these other species. The presence of what appears to be a six-month-old piglet buried in the monastic enclosure could reflect the breeding of, or at least fattening, of pigs by the community itself, which could be extended to other species and especially to poultry, which in turn would have enabled the community to have its own supply of eggs. It is more difficult to propose the same for goats and sheep, unless it took place on a farm owned by the community outside the urban centre, which would have facilitated the necessary grazing of these animals.

The other resources recorded are related to the exploitation of the marine environment, either by fishing (hake and conger eel) or shellfish gathering (molluscs and, perhaps, barnacles). The consumption of these fish species was common in the north-west throughout the Middle Ages and Early Modern period, in addition to assuming a very important role in the diet of monastic centres, as we have been able to detect archaeologically in Acibeiro or to document in Sobrado and Santa María de Oseira (Vázquez Lijó 2009; Juega Puig 2012, 241). Perhaps the most striking element in this assemblage is the presence of young individuals that could be associated with whiting, compared to the large specimens identified at other Galician sites (Llorente Rodríguez *et al.* 2022). This could be linked to the consumption of fresh fish from port markets, more often associated with small-scale fishermen (Ferreira Priegue 1998; Juega Puig 2012, 317), than with catches destined for long-distance export, where specimens of much larger sizes are recorded.

For chronological periods as recent as the present one, written sources are essential. Those available in the case of the convent of Santa Clara (Otero Piñeiro 1999) do not provide information regarding the possible raising of animals or the consumption of meat or other animal-derived products, such as eggs. The analysis of the convent's income, as far as the income in kind is concerned, centres on cereals, while when analysing the food consumption of the community, cereal and wine are also mentioned. In short, it is not this information that resolves the questions that have arisen from the study of the faunal evidence from this site.

Death in the convent of Santa Clara

The main data related to funerary contexts has been provided by the analysis of wood and textile remains, and it is related to the shrouds and coffins used for burial in the convent of Santa Clara, during the time the cloister area was used as a cemetery. In all the cases that the remains of wood attached to iron nails were preserved by mineralisation, the species selected was chestnut (*Castanea sativa*). The use of chestnut wood was predominant from the Middle Ages onwards in Pontevedra. Its use as raw material for wooden coffins has also been attested at the convent of San Domingos

(Martín-Seijo unpublished report). In the port of Pontevedra, in contexts dating from the 15th to 16th century, chestnut trunks were the preferred timber for crafting planks (Martín-Seijo *et al.* 2021). This wood arrived in Pontevedra by cart or by it being floated downstream along the River Lérez to fulfil the demands of the city; the documentary sources also describe shipments of chestnut planks to the shipyards of Lisbon and Seville (Armas Castro 1992).

The exceptional preservation of a shroud has provided a unique opportunity to identify the kind of textiles used for wrapping the bodies of deceased nuns. The use of a fine tabby fabric made of bast fibres, probably linen, is the first archaeological evidence of this kind in the region.

Conclusions

This multidisciplinary approach to the study of different kinds of archaeological evidence has demonstrated its potential for producing significant information about key aspects of life and death in a cloistered female community that are usually not reflected in the written sources. The analysis of plant and animal remains has provided information about the management of supplies from outside the convent, such as firewood and food from terrestrial and marine animals. Moreover, it has offered an exceptional opportunity to catch a glimpse of the material nature of nuns' burials, such as the wooden coffins and the shrouds. Finally, this research stresses the relevance of addressing archaeobotanical and zooarchaeological studies in contexts dating from the Middle Ages onwards, for producing complementary information to the written sources.

Acknowledgements

We would like to acknowledge the support provided from the Museo de Pontevedra, especially to José Manuel Rey García and Sonia Mateos Escudero.

References

Armas Castro, J. (1992) *Pontevedra en los siglos XII a XV. Configuración y desarrollo de una villa marinera en la Galicia medieval*. Colección Galicia Histórica, Instituto P. Sarmiento de Estudios Gallegos. Pontevedra, Fundación 'Pedro Barrié de la Maza Conde de Fenosa'.

Barone, R. (1999) *Anatomie comparée des mammifères domestiques. Tome I Ostéologie*. Paris, Vigot Frères.

Bender Jørgensen, L. and Grömer, K. (2012) The archaeology of textiles–recent advances and new methods. *Portal: godišnjak Hrvatskoga restauratorskog zavoda* (3), 45–68.

Boessneck, J. (1969). Osteological differences between sheep (*Ovis aries* Linné) and goat (*Capra hircus* Linné). In D. Brothwell and E. Higgs (eds) *Science in Archaeology. A Survey of Progress and Research*, 331–58. London, Thames and Hudson.

Carlé, M.C. (1976) El bosque en la Edad Media (Asturias-León-Castilla). *Cuadernos de Historia de España* 59, 297–375.

Cartwright, C.R. (2015) The principles, procedures and pitfalls in identifying archaeological and historical wood samples. *Annals of Botany* 116(1), 1–13.

Cohen, A. and Serjeanston, D. (1996) *A Manual for the Identification of Bird Bones from Archaeological Sites*. 2nd ed. London, Achetype Publications.

Crader, D.C. (1990) Slave diet at Monticello. *American Antiquity* 55(4), 690–717.

Cybulska, M., Jedraszek-Bomba, A., Kuberski, S. and Wrzosek, H. (2008) Methods of chemical and physicochemical analysis in the identification of archaeological and historical textiles. *Fibres and Textiles in Eastern Europe* 5 (70), 67–73.

Fernández-Jalvo, Y. and Andrews, P. (2016) *Atlas of Taphonomic Identifications*. Vertebrate Paleobiology and Paleoanthropology Series. London, Springer.

Ferreira Priegue, E. (1998) O desenvolvemento da actividade pesqueira desde a Alta Idade Media ó século XVII. In C. Fernández Casanova (coord.) *Historia da pesca en Galicia*, 51–86. Biblioteca de Divulgación, Serie Galicia, 24. Santiago de Compostela, Universidade de Santiago de Compostela.

Fuller, D.Q., Stevens, C. and McClatchie, M. (2014) Routine activities, tertiary refuse and labor organization: social inferences from everyday archaeobotany. In M. Madella, C. Lancelotti and M. Savard (eds) *Ancient Plants and People: Contemporary Trends in Archaeobotany*, 174–217. University of Arizona Press.

Gale, R. and Cutler, D. (2000) *Plants in Archaeology. Identification Manual of Vegetative Plant Materials Used in Europe and the Southern Mediterranean to c. 1500*. Kew, Westbury and Royal Botanic Gardens.

Gautier, A. (1987) Taphonomic groups: how and why? *Archaeozoologia* 1, 47–52.

Gleba, M. and Mannering, U. (2012) Introduction: textile preservation, analysis and technology. In M. Gleba and U. Mannering (eds) *Textiles and Textile Production in Europe from Prehistory to AD 400*, 1–26. Oxford, Oxbow Books.

Gutiérrez Zugasti, I. (2009) *La explotación de moluscos y otros recursos litorales en la región cantábrica durante el Pleistoceno Final y el Holoceno Inicial*. Santander, Publican, Ediciones de la Universidad de Cantabria.

Habermehl, K.-H. (1975) *Die Altersbestimmung bie Haus- und Labortieren*. Berlin and Hamburg, Verlag Paul Parey.

Hather, J.G. (2000) *The Identification of the Northern European Woods. A Guide for Archaeologists and Conservators*. London, Archetype Publications.

Hausman, L.A. (1920) Structural characteristics of the hair of mammals. *The American Naturalist* 54(635), 496–523.

Juega Puig, J. (2012) *El comercio marítimo de Galicia (1525–1640)*. PhD thesis, Universidade de Santiago de Compostela.

Klein, R.G. and Cruz-Uribe, K. (1984) *The Analysis of Animal Bones from Archaeological Sites*. Prehistoric Archeology and Ecology Series. Chicago and London, University of Chicago.

Llorente-Rodríguez, L., Craig, O.E., Colonese, A.C., von Tersch, M., Roselló-Izquierdo, E., González Gómez de Agüero, E.,

Fernández-Rodríguez, C., Quirós-Castillo, J.A., López-Arias, B., Marlasca-Martín, R., Nottingham, J. and Morales Muñiz, A. (2022) Elucidating historical fisheries' networks in the Iberian Peninsula using stable isotopes. *Fish and Fisheries* 23(4), 86273. https://doi.org/10.1111/faf.12655

Lyman, R.L. (2008) *Quantitative Paleozoology*. Cambridge Manuals in Archaeology. Cambridge and New York, Cambridge University Press.

Marguerie, D. and Hunot, J.Y. (2007) Charcoal analysis and dendrology: data from archaeological sites in North-western France. *Journal of Archaeological Science* 34, 1417–33.

Martín-Seijo, M., Sartal Lorenzo, M., Kaal, J. and Teira-Brión, A. (2021) A multi-disciplinary study of woodcrafts and plant remains that reveals the history of Pontevedra's harbour (Northwest Iberia) between the 13th and 19th centuries AD. *Environmental Archaeology* 26(2), 192–208.

Martín-Seijo, M. and Vázquez Collazo, S. (2020) *Ad comburendum.* El uso de la leña en el Castillo de Pambre (Palas de Rei, Lugo) durante los siglos XV–XVII. *SPAL-Revista de Prehistoria y Arqueología* 29(1), 321–341.

Morales Muñiz, A. and Roselund, K. (1979) *Fish Bone Measurement. An Attempt to Standardize the Measuring of Fish Bones from Archaeological Sites*. Steenstrupia. Copenhagen, Zoologisk Museum.

Otero Piñeiro, G. (1999) Santa Clara de Pontevedra en la Edad Moderna. Estructura económica del convento (1640–1834). *Obradoiro de Historia Moderna* 8, 270–300.

Rast-Eicher, A. (2016). *Fibres: Microscopy of Archaeological Textiles and Furs*. Budapest, Archaeolingua.

Ruscillo, D. (2014) Zooarchaeology: methods and collecting age and sex data. In C. Smith (ed.) *Encyclopedia of Global Archaeology*, 8000–10. New York, Springer.

Schmid, E. (1972) *Atlas of Animal Bones*. Amsterdam–London, Elsevier.

Schramm, Z. (1967) Long bones and height in withers of goat. *Roczniki Wyzszej Szkoly Rolniczeyw Poznani* 36, 89–105.

Schweingruber, F.H. (1990) *Anatomy of European Woods. An Atlas for the Identification of European Trees, Shrubs and Dwarf Shrubs*. Stuttgart, Paul Haupt.

Schweingruber, F.H., Börner, A. and Schulze, E.D. (2008) *Atlas of Woody Plant Stems. Evolution, Structure and Environmental Modifications*. Berlin-Heidelberg, Springer Verlag.

Shigo, A.L. (1979) Tree decay an expanded concept. *Agriculture Information Bulletin* 419. Washington, DC, Forest Service, United States Department of Agriculture Forest Service.

Shigo, A.L. (1984a) Compartmentalization: a conceptual framework for understanding how trees grow and defend themselves. *Annual Review of Phytopathology* 22(1), 189–214.

Shigo, A.L. (1984b) Tree decay and pruning. *Arboricultural Journal* 8(1), 1–12.

Shigo, A.L. and Marx, H.G. (1977) Compartmentalization of decay in trees. *Agriculture Information Bulletin* 405, 1–73. Washington, DC, Forest Service, United States Department of Agriculture Forest Service.

Silver, I.A. (1969) The ageing of domestic animals. In D. Brothwell and E. Higgs (eds) *Science in Archaeology. A Survey of Progress and Research*, 283–302. London, Thames and Hudson.

Teichert, M. (1975) Osteometrische Untersuchungen zur Berechnung der Widerristhöhe bei Schafen. In A.T. Clason (ed.) *Archaeozoological Studies*, 51–69. Amsterdam, North-Holland Publishing Company.

Trigo, J.E., Díaz-Agras, G.J., García Álvarez, O.L., Guerra Sierra, A., Moreira da Rocha, J., Pérez Dieste, J., Rolán Mosquera, E., Troncoso, J.S. and Urgorri, V. (2018) *Guía de los moluscos marinos de Galicia*. Vigo, Universidade de Vigo.

Vázquez Lijo, J.M. (2009) Sabrosa abstinencia. Abasto y consumo de pescado en el monasterio de Sobrado (ss. XVII–XVIII). *Obradoiro de Historia Moderna* 18, 151–79.

von den Driesch, A. (1976) *A Guide to the Measurement of Animal Bones from Archaeological Sites*. Peabody Museum Bulletin 1. Cambridge, Harvard University.

Zeder, M.A. and Lapham, H.A. (2010) Assessing the reliability of criteria used to identify postcranial bones in sheep, *Ovis*, and goats, *Capra*. *Journal of Archaeological Science* 37(11), 2887–905.

Zeder, M.A. and Pilaar, S.E. (2010) Assessing the reliability of criteria used to identify mandibles and mandibular teeth in sheep, *Ovis*, and goats, *Capra*. *Journal of Archaeological Science* 37(2), 225–42.

Part 3

Animals in the monastic environment

13

Paw prints in the cloister: the study of ichnites on tiles from the Monastery of Santa Maria de Pedralbes (Barcelona, Spain)

Jordi Nadal, Philip Banks, Anna Castellano-Tresserra, Santiago Riera, Marina Fernández-Liarte and Lluís Lloveras

This study presents the results of the analysis of a minimum of 136 animal tracks located on the floors of different buildings forming part of the Monastery of Santa Maria de Pedralbes (Barcelona, north-eastern Spain), especially on the ground floor and first floor of the Gothic cloister. The taxonomical identification of the footprints demonstrates that the main species represented are dogs, cats and fowl. With the exception of this last species, these taxa were not considered to be food animals in medieval Europe. The results are compared with those of the zooarchaeological studies carried out on the material recovered at the site and with evidence from documents. The latter provide information about the origin of the tiles and other clay building materials used in the construction and successive restorations of the conventual buildings from the 14th century onwards and indicate that tiles were produced in suburban brickyards located in the former village of Sarrià (nowadays a district of Barcelona) and the city of Barcelona itself. Our results demonstrate that the area of tile production for the community was not the same as that of meat production, except perhaps in the case of poultry.

Keywords: Santa Maria de Pedralbes, ichnology, footprints, tile production.

Introduction: geographical and historical background

The close relationship between ichnology, palaeontology and taphonomy has a long tradition (Frey 1975; Bromley 1990). This discipline is applicable to the study of any motile organism, whether marine or terrestrial, vertebrate or invertebrate, and has been used for both palaeobiological and taphonomic reconstructions (Seilacher 2007). Despite this, the application of vertebrate ichnology in anthropogenic contexts (archaeological sites) has primarily focused on the study of human footprints, generally in Pleistocene chronological contexts and normally to demonstrate the presence of hominids at a site or in the study of the evolution of bipedalism (Pastoors and Lenssen-Erz 2021). It is only recently that ichnology has been applied to periods in which the use of clay is widespread, especially as a construction material, in the manufacture of adobe or fired bricks and tiles. This production grew exponentially with the expansion of the Roman world in the Mediterranean context, and this has been the focus of interest of several studies about the presence of animals on archaeological sites from the analysis of footprints (Bennet 2012; Dobosi 2016, among others).

The aims of our paper are: (1) to present the results of an ichnological study carried out in the Monastery of Santa Maria de Pedralbes (nowadays in the district of Sarrià, in Barcelona, north-eastern Spain) since, as a result of a preliminary examination, the monastery revealed a considerable concentration of this kind of zooarchaeological-taphonomic evidence, and (2) to establish the similarities and differences between the ichnological results and the osteological ones from the same monastery (which are presented in another chapter of this publication). Moreover, the origin of such differences or similarities will be discussed and evaluated in historical and social terms.

The Monastery of Santa Maria de Pedralbes is located in the neighbourhood of Sarrià, formerly a village that was progressively incorporated within Barcelona as a result of the city's growth in the late 19th century and early 20th century, until it finally became part of Barcelona for administrative purposes in 1921. The monastery was founded by Queen Elisenda de Montcada, the fourth and last wife of King Jaume II, called 'the Just', in 1327, in order to spend the rest of her life, once she was widowed, which occurred in the same year. The Queen lived in accommodation close to the monastery until her own death in 1364. Her remains still lie buried in the monastic church (Castellano-Tressera 1998; Sanjust 2010).

The monastery belongs to the order of the Poor Clares and is still used by the present-day community, even though a large part of the complex is accessible to visitors. A number of archaeological excavations and also restoration projects have been carried out in certain areas of the monastery in recent times. Besides, over the years, the monastery has been expanded with new buildings, reformed and rebuilt. Some of these reconstructions involved the use of new bricks and tiles that potentially had footprint marks. In fact, for the dating of the tiles with footprints, the documentation relating to the aforementioned renovations has been used (Anzizu 1897).

Materials and methods

A hundred and thirty-six individual footprints, complete or incomplete, isolated or clustered were recorded and analysed. These are located on floors of different parts of the monastery: the infirmary, the garden and the so-called 'claustre dels gats' (literally 'cats' cloister'), but the majority of them are to be found on tiles from the ground floor and the first-floor galleries of the cloister.

The chronology of these tiles ranges from the 15th to the 17th century, according to different contemporary records and the interpretation of data from the previously mentioned study by Anzizu. Nevertheless, some floors were restored in the 19th and 20th centuries and earlier tiles were reused in these modern restorations, according to the research undertaken by one of the authors (A.C-T.).

The methodology has been divided into a first part comprising the taxonomical and anatomical description of the footprints, in accordance with previous works (Bang and Dahlström 1992; Bennet 2012): the taxonomic identification by means of diverse diagnostic clues, the identification of fore or hind paws, and right or left paws, depending on the species and only in the case of related and clustered footprints (trackways); and metric analysis, in order to establish size or breed within the species (measured in millimetres and taken by means of callipers). The measurements registered, if possible, were: the total length of the footprint, including the claws; the length of the fore claws; the length of the footprint from the distal part of the heel pad to the proximal end of the front two toes excluding the claws; and the greatest width, between the lateral sides of the outer toes. These measurements were taken in carnivore mammals. In the case of birds, the measurements were limited to the greatest length and the greatest width. Although other measurements were designed for other species (two- or four-toed Artiodactyla and Perissodactyla, among others), in the end, they were not used because such taxonomic groups were not identified at the site. The relationship among diverse footprints when more than one mark was present on a sole tile has also been established. Prints are classified as isolated (just one print on a tile) or clustered (two or more prints), which in turn can be classified as forming a track cluster (the different impressions representing the same displacement of a single individual) or randomly clustered (when prints do not correspond to the same track or even the same animal). The second part of the methodology focuses on the taphonomic modifications of the ichnites. The footprints were evaluated in terms of preservation and completeness or integrity. The integrity of an ichnite was classified in two categories, as a complete or incomplete footprint. In the case of incomplete prints, which part was present was registered: anterior/posterior and medial/lateral as well. Footprint preservation was classified as good, when all the anatomical parts of the foot could be observed clearly (in some cases even fur marks); moderate, if the print was not clear enough to take any measurements; or poor, if some parts of the anatomy were not identifiable and no measurements at all were taken. An attempt must be made to distinguish the different causes of the completeness of the impression, depending on whether it was a result of finished or unfinished trampling when the animal walked – and therefore data related to the moment when the fossil entity was produced – or a result of preservation, due to biostratinomic agents after the ichnite had been produced: breakage, weathering agents or wear over the course of time, once floor tiles were already fired. Even though it may seem obvious, it should be pointed out that the completeness of the footprints is related to events prior to the firing of the tile or brick, and agents that affect the preservation reflect events after their removal from the kiln.

By defining completeness on the basis of the association between footprints, the patterns of movement for quadrupeds have been determined as falling into four categories, namely stalking, trotting, loping or galloping.

Results

Of the 136 footprints analysed, 124 have been taxonomically identified at the species level (91.18%), and only 12 (8.82%) have not been identified. Nevertheless, among the unidentified footprints, carnivore prints can be recognised in 10 cases and bird ones in the remaining 2 cases. It must be considered that they were left by the same species as those identified among the species detected.

Only three species can be identified in the assemblage, which demonstrates a very limited zoological diversity: dogs (*Canis familiaris*), cats (*Felis catus*) and domestic fowl (*Gallus gallus*). Thus, it might be supposed that the unidentified bird footprints were of the genus *Gallus*, but in such a poor state of preservation that criteria for a more precise identification could not be found. It is more difficult to relate the unidentified carnivore prints to small dogs or, alternatively, to cats. Be that as it may, no other species that it would not be unusual to find on other sites, generally in rural contexts, have been identified: for instance, pigs, sheep or goats, cattle or even equids (Eliot 1991; Bar-Oz and Tepper 2010; Muscolino 2018, among others).

These footprints appear as isolated impressions in 60 cases (44.1%), a percentage that represents 37.3% in the case of dog prints (25), 11.8% in cats (2 cases) and 70% for chickens (28 cases). In 68 cases (50%), the prints formed a track (in general of no more than 2 prints due to the size of the tiles), a figure that represents 58.2% in the case of dogs (39 cases), 82.4% in cats (14 cases) and 27.5% among domestic fowl (11 cases). The reason why the presence of tracks among cats is substantially higher than in the other taxa may be a consequence of two factors: the number of cat prints is smaller than that for the other species, and this may result in a noticeable bias as the sample is less statistically significant. On the other hand, the size of the footprints, the smallest among the three species represented, makes it possible to record more evidence for a track on a sole tile. Finally, there is a substantial number of footprint specimens that are not structured in an orderly way as a track but randomly clustered, since several animals trampled on the tile before it was dried and fired. In some cases, the prints could correspond to different individuals of the same species. These random clusters represent 5.9% of the total (8 footprints), which are distributed as follows: 4.5% among dogs (3 cases), 5.9% among cats (1 case) and 2.5% among chicken (1 case) (Fig. 13.1).

As regards the preservation of the footprints, 37 cases can be considered as well preserved (27.2%), 71 cases as moderately preserved (52.2%), and 28 cases as poorly preserved (20.6%) (Fig. 13.2).

Dogs (Canis familiaris)

As concerns the degree of completeness of the footprints of this species, 30 specimens are complete (44.8% among this species). Among the incomplete prints, 36 cases represent just the anterior part of the foot (53.7%) and 1 just the medial part (1.5%). It seems likely that in many cases the animals must have been moving at a moderate speed, probably at a trot.

Regarding the measurements obtained, especially the maximum length of the footprint, without taking into

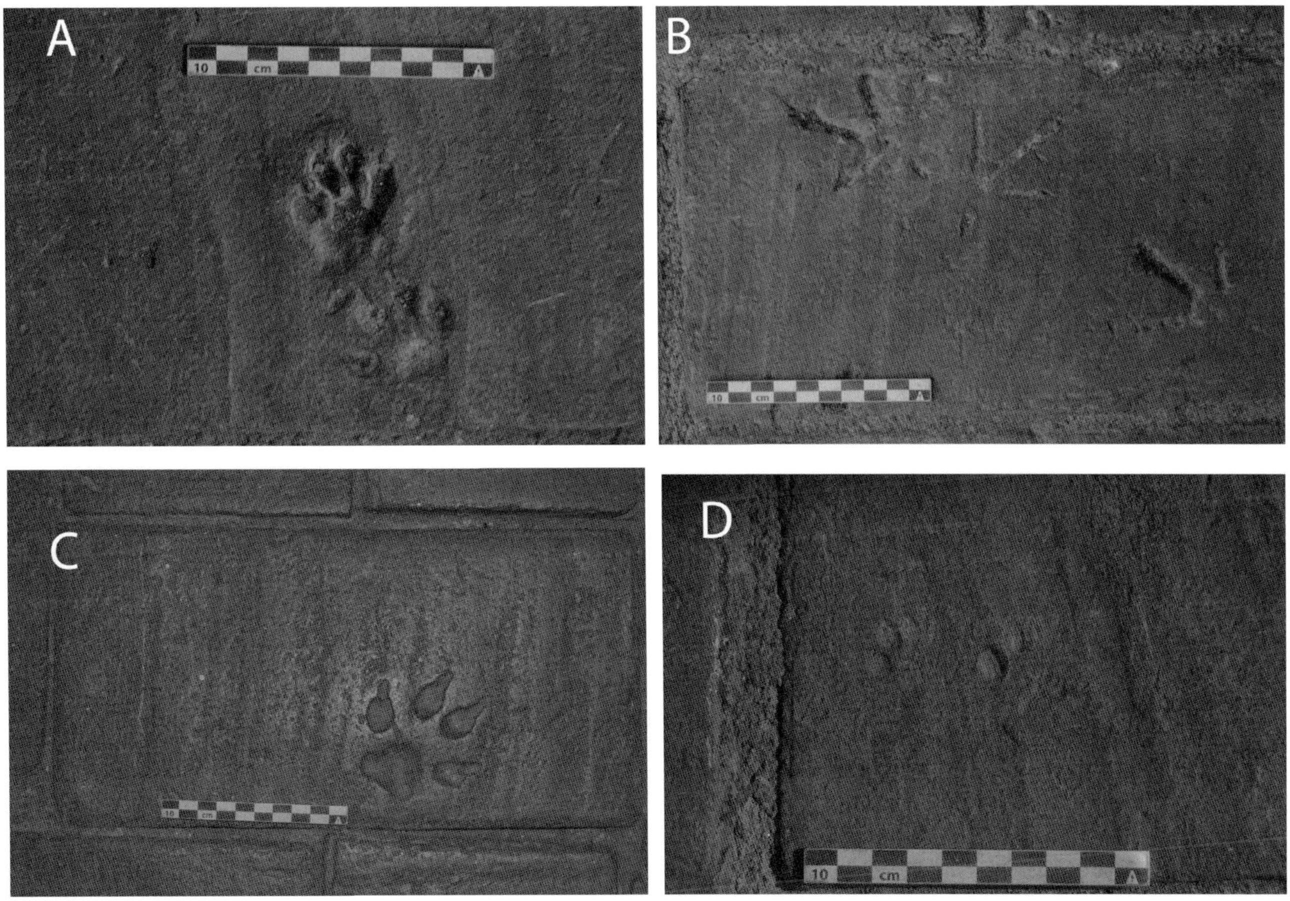

Figure 13.1. A: two dog footprints related in a track; B: three chicken footprints, two of them related in a track and one randomly clustered to the others; C: isolated dog footprint in a very good state of preservation; D: incomplete dog paws.

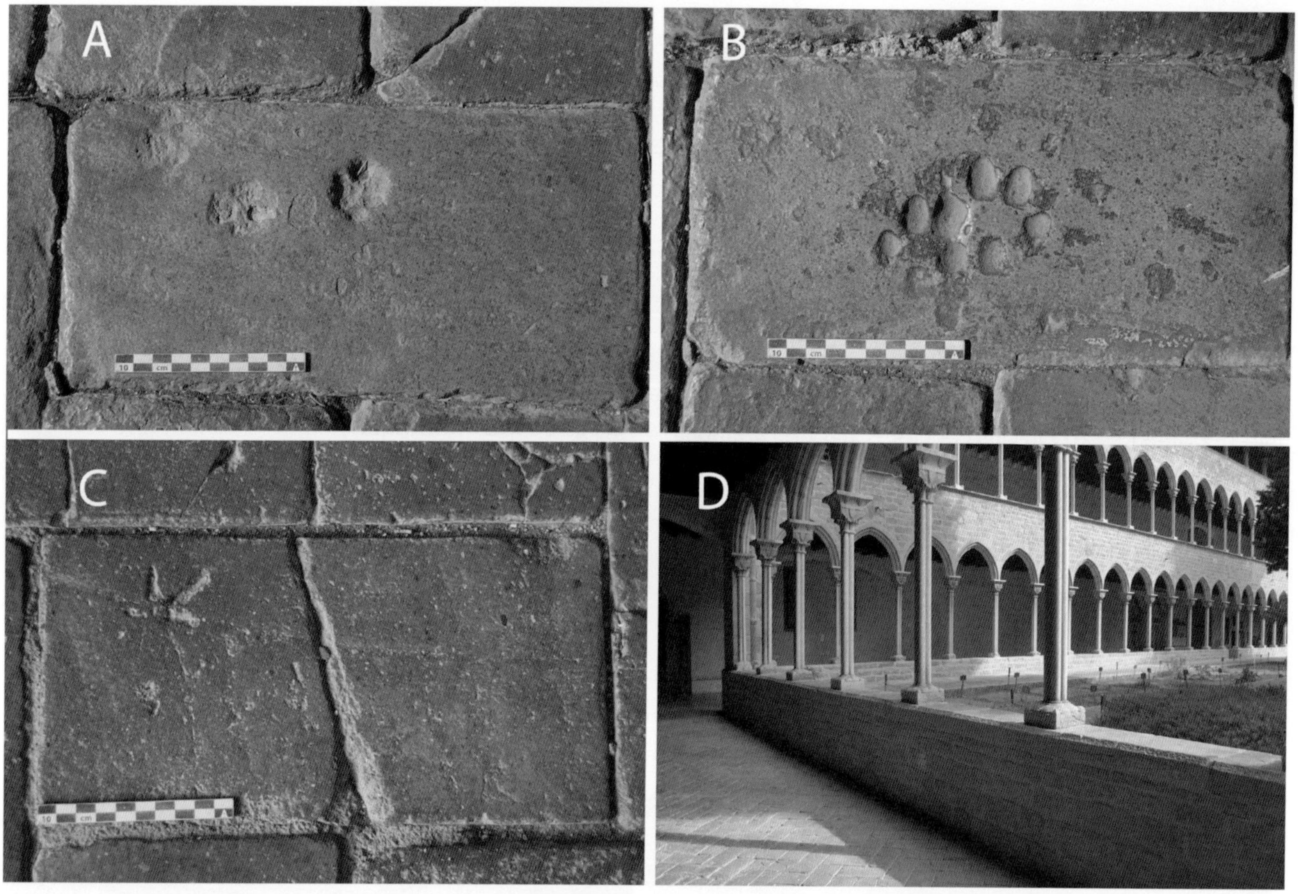

Figure 13.2. A: two cat paws exhibiting very poor preservation; B: dog footprints with moderate preservation; C: broken tile with a chicken print affected by the scar; D: an overview of the ground floor and first-floor cloister galleries of the monastery, where most of the tiles with footprints are located.

account the impression of the claw, dogs were represented by different sizes. The possibility of individuals of different ages cannot be discarded, even though the characteristics in the morphology and development of claws can help us to distinguish young and adult animals with a certain degree of precision. In any case, using the results of this measurement provides evidence that may point to the presence of diverse breeds of different heights at their withers. In accordance with these results and if they are compared with the measurements of footprints of different modern breeds, at least one print has a maximum length of over 8 cm. This measurement is reconcilable with measurements of large breeds, such as modern retrievers. Nevertheless, the majority of the results form a group with measurements ranging between 5 and 7 cm, which can be placed alongside modern medium-sized and small breeds, such as the border collie as a model of a medium-sized breed and the Shetland sheepdog as an example of small breeds. One or two cases present measurements under 5 cm. These animals can be related to very small breeds, such as the pug (Fig. 13.3). This diversity of measurements has been identified by osteoarchaeological evidence in Barcelona at least since the Roman period (Colominas 2016). In a recent study covering the medieval period in the same city, many bones of dwarf dog breeds, both gracile and robust examples, have been identified. In general, these dwarf animal bones are related to the presence of different pathologies arising from their distorted morphology (brachymelia), such as *osteochondritis dissecans* or *spondylosis deformans* (Fernández *et al.* 2024).

Cats (Felis catus)

Cat footprints are the least frequent among the species identified. The degree of completeness of the prints is total in 12 cases (70.58%) while incomplete prints are always represented by the anterior part of the foot, in 5 cases (29.4%). In 2 cases (11.8%) the prints are isolated on the tile. One case is randomly clustered (5.9%), and most prints appear to form a track (82.3%). As mentioned above, the cat prints are clearly smaller than dog ones, and that trait could facilitate the presence of more than one footprint on a sole tile, and tracks would thus be more frequent in this species. As in the case of dogs, the cats must have been moving at moderate speed, probably trotting on the surface of the new tiles.

Measurements of cat prints exhibit less metric variability within the species. It is true that no great size variability has so far been noticed among the different breeds of cats. No one print is larger than 4 cm in length and width, while the smallest examples are around 3.4 cm in both length and width. Only two cases reveal smaller measurements (one with results of less than 3 cm and the other slightly more than 2 cm). Even though the fact that these prints, of relatively poor quality in both cases, were preserved might suggest that the presence of infantile individuals cannot be discarded, we consider that there is some taphonomic bias in the measurements of these two last footprints (Fig. 13.4).

Domestic fowl (Gallus gallus)

Chicken footprints (including under the heading 'chickens' all the variabilities of age and sex among domestic *Gallus gallus* species) are complete in 23 cases (57.5%), only displaying the anterior part of the foot in 14 cases (35%), and the posterior part in just one case (2.5%). There are two incomplete examples (5%), but no specific part of the foot is represented (just parts of the toes, whether anterior or posterior). Most of the incomplete cases in this species are a consequence not of the effect of trampling, but because of the poor state of preservation of the tiles. Two unidentified bird footprints, which surely have to be considered as being related to chickens as well, should be added, but no diagnostic evidence is preserved.

As regards the interpretation of the gait, these footprints are isolated in 28 cases (70%), 1 case is randomly clustered (2.5%), and in 11 cases the prints are clustered in a track (27.5%), a figure that is far lower than in the other species studied. This is likely to be a consequence of the overall dimensions of the *Gallus* print. As the toes are extremely large, it is quite difficult for more than one print to have been present on a sole tile. Nevertheless, the tracks identified are indicative of the relatively slow progress of the animals over the fresh surface of the tiles.

Regarding the measurements, two more or less clear groups can be identified. The first one, with a substantial spread of results, but in general representing large animals, ranges from 7 cm to slightly more than 10 cm in length, and from slightly more than 6.5 to 8 cm in width. These measurements are most likely indicative of both females – the smaller specimens – and males – the larger ones – but without a clear break in continuity. On the other hand, there are a few prints with a variability of measurements that ranges between 7 cm and 8 cm in terms of greatest length and between 5.5 and 6 cm in width. Even though we cannot rule out the presence of young individuals, these results can be interpreted in terms of the presence of two breeds. The larger prints could correspond to a common large chicken breed and the smaller ones to bantams (Fig. 13.5). In fact, among Catalan chicken breeds, there is the relatively large 'El Prat' breed (which originated in the municipality of El

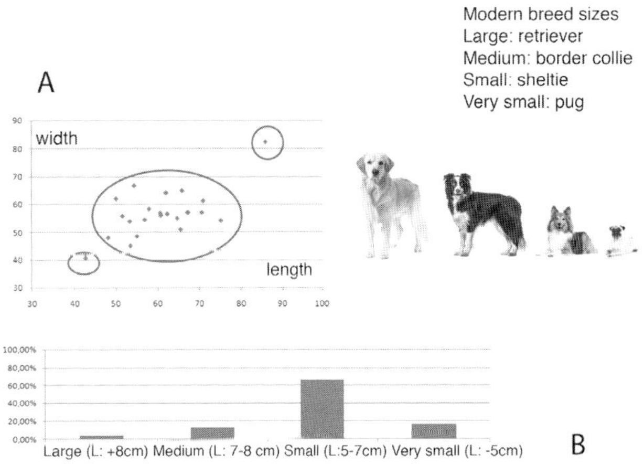

Figure 13.3. A: scatterplot with the measurements of dog paws (measurements in millimetres). At least three groups can be seen, the very large and very small animals, represented at least by one individual. And a very important group of medium-sized dogs; B: percentual barplot of the size results, in millimetres, according to the greatest length of the foot.

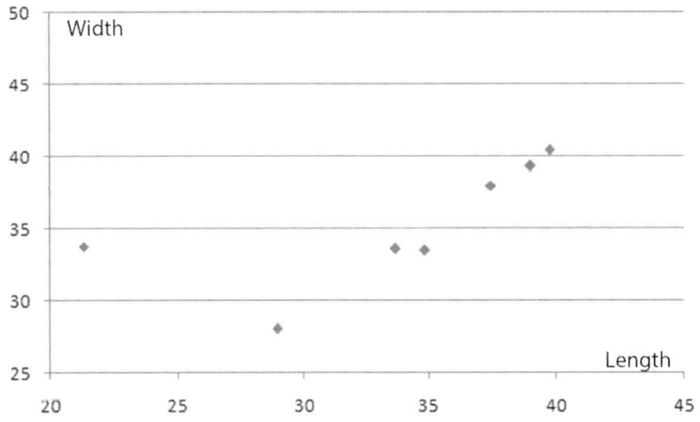

Figure 13.4. Scatterplot with the measurements of cat paws (measurements in millimetres).

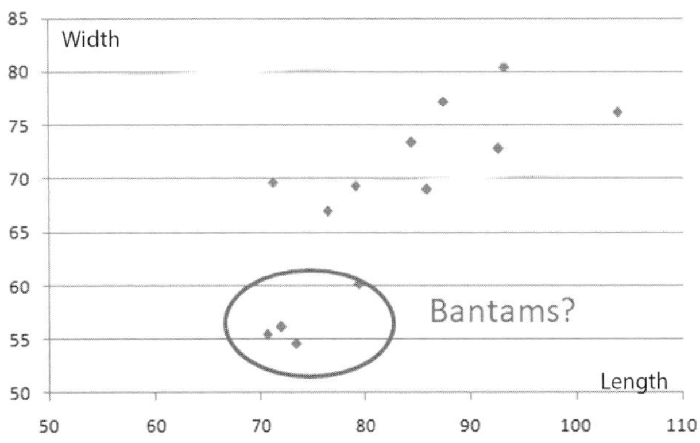

Figure 13.5. Scatterplot with the measurements of chicken prints (measurements in millimetres). Several individuals could be considered as bantams.

Prat del Llobregat, in the province of Barcelona), in which some males can reach 2.2 kg in weight, and bantam breeds such as 'Flor d'ametller' (literally 'Almond flower' in Catalan, because of the colour of the feathers, white with some brown spots, which resemble almond tree flowers), which reach no more than 0.8–1.2 kg in weight (Parés *et al.* 2006).

Discussion

The most noteworthy aspect of the study of the footprints at Santa Maria de Pedralbes is that the taxonomic results cannot be related to the taxonomic distribution obtained in previous zooarchaeology studies on the osteological remains from the monastery (see Lloveras *et al.* 2023, and Lloveras *et al.* (forthcoming) in volume 2 of this publication). Among them, mammal remains are the most abundant, with more than 86% of the sample in terms of NISP, followed by fish bones, slightly more than 9%, and finally birds, which fail to reach 4%. The best represented mammal species is *Ovis aries*, and the sample is completed by other food species, which, however, are far less noticeable in the sample: *Bos taurus* and *Sus domesticus*. The only mammal species represented among the faunal remains that is considered in our study is the cat (6% of mammal bones). In spite of its relatively abundant presence, it cannot be considered to fall into the category of food remains. In fact, all those remains represent just two animals, one of them almost anatomically complete. They must be considered to be the bone remains of pets that lived in the monastery whose bodies were abandoned in rubbish dumps once the animals had died. Other evidence indicates the presence of cats as commensal animals within the monastery: bird bones with cat toothmarks; some parts of the complex with references to cats (the above-mentioned 'Cats' cloister'); or the presence of cat flaps in some monastery doors. In contrast, there is no evidence for the presence of dogs. Among bird remains from the kitchen waste dump, *Gallus gallus* is clearly the most abundant species, with 90.3% NISP (Lloveras *et al.* 2023). Therefore, with the exception of chickens, no relationship can be established between the footprints and faunal remains from the monastery. These differences between the results derived from the zooarchaeological study and the ichnological study should not come as a surprise. The majority of the food animals were slaughtered or even butchered before they reached the monastery, so individuals were never alive near or inside the building. Poultry may have been slaughtered in the kitchens or in other domestic spaces around the monastery since texts mention there was a henhouse or a chicken coop near the monastic garden.

On the other hand, it must be accepted that tiles were not produced inside the monastery. Two hypotheses could be proposed to explain the origin of the building materials used in the monastery: a local origin, maybe a brickyard associated with a farmhouse in the municipality of Sarrià (it should be recalled that Sarrià was, at that time, a small rural village and Santa Maria de Pedralbes was located within the limits of the corresponding parish) or alternatively an origin that was not so close, for example any brickyard located in the municipality of Barcelona. The city of Barcelona, during medieval and post-medieval times was approximately 5 km from the monastery itself. Whatever the case, the literature (Anzizu 1897) informs us that building material, including tiles and bricks, was purchased from craftsmen in both Sarrià and Barcelona. It is through such an approach that taxonomic identification of the footprints can help us to elucidate where the tiles came from. Even though the most frequent species identified among the footprints on clay building materials is the dog, it seems that there is little evidence of rural animals on tiles found in cities, while the prints of typical farm animals, such as cattle, equids or sheep and goats are not unusual on tiles or bricks from rural sites (as can be seen in some examples cited previously, or other cases such as Castro and Gimeno 1996; Sabio and Detry 2019). In fact, the zoological diversity detectable on the tiles from Pedralbes is extremely limited and related to a small number of species (domestic carnivores) that are very common in urban contexts, rather than food animals: to be more precise they are restricted to cats and dogs, as might be expected at an urban tile production centre. In the case of dogs, the variability in foot measurements recorded on the Pedralbes tiles could reinforce the hypothesis of an urban origin as has already been mentioned (Fernández *et al.* 2024), with small, medium-sized and very large breeds. The fowl prints can be related to both rural and urban contexts. In Barcelona, some evidence shows that people could have had hen-coops on the terraced roofs of urban buildings in order to obtain eggs and also so that the animals could be eaten when egg production dropped as a result of the age of the hens (Lloveras *et al.* 2021).

The most important nearby area of brickyards lay on the outskirts of the city of Barcelona, along the present-day carrer Tallers, which has sometimes been understood as the 'street of workshops'; however, there is evidence that suggests that the word *tallers* should, in fact, be related to the medieval Catalan form 'teyers' or 'teulers', that is to say tile-makers, where those who produced bricks and tiles are recorded as having kilns from the 14th century onwards, and where production using semi-industrial methods was still taking place in the 19th century. A document of the year 1365 in the Cathedral archives (Arxiu Capitular de Barcelona, parchment 4-41-659) refers to a *furno de coquendi rajolas*, that is a kiln for firing floor tiles and bricks, located at the *capite superiori vici vocati dels Tayers*, the upper end of the street called Tayers (or Tallers), a point some distance from the densely inhabited core of the city, thus minimising the fire risk inherent in the process (Vila 2005; Sánchez and Gurt 2018; Sales 2021). It might be pointed out that this area adjoined the road leading from the city to the district of Sarrià, where the monastery stands. The purchase of building materials for the Monastery of Pedralbes in Barcelona would be explained by the existence of a set of commitments and agreements between the government of the city and Santa Maria de Pedralbes, which have been maintained down to the present day.

Conclusions

Even though the presence of animal footprints from the monastery of Santa Maria de Pedralbes cannot be related to either day-to-day life in the monastery or to the animals used to fulfil the dietary requirements of the nuns, the study of this kind of zooarchaeological data enables us to increase the information available about the commercial networks, in this case of clay building material (tiles), involving the city of Barcelona. The taxonomic distributions (with the presence of dogs, cats and fowl, but the total absence of footprints of more rural animals, such artiodactyls and perissodactyls as goats, sheep, pigs, cattle, horses and so on) and the diversity of dog breeds inferred from the size of the prints, suggest as the most parsimonious hypothesis that the tiles used as building material in different dependences of the monastery came from Barcelona brickyards. This hypothesis can be supported by texts that mention commercial contacts between Santa Maria de Pedralbes and the city.

Acknowledgements

This work is supported by research projects PID2020-118194RJ-100 and PID2023-146276NB-100 funded by AEI/MINECO, and 2021SGR00337 funded by the Generalitat de Catalunya.

References

Anzizu, E. (1897) *Fulles històriques del Real Monestir de Santa María de Pedralbes*. Barcelona, Estampa F. Xavier Altés (Facsimile edition 2007).

Bang, P. and Dahlström, P. (1992) *Huellas y señales de los animales de Europa*. Barcelona, Omega.

Bar-Oz, G. and Tepper, Y. (2010) Out on the tiles. Animal footprints from the Roman site of Kefar 'Othnay (Legio), Israel. *Near Eastern Archaeology* 73(4), 244–7.

Bennet, D. (2012) Life history information from tracks of domestic dogs (*Canis familiaris*) in ceramic building materials from a Roman bath-house at Vindolanda, Northumberland, England. *Archaeofauna* 21, 7–36.

Bromley, R.G. (1990) *Trace Fossils. Biology and Taphonomy*. London, Unwin Hyman.

Castellano-Tresserra, A. (1998) *Pedralbes a l'edat mitjana: història d'un monestir femení*. Barcelona, Publicacions de l'Abadia de Montserrat.

Castro, O. and Gimeno, R. (1996) Tégulas con huellas de animales en el Castro de Viladonga. *Boletín CROA* 6, 9–11.

Colominas, L. (2016) Morphometric variability of Roman dogs in Hispania Tarraconensis. The case study of the vila de Madrid Necropolis. *International Journal of Osteoarchaeology* 26(5), 897–905.

Dobosi, L. (2016) Animal and human footprints on Roman tiles from Brigetio. *Disertationes Archaeologicae ex Instituto Archaeologico Universitatis de Rolando Eötvös Nominate* 3(4), 117–33.

Eliot, W. (1991) Animal footprints on Roman bricks from Newstead. *Proceedings of the Society of Antiquaries of Scotland* 121, 223–6.

Fernández, M., Lloveras, L., Tarragó, A.. Banks, P. Bordas, A., Matas, O., Ramos, J., Triay, V., Albizuri, S. and Nadal, J. (2024) Els gossos a la Barcelona medieval. Registre arqueològic, icnològic i textual. In A. Sanchis (ed.) *Cànids i grups humans en el passat. VI Jornades d'Arqueozoologia*, 197–218. València, Museu de Prehistòria de València-Diputació de València.

Frey, R.W. (ed.) (1975) *The Study of Trace Fossils*. New York, Springer.

Lloveras, L., Fernández, A., García, L., Riera, S., Miró, C. and Nadal, J. (2021) Les restes d'aus recuperades a la Casa Corrales (El Born). Reflexions al voltant del consum d'aviram a Barcelona entre els segles XIV i XVIII. *Quarhis: Quaderns d'Arqueologia i Història de la Ciutat de Barcelona* Època II 15, 148–59.

Lloveras, L., Cruells, J., Castellano-Tresserra, A., Riera, S. and Nadal, J. (2023) Monastic meals: The role of birds in the diet of Poor Clares at the Monastery of Santa Maria de Pedralbes (Barcelona, Spain) from medieval to modern periods (14th to 19th centuries AD). *International Journal of Osteoarchaeology* 33, 701–16.

Lloveras, L., Nadal, J., Castellano, A., Cruells, J. and Riera, S. (forthcoming) The importance of mammalian meat in the diet of the Poor Clare nuns of the monastery of Santa Maria de Pedralbes (Barcelona, Spain) according to zooarchaeological and documentary data. In L. Lloveras, J. Nadal, C. Rissech and P. Banks (eds) *Reconstructing Past Monastic Life. New Trends from Archaeological, Bioanthropological and Documentary* Perspectives, vol. 2. Oxford, Oxbow Books.

Muscolino, F. (2018) Bolli e impronte su laterizi. In L. Arslan (ed.) *Amoesnissimis... Aedificiis. Gli scavi di Piazza Marconi a Cremona. Vol. II. I Materiali*, 349–602. Mantova, SAP Società Archeologica S.R.L.

Parés, P.M, Francesch, A., Jordana, J. and Such, X. (2006) *Catalans de pèl i ploma. Races domèstiques autòctones de Catalunya*. Barcelona, Lynx ed.

Pastoors, A. and Lenssen-Erz, T. (eds) (2021) *Reading Prehistoric Human Tracks*. New York, Springer.

Sabio, R. and Detry, C. (2019) Ladrillos con marcas de animales. In R. Sabio. (ed.) *Animalia inter Emeritenses*, 87–8. Madrid, Gobierno de España. Ministerio de Cultura y Deporte.

Sales, L. (2021) *La ciutat de Barcelona segons el fogatjament de 1496*. Barcelona, MUHBA, Ajuntament de Barcelona.

Sánchez, J. and Gurt, J.M. (2018) La fàbrica de ceràmica ornamental d'Antoni Tarrés, segles XVIII–XIX. *Quarhis: Quaderns d'Arqueologia i Història de la Ciutat de Barcelona* Època II 14, 150–65.

Sanjust, C. (2010) *L'obra del Reial Monestir de Santa Maria de Pedralbes des de la seva fundació fins al segle XVI. Un monestir reial per a l'orde de les clarisses a Catalunya*. Cerdanyola del Vallès, Edicions de la UAB.

Seilacher, A. (2007) *Trace Fossil Analysis*. New York, Springer.

Vila, J.M. (2005) Política municipal sobre la producció de ceràmica i l'organització corporativa dels terrissaires a Barcelona, segles XIV–XVI. *Arqueologia Medieval* 1, 104–26.

14

Bilateral skull asymmetries in two ancient semi-feral horse breeds and their implications for archaeological equine studies: how can they help us in monastic contexts?

Pere M. Parés-Casanova, Nuno Carolino, José V. Leite, Ruy Dantas, Susana Lopes and Abu B. Siddiq

This study aims to assess bilateral asymmetries (directional and fluctuating asymmetries) in the skulls of semi-feral horse populations in the Iberian Peninsula. For this purpose, a sample of 31 skulls with no bone pathologies was analysed using geometric morphometric methods. Of these, 4 specimens belonged to the Galego breed (Galicia, north-western Spain) and 27 to the Garrano breed (northern Portugal), both ancient horse breeds native to the Iberian Peninsula maintained in semi-feral conditions. A set of 15 homologous landmarks was identified on the ventral aspect of each skull for analysis. Both types of asymmetries were found to be statistically significant. Although directional asymmetry (DA) contributed most to the asymmetries detected, it does not seem to have any adaptive significance. Fluctuating asymmetry (FA) presented lower values. Given that inbreeding and hybridisation with other horse breeds can be ruled out for these two breeds and considering that horses can thrive in very harsh habitats, the asymmetries observed are likely to be the result of adaptation to challenging conditions in which these breeds have lived for millennia. This resilience can be interpreted as high degree of hardiness, indicating their strong adaptive capacities to challenging environmental conditions. Consequently, the observation of asymmetric features in both equine breeds can help us to establish the sturdiness of individuals and infer their use in different human contexts, such as, for example, in monasteries, friaries and nunneries, where horses, donkeys and mules were kept for different purposes.

Keywords: Adaptation, developmental instability, directional asymmetry, *Equus caballus*, fluctuating asymmetry.

Introduction

Horses played a significant role in ancient societies, serving various essential functions that had a profound impact on social, economic, military and cultural development (Hyland 2003). Following their domestication, horses revolutionised transportation, enabling faster travel over long distances compared to travel on foot or in vehicles drawn by oxen. In rural societies, horses were used for heavy labour such as ploughing fields and pulling carts. With the emergence of states, horses became vital in warfare, providing speed and mobility. Horse-drawn chariots were employed by many ancient civilisations, including the Egyptians, Hittites and Assyrians (Hyland 2003, 71–107). Throughout various empires across the Old World, horse production and specialised breeding became well-developed practices, playing crucial roles in military, economic and social structures (Hyland 2003, 108–63; Donaghy 2014).

In this regard, archaeological equine studies within the field of zooarchaeology are essential for understanding the multifaceted horse–human relationships in ancient societies, including the various roles of horses in past economies, social hierarchies, technological advancements, military strategies and environmental adaptations. This is also applicable to the study of daily life and economies in Christian monastic contexts (even in the warfare waged by military orders), especially in rural regions.

In zooarchaeology, geometric morphometrics is widely employed to identify variations in shape and size, thus providing the opportunity to extract additional information from animal bones (Siddiq *et al.* 2021; Seguchi *et al.* 2023). While bilateral asymmetry has been extensively studied in palaeoanthropology and archaeological anthropology, its investigation in animal remains, particularly from domestic animal taxa in archaeological contexts, has not been as thoroughly explored (Albert and Greene 1999). However, studying asymmetries can contribute significantly to some of the key recurring themes in zooarchaeology, such as shifts in domestication, herd management, species health, as well as the effects of environmental changes or adverse ecological pressures on archaeological populations (Siddiq *et al.* 2021; Seguchi *et al.* 2023).

Directional asymmetry (DA) occurs when one side of a paired structure is more developed than the other (Rivero-Aragón *et al.* 2018). DA is characterised by a distribution skewed to one side (right or left) at the population level, originating as a response to external stimuli that differentially affect both sides of the organism (Graham *et al.* 1993; Klingenberg *et al.* 2002).

Fluctuating asymmetry (FA) refers to the random developmental variation in traits that are, on average, perfectly symmetrical across a population (Valen 1962). Deviations from perfect symmetry in normally bilaterally symmetrical characters occur due to the influence of multiple factors, including environmental stress (Alibert *et al.* 1994; Palmer 1996; Albert and Greene 1999; Auffray *et al.* 1999). Recent studies have shown that high levels of FA are associated with high levels of stress, as observed among Byzantine dogs in Istanbul, probably because they were stray animals (Siddiq *et al.* 2021). Additionally, high levels of FA have been recorded among domestic animals subjected to intensive farming and labour activities (Parés-Casanova and Esteve-Puig 2014). Conversely, FA among ancient camel hybrids was lower than in pure species, possibly indicating higher stress resistance (Parés-Casanova, Siddiq *et al.* 2020).

In archaeological contexts, therefore, the study of FA may reflect factors such as: a) environmental or feeding stress; b) stresses of overhunting, which may have caused high levels of natural (in the wild) inbreeding; and c) animal labour, as it is assumed that the use of animals in farming and other laborious activities may result in higher levels of stress. This last factor must have been of great importance in horses used for transport and burden, such as those found in monastic contexts.

When inferring extensive husbandry practices, it is crucial to consider the suitability of a breed to a particular management system. Morphological measures of asymmetries can serve as proxies for an animal's performance under specific management conditions. Here, we conduct tests to determine whether skull asymmetries (DA and FA) in an ancient horse population managed under semi-feral conditions would reflect environmental stress and morphological variance. The aim is to contribute to the limited studies of skull shape asymmetries that can be applied to archaeological animals using geometric morphometric techniques. Addressing this question could incentivise further research into skull asymmetries in domestic mammals, particularly equine species, from archaeological sites – a subject which has only recently begun to receive attention.

Materials and methods

The Galego *and* Garrano *breeds*

The rugged mountainous regions of north-western Spain and Portugal, within the Iberian Peninsula, have long been home to the *Galego* and *Garrano* horse populations, respectively. These are two ancient breeds of small horses, referred to as 'facas', not ponies. For example, *Galego* horses stand between 1.20 and 1.47 m at the withers (Devimeux 1988). The diminutive stature of these horses is likely to be an adaptation to the harsh environment (Devimeux 1988).

The *Galego* breed, officially known as 'Cabalo de Pura Raza Galega', originated in Galicia in north-western Spain. Historically, these horses were utilised as war horses, for agricultural work, and for meat production (Devimeux 1988). They typically have bay or black coats (Fig. 14.1A). Nearly all *Galego* horses are managed extensively, living in semi-feral conditions in the mountainous regions of Galicia. It is believed that these horses have existed in Galicia since at least 600 BC (Devimeux 1988). These small animals require less energy and can survive periods of famine.

The *Garrano* breed (Fig. 14.1B), found in north-western Portugal, is genetically very similar to the *Galego* horse. It is primarily used as a pack horse, for riding and for light farm work. This ancient breed has remained largely unchanged for thousands of years, although it is currently in decline due to wolf predation and decreased interest in breeding. Despite their close relationship, both the *Galego* and *Garrano* belong to a larger group of small horses, which includes the Asturcón (Asturias, Spain), the Monchino (Cantabria, Spain), the Basque Pottok, the Mérens horse (Ariège, France), the Landaise (Landes, France), the Connemara (Ireland), the Bardigiano (Italy), and the Dartmoor, Exmoor, Welsh Mountain, Fell, and New Forest ponies (Great Britain) (Martínez Saïz *et al.* 1986; Sánchez García *et al.* 1996; Jordana and Parés-Casanova 1999).

Given the close phenotypic identity and continuous genetic flow between the *Galego* and *Garrano* breeds, as they graze year-round freely and often share mountain grasslands, we considered the sample studied as belonging to the same population.

Sample

A sample of skulls without bone pathologies was analysed using geometric morphometrics (GM) methods. The skulls belonged to animals over one year of age that had died naturally in the countryside, and all were well-preserved. Skulls with pathological lesions (assessed via macroscopic examination) or incomplete specimens were excluded from the study.

The final sample was 31 complete skulls, which comprised 4 skulls from the *Galego* breed and 27 from the *Garrano* breed. The *Galego* specimens are curated at the *Asociación Pura Raza Cabalo Galego* in Tui, Galicia, Spain, while the *Garrano* specimens are curated at the *Associaçao de Criadores de Equinos de Raça Garrana* in Vieira do Minho, Portugal.

Figure 14.1. Two mares (A) and one foal (B) of Garrano *horses in their natural habitat. Photo: Pere M. Parés-Casanova.*

Data acquisition

For image capture each skull was positioned on a horizontal plane, ventral side facing upward. A Nikon® D70 digital camera (image resolution of 2240 × 1488 pixels) equipped with a Nikon AF Nikkor® 28-200 mm telephoto lens was used to photograph the ventral aspect of each skull. The camera's focal axis was kept parallel to the horizontal plane and centred on the ventral aspect of each skull. A scale was included in each image for reference.

TPSUtil v. 1.50 software was employed to prepare and organise the images (Rohlf 2015). Subsequently, 15 homologous landmarks (Type 1: 5 on the midsagittal line and 5 paired on right-left sides) were digitised twice using TPSDig v. 2.16. These cranial variables were selected based on their anatomical relevance in similar studies (Parés-Casanova, Salamanca-Carreño *et al.* 2020). The description of the localisation of the 15 landmarks can be found in Figure 14.2, and they are detailed in Table 14.1.

Shape asymmetry

Shape asymmetry of the skulls was analysed by superimposing the configurations of landmarks from each side of the skull using Procrustes superimposition (Klingenberg *et al.* 1998). A generalised full Procrustes fit was performed on the two-dimensional landmark coordinates to extract shape information (Zelditch *et al.* 2004). After the configurations were scaled to unit Centroid Size (CS) – computed as the square root of the sum of squared distances of all landmarks from the centroid (Bookstein 1991) – they were rotated around their centroid, the point with average coordinates. Asymmetry was then measured as the deviations between the bilateral pairs of the corresponding superimposed landmarks (Webster and Sheets 2010).

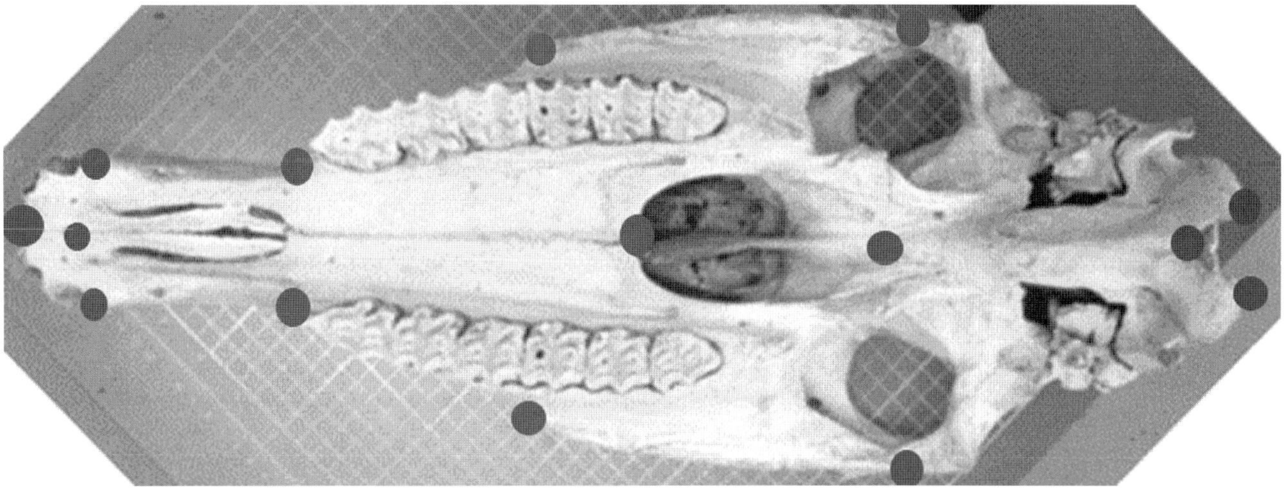

Figure 14.2. Location of landmarks used for the study of asymmetries in horse skulls (ventral aspect). In total, 15 two-dimensional landmarks were used: five of them (1 to 5) were midline and 10 (6 to 15) were bilateral paired landmarks. The set was considered to encompass elements of both neurocranium (4 to 9) and viscerocranium (1 to 3; 10 to 15).

Table 14.1. Anatomical reference points used in the study of skull asymmetries

No. of landmark	*Anatomical reference points*
1	Most rostral part of *os incisivum* on mid-sagittal line
2	*Canalis interincisivus*
3	Most caudal part of *lamina horizontalis ossis palatini* on mid-sagittal line
4	Most rostral part of *vomer* on mid-sagittal line
5	Most rostral part of *foramen magnum* on mid-sagittal line
6–7	Most lateral part of both *condylus occipitalis*
8–9	Most lateral part of both *tuberculum articulare ossis temporalis*
10–11	Most latero-rostral part of both *os zygomaticum*
12–13	Narrowest points on *ossis incisive*
14–15	Widest points on *ossis incisive* on its rostral part

Intra-observer error

To establish the degree of error in acquiring the landmark series, we repeated the measurements twice on different days for all specimens. The measurement error was tested to verify whether asymmetry estimates were significantly larger than those predicted due to intra-observer error alone (Fruciano 2016).

Statistical analysis

The effect of allometry was verified using multivariate regression of shape (Procrustes coordinates) on the Centroid Size (CS), which was $\log_{10}$-transformed (Mosimann 1970). Here, CS served as a proxy for general skull size. A two-way, mixed-model analysis of variance (ANOVA) was performed separately on each of the characters, including two replicas. To assess directional asymmetry (DA) and fluctuating asymmetry (FA), we conducted a Procrustes ANOVA, evaluating FA as the 'individual * side effect' and DA as the 'side effect' (Klingenberg *et al.* 2002). In this analysis, 'sides' was treated as a fixed effect, while 'individual' was considered a random effect. The individual variation of each character was partitioned into DA, the main effect due to 'sides' at a population level, and individual variation in size and shape (the main effect due to 'individual'), into FA, the 'sides-by-individual' interaction (Klingenberg and McIntyre 1998). Degrees of freedom for the ANOVA shape were calculated as the degrees of freedom for each of the effects multiplied by the number of landmark coordinates, minus four.

Analyses were conducted using the PAST v. 4.17c (Hammer *et al.* 2001) and MorphoJ v. 1.06c (Klingenberg 2011) packages. Statistical significance was determined at the 95% level for all tests.

Results

Asymmetry detection and measurement error

The results of the ANOVA analysis indicated a significant presence of both directional asymmetry (DA, 'side effect') and fluctuating asymmetry (FA, 'individual * side effect') ($p<0.05$), but not measurement error ('error effect').

Results showed distinct patterns of mean square values for individual variation, DA and FA. DA accounted for 16.0% of the total variance and was centred on the neurocranium (Fig. 14.3), while FA accounted for 5.6% of the total variance (Table 14.2). The measurement error was significantly lower, accounting for only 1.93% of the total variance (Table 14.2). This suggests that the FA observed is not due to intra-observer measurement error.

Allometry

The results did not show a correlation between asymmetries and size (p=0.307), indicating that size differences did not significantly affect asymmetry. This implies that regardless of age, the same right-left cranial configuration is found in the sample studied.

Discussion and conclusions

This study assesses bilateral asymmetries in the skulls of semi-feral *Galego* and *Garrano* horse populations and their possible implications for understanding environmental adaptability and resilience among archaeological horses.

The directional asymmetry (DA) observed in the sample, which was focused mainly on the neurocranium, suggests that the asymmetries in the basal skull are not functional. Therefore, it can be argued that the DA observed in *Galego* and *Garrano* horse skulls is not an adaptive consequence of developmental physiological constraints. We deduce that the genes involved in DA, which are conspicuously expressed in other non-bilateral structures of the body (such as lungs, heart, kidneys, adrenal glands and thyroid) and related to anatomical variation, are also expressed physiologically in the skull, as has been described in other domestic mammals (Parés-Casanova and Bravi 2014; Parés-Casanova 2016; Parés-Casanova and Medina 2019; Parés-Casanova, Salamanca-Carreño *et al.* 2020). While DA may obscure the effects of fluctuating asymmetry (FA), the latter, although lower, was also statistically significant.

Although *Galego* and *Garrano* horses are managed in semi-feral conditions, both breeds have solely inhabited the mountainous landscapes of north-western Iberia since ancient times (Devimeux 1988). These breeds are very similar, unlikely to have hybridised with other horse breeds and are likely to have remained unchanged for thousands of years. Considering that horses can thrive in

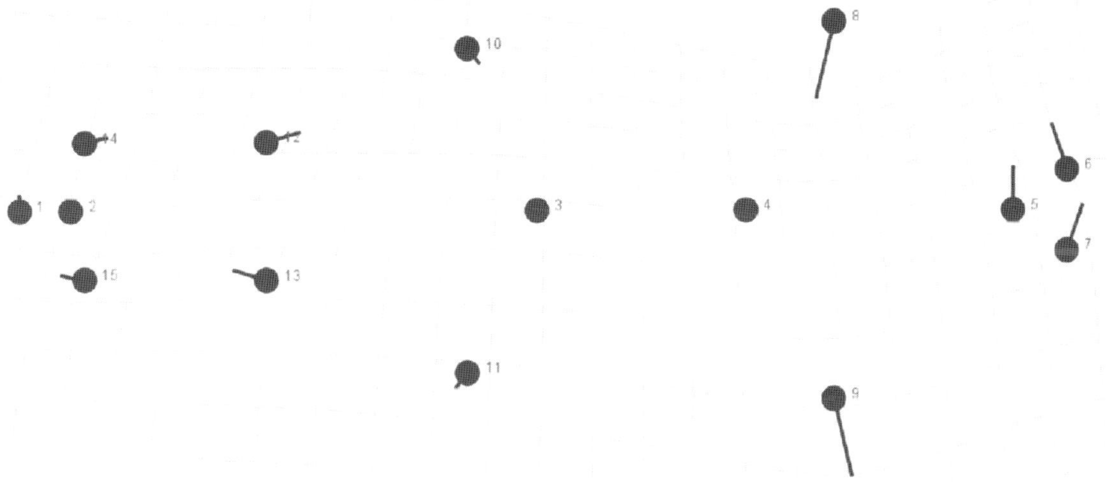

Figure 14.3. Deformation grid in horse skulls (ventral aspect) according to 15 two-dimensional landmarks.

Table 14.2. Results of Procrustes ANOVA. Fluctuating Asymmetry is expressed as 'individual side' effect and Directional Asymmetry as 'side' effect*

Effect	*Sum of Squares*	*Mean Squares*	*Degrees of freedom*	*F*	*p*
Individual	0.00098	7.5E-05	13	9.95	<.0001
Side	0.00028	2.2E-05	13	2.84	0.035
Individual*side	9.8E-05	7.6E-06	13	2.93	0.003
Error	0.00013	2.6E-06	52	0.10	1
Residual	0.04023	2.7E-05	1508		

harsh environmental conditions such as extreme climates, steep orography and wolf predation, we hypothesise that the fluctuating asymmetry (FA) in *Galego* and *Garrano* horses is an adaptive response to the adverse geo-climatic conditions of their habitats. In other words, the low FA in these horses may indicate their high level of hardiness and fitness in the face of the challenging natural environments they have lived in for thousands of years.

In conclusion, the functional significance of directional asymmetry (DA) is apparent among the *Galego* and *Garrano* horse skulls studied. Although fluctuating asymmetry (FA) appears low, it is still significant, suggesting an expression of buffered developmental stability. Given the resilience of *Galego* and *Garrano* horses to adverse conditions, these breeds are expected to have high adaptive capacity, or a high level of sturdiness. Therefore, based on the results of bilateral asymmetries observed in the skulls of these two ancient horse breeds, we can argue that developmental stress is not increased in highly rustic breeds.

Recently, geometric morphometrics has been widely introduced in zooarchaeology (*e.g.* Lloveras *et al.* 2022; Parés-Casanova *et al.* 2023). However, the study of asymmetries can significantly contribute to recurring themes in zooarchaeology, such as shifts in domestication, herd management, species health and the effects of environmental changes or adverse ecological pressures on archaeological populations (Siddiq *et al.* 2021; Seguchi *et al.* 2023). Notably, the application of geometric morphometrics for studying asymmetries in archaeological horse populations has not yet received considerable attention.

Morphometric analysis of skull remains offers valuable insights into various aspects, including stress responses experienced during individuals' lifetimes (Albert and Greene 1999; Parés-Casanova, Siddiq *et al.* 2020; Siddiq *et al.* 2021). While geometric morphometric (GM) techniques are powerful, they come with certain constraints. These techniques require relatively large sample sizes to ensure statistical robustness and are sensitive to measurement errors, as the differences in bilateral trait values are often minimal. Therefore, it is crucial that the samples be well preserved to obtain accurate and reliable measurements. Despite the moderate sample size, our observations of bilateral asymmetries – both DA and FA – in the skulls of *Galego* and *Garrano* horses provide valuable insights into the resilience and adaptive characteristics of these two ancient semi-feral horse breeds. Acquiring a large number of complete horse skulls from archaeological assemblages is nearly impossible. However, the GM data on DA and FA from *Galego* and *Garrano* horse skulls may offer important implications for archaeological equine studies.

In the archaeofaunal assemblages recovered in monastic houses, equine bones are not as frequent as other species dedicated to human consumption, especially sheep, birds and fish. Nor are they comparable to the evidence of species that must have been treated as pets, particularly cats. Despite this, documentary references pointing to the presence of equines (horses, donkeys or mules) kept on monastic premises, often outside the enclosure areas, are not uncommon. Some examples are the convent of Sant Domenech in Girona (Spain) where documents indicate that friars used a mule to transport purchases and to help in the grape harvest (Vila 2016), or the monastery of Sant Benet de Bages (Sant Fruitós de Bages, Spain) where textual evidence shows the use of mules for agricultural activities (Serra de Manresa 1993). Another interesting case is that of Santa María de Pedralbes in Barcelona, where medieval documentation tells us about the presence of a mule in the monastery (Castellano 1998). Among the large archaeological faunal assemblage studied from this monastery (Lloveras *et al.* 2023), coming from the kitchen rubbish dumps, no equine skeletal remains were identified. However, in the archaeological excavations that are currently being carried out, the complete skeleton of a buried equine has been found in a pit located within the monastic complex (Fig. 14.4; A. Bordas, personal communication). Similarly, in other monasteries in the area of study of our project the presence of horses or other equines is always present although represented by only a few elements. This is the case of Sant Agustí Vell (Barcelona) with a single example, or Nostra Senyora dels Àngels d'Avinganya (Lleida) with two specimens.

The characterisation of these animals is essential to be able to know their exact taxonomy as well as their functionality within the monastic complex (agricultural work, transport of goods, human transport, etc.). In this sense, metric

Figure 14.4. Horse skeleton excavated in the monastery of Santa Maria de Pedralbes in Barcelona. Courtesy of A. Bordas from ATICS S.L.

studies and, in our opinion, the analysis of aspects such as asymmetry, can bring us closer to the study of robustness and other intrinsic characteristics of the different domestic breeds, and particularly, on horses.

The observation of asymmetric features in both equine breeds can help us to establish the hardiness of individuals and infer their use in different human contexts, such as, for example, in friaries and nunneries, where horses, donkeys and mules were kept for different purposes.

Acknowledgements

We want to express our gratitude to Anna Bordas (ATICS SL) for allowing us to cite references to the unpublished data from new excavations at Santa Maria de Pedralbes.

This work was supported by research project PID2020-118194RJ-100 funded by AEI/MINECO.

References

Albert, A.M. and Greene, D.L. (1999) Bilateral asymmetry in skeletal growth and maturation as an indicator of environmental stress. *American Journal of Physical Anthropology* 110(3), 341–9. https://doi.org/10.1002/(SICI)1096-8644(199911)110:3<341:AID-AJPA6>3.0.CO;2-8

Alibert, P., Renaud, S., Dod, B., Bonhomme, F. and Auffray, J.C. (1994) Fluctuating asymmetry in the Mus musculus hybrid zone. A heterotic effect in disrupted co-adapted genomes. *Proceedings of the Royal Society B: Biological Sciences*, 258 (1351), 53–9. https://doi.org/10.1098/rspb.1994.0141

Auffray, J.C., Debat, V. and Alibert, P. (1999) Shape asymmetry and developmental stability. In J.C.M. Mark, A.J. Chaplain and G.D. Singh (eds) *On Growth and Form: Spatio-Temporal Pattern Formation in Biology*, 309–24. New York, John Wiley and Sons Ltd.

Bookstein, F.L. (1991) *Morphometric Tools for Landmark Data: Geometry and Biology*. Cambridge, Cambridge University Press. https://doi.org/10.1002/bimj.4710350416

Castellano, A. (1998) *Pedralbes a l'Edat Mitjana. Història d'un monestir femení*. Barcelona, Publicacions de l'Abadia de Montserrat.

Devimeux, Th. (1988) *Les Equides en Espagne*. Paris, CEREOPA.

Donaghy, T. (2014) *Horse Breeds and Breeding in the Greco Persian World: 1st and 2nd Millennium BC*. Newcastle upon Tyne, Cambridge Scholars Publishing.

Fruciano, C. (2016) Measurement error in geometric morphometrics. *Development Genes and Evolution* 226(3), 139–58. https://doi.org/10.1007/s00427-016-0537-4

Graham, J.H., Freeman, D.C. and Emlen, J.M. (1993) Antisymmetry, directional asymmetry, and dynamic morphogenesis. *Genetica* 89(1–3), 121–37. https://doi.org/10.1007/BF02424509

Hammer, Ø., Harper, D.A.T. and Ryan, P.D. (2001) PAST v. 2.17c. *Palaeontologia Electronica* 4(1), 1–229.

Hyland, A. (2003) *The Horse in the Ancient World*. Stroud, Sutton.

Jordana, J. and Parés-Casanova, P.M. (1999) Relaciones genéticas entre razas ibéricas de caballos utilizando caracteres morfológicos (prototipos raciales). *Animal Genetic Resources Information* 26, 75–94.

Klingenberg, C.P. (2011) MorphoJ: an integrated software package for geometric morphometrics. *Molecular Ecology Resources* 11 (2), 353–7. https://doi.org/10.1111/j.1755-0998.2010.02924.x

Klingenberg, C.P., Barluenga, M. and Meyer, A. (2002) Shape analysis of symmetric structures: quantifying variation among individuals and asymmetry. *Evolution* 56(10), 1909–20. https://doi.org/10.1111/j.0014-3820.2002.tb00117.x

Klingenberg, C.P. and McIntyre, G.S. (1998) Geometric morphometrics of developmental instability: analyzing patterns of fluctuating asymmetry with procrustes methods. *Evolution* 52(5), 1363. https://doi.org/10.2307/2411306

Klingenberg, C.P., McIntyre, G.S. and Zaklan, S.D. (1998) Left-right asymmetry of fly wings and the evolution of body axes. *Proceedings of the Royal Society B: Biological Sciences* 265 (February), 1255–9. https://doi.org/10.1098/rspb.1998.0427

Lloveras, L., Rissech, C., Davis, S. and Parés-Casanova, P.M.(2022) Morphological differences between sheep and goat calcanea using two-dimensional geometric morphometrics. *Animals* 12(21), 2945. https://doi.org/10.3390/ani12212945

Lloveras Ll., Rissech C., Nadal J., Asensio V., Miró C. and Riera S. (2023) MonBones: Dieta, salut, economia i societat en contextos monàstics de Barcelona (s. XIV - XIX) des d'una perspectiva arqueològica. Primeres dades aportades per la fauna. *Rodis. Journal of Medieval and Post-Medieval Archaeology* 6, 123–42.

Martínez Saïz, J., Valera Córdoba, M. and Molina Alcalá, A. (1986) El caballo losino: estudio morfológico de la raza. *Boletín de la Sociedad Española para los Recursos Genéticos Animales* 3(1), 34–46.

Mosimann, J.E. (1970) Size allometry: size and shape variables with characterizations of the lognormal and generalized gamma distribution. *Journal of the American Statistical Association* 65(330), 930–45.

Palmer, A.R. (1996) From symmetry to asymmetry: phylogenetic patterns of asymmetry variation in animals and their evolutionary significance. *Proceedings of the National Academy of Sciences of the United States of America* 93(25), 14279–86. https://doi.org/10.1073/pnas.93.25.14279

Parés-Casanova, P.M. (2016) Skull asymmetry in sheep is dominated by right side. *Journal of Morphology and Anatomy* 3(2), 1–5.

Parés-Casanova, P.M. and Bravi, R. (2014) Directional and fluctuating asymmetries in domestic sheep skulls. *Journal of Zoological and Bioscience Research* 2(3), 11–17.

Parés-Casanova, P.M. and Esteve-Puig, C. (2014) Directional and fluctuating asymmetries in domestic pig skulls. *Research* 1(828), 1–5. http://dx.doi.org/10.13070/rs.en.1.828

Parés-Casanova, P.M. and Medina, A. (2019) Skull directional asymmetry in companion rabbit. *Revista de Investigaciones Veterinarias del Peru* 30(3), 1003–8. https://doi.org/10.15381/rivep.v30i3.15428

Parés-Casanova, P.M., Rissech, C., Davis, S. and Lloveras, L. (2023) Morphological integration on the calcaneum of domestic sheep (Ovis aries Linnaeus, 1758) – a geometric morphometric study. *Applied Sciences* 13(7), 4177. https://doi.org/10.3390/app13074177

Parés-Casanova, P.M., Salamanca-Carreño, A., Crosby-Granados, R.A. and Bentez-Molano, J. A. (2020) A comparison of traditional and geometric morphometric techniques for the study of basicranial morphology in horses: A case study of the Araucanian horse from Colombia. *Animals* 10(1), 118. https://doi.org/10.3390/ani10010118

Parés-Casanova, P.M., Siddiq, A.B., Morilla-Gorgot, C. and Onar, V. (2020) Lower palatine developmental instability in hybrid Old World camelids. *Journal of Advanced Veterinary and Animal Research* 7(4). https://doi.org/10.5455/javar.2020.g465

Rivero-Aragón, A., Suárez-Hernández, A.M. and Alvaré-Jaramillo, A. (2018) Evolution of population level lateralization in Dysdercus andreae L. (Heteroptera: Pyrrhocoridae). *Centro Agrícola* 45(4), 75–84.

Rohlf, F.J. (2015) The tps series of software. *Hystrix* 26(1), 9–12. https://doi.org/doi:http://dx.doi.org/10.4404/hystrix-26.1-11264

Sánchez García, I., Iglesias, A., Fernández, A. and Viana, J.L. (1996) Caballo gallego de monte (Poney Gallego). *Animal Genetic Resources Information* 19, 45–56. https://doi.org/10.1017/S101423390000078X

Seguchi, N., Loftus, J.F. III, Yonemoto, S and Murphy, M.M. (2023) Investigating intentional cranial modification: a hybridized two-dimensional/three-dimensional study of the Hirota site, Tanegashima, Japan. *PLOS ONE* 18 (8), p. e0289219. https://doi.org/10.1371/journal.pone.0289219

Serra de Manresa, F. V. (1993) La vida quotidiana de les institucions conventuals i monàstiques, a través dels llibres de comptabilitat (Segles XVI i XVII). *Pedralbes. Revista d'Història Moderna* 13(2), 181–3.

Siddiq, A.B., Parés-Casanova, P.M., Öncü, Ö., Kar, H. and Onar, V. (2021) High level of fluctuating asymmetry in the Byzantine dogs from the Theodosius Harbor, Istanbul, Turkey. *Turkish Journal of Veterinary and Animal Sciences* 45(2), 248–56. https://doi.org/10.3906/vet-2011-79

Valen, L. van (1962) A study of fluctuating asymmetry. *Evolution* 16, 125–42.

Vila, P. (2016) Aspectes de la cuina i l'alimentació, a l'època del Barroc, al convent de Sant Domènec de Girona (1600-1612). *Annals de l'Institut d'Estudis Gironins* 57,187–234.

Webster, M. and Sheets, H.D. (2010) A practical introduction to landmark-based geometric morphometrics. In J.A. and G. Hunt (eds) *Quantitative Methods in Paleobiology*, 163–88. Chicago, The Paleontological Society.

Zelditch, M.L., Swiderski, D.L. and Sheets, H.D. (2004) *Geometric Morphometrics for Biologists: A Primer*. Boston, Elsevier Academic Press. https://doi.org/10.1016/B978-0-12-778460-1.X5000-5

15

Estimating the withers height and body length of domestic cattle from head values: its applications to archaeological samples

Arcesio Salamanca-Carreño, Pere M. Parés-Casanova, Germán Martínez Correal, Mauricio Vélez-Terranova, David Eduardo Rangel and Abu B. Siddiq

Few equations exist to estimate body measurements from the post cranial skeletal elements of archaeological cattle. Additionally, no study has yet formulated equations for estimating cattle body measurements from head values. This study aims to investigate the reliability of predictive formulae based on simple cephalic measurements to estimate withers height and body length in cattle. Six cephalic measurements (length and breadth of skull, face and head) were collected from 79 Sanmartinero breed females, an ancient creole cattle breed from Colombia, aged between 2 months and 10 years. Using these measurements, we generated simple linear Ordinary Least Squares equations y=a+bx to estimate body length and withers height. Except for head breadth, all skull measurements provided good estimations of withers height and body length. Therefore, it is proposed that linear measurements of the face and skull serve as reliable estimators for body length and withers height in cows. These measurements are crucial for assessing both the visual morphology and body weight of animals. Zooarchaeologists can benefit from using the formulae proposed, given the rarity of complete skeletal elements in archaeological cattle assemblages.

Keywords: Body length, *Bos taurus*, osteometry, Sanmartinero breed, withers height.

Introduction

Body mass estimates are crucial for understanding adaptive strategies, comparing features across animal groups, deducing husbandry practices, and reconstructing ancient phenotypes. As a significant individualising feature, body size can also influence taphonomic processes. Therefore, accurate methods for estimating animal body size are essential in zooarchaeological research (Von den Driesch and Boessneck 1974; Kyselý 2016). Specifically, body measurements such as body length and withers height are vital for assessing the visual morphology of archaeological cattle populations, as there is a strong correlation between body measurements, body weight and productivity in domestic cattle (Sieber *et al.* 1988). Thus, estimating body size also plays an important role in zooarchaeological analyses of cattle remains (Boessneck 1956; Matolcsi 1970; Kyselý 2016; Davis and Sendim 2020).

In relation to the monastic environment, the topic that this volume deals with, monasteries may be a focus of attraction for external food resources, especially of animal origin, whether through the purchase of meat in the surrounding markets or by the acquisition of animals through tithes from the subject populations (Fig. 15.1). The metric homogeneity of measurements among different individuals can reflect the presence of different breeds and possibly the diverse origins of these resources. For example, withers heights obtained from measurements of different cattle post-cranial bones from excavations at the Charité-sur-Loire monastery (France) are very homogeneous, with a range width from 115 cm to 120 cm (Audoin-Rozeau 1986). In contrast, at the monastery of Santa Maria de Pedralbes (Barcelona, Spain), in an unpublished study where a collection of 11 metapodial bones (both metacarpi and metatarsi) were measured, the results are clearly heterogeneous, with a range width varying

Figure 15.1. Medieval monks and ox-drawn ploughing. A miniature from the period (c. *1300) (source: Wikimedia Commons).*

from 116 cm to 138 cm on withers height. At La Charité-sur-Loire it can be assumed that there was little breed variability among the cattle, while at Santa Maria de Pedralbes variability must have been considerable, surely representing the presence of different breeds and diverse origins.

The relationship between skeletal parameters and body mass is generally allometric. The body size of an archaeological population can be calculated using the relationship between body mass and the dimensions of skeletal parts in their living relatives. A common approach to generating a predictive equation involves regressing body mass on a skeletal variable (Elliott *et al.* 2016). However, obtaining accurate estimates can be challenging because the mass of extant breeds is only approximately reflected in the size of their bones and teeth. For example, tooth measurements correlate with body mass due to the relationship between the size of the animal and the amount of food it needs to process for survival (Wimberly 2023). However, significant differences can exist among these variables due to different husbandry practices.

In addition to teeth, various skeletal elements commonly found in faunal assemblages, such as skulls and long bones, can be used to estimate body mass. The choice of variables for estimating conformation is crucial, as different proxies can yield a wide variety of results (Bover *et al.* 2010). The estimation of body mass from skeletal remains typically relies on regression analysis. Most equations for estimating body mass found in the literature are based on post-cranial variables, with few derived from cranial variables (Elliott *et al.* 2016).

Despite their widespread use, the currently available predictive equations for estimating cattle body size may not be as reliable as commonly assumed (Von den Driesch and Boessneck 1974; Wimberly 2023). One risk is that sample sizes are often too small, or the procedures are based on modern samples from highly selected breeds or breeds from temperate zones with low degrees of resilience, which can lead to unreliable estimations. Additionally, regional cattle populations exhibit management differences in body size, which means that the true relationship between post-cranial skeletal variables and body size conformation has often not been adequately captured. In contrast, the spectrum of head conformations is less affected by husbandry practices.

Harmonies, or correlations between morphological traits, can vary between breeds (Silva-Jarquin *et al.* 2019). Using morphological values obtained from nine modern cattle breeds, namely Hungarian Simmental, Hereford, Aberdeen Angus, Red Angus, Lincoln Red, Shaver, Charolais, Limousin and Blonde d'Aquitaine (Bene *et al.* 2007), we performed a bivariate regression of head breadth (HdB) against body length (BL) and withers height (WH). The results showed no significant changes according to that cephalic trait (r=0.277, p=0.469 for BL; r=0.171, p=0.659 for WH). Therefore, we hypothesise that formulae based on head measurements can be considered good predictors of cattle body measurements, specifically body length and withers height.

To obtain reliable inferences, databases of linear measurements must be compiled from the closest living populations, a procedure that minimises bias. Ideally, these living breeds should be phylogenetically close to extinct breeds and exhibit low specific diversity and a narrow range of body masses. The best modern analogues to medieval domestic cattle would be those that have undergone the least morphological changes since the 15th century AD, meaning they have low phylogenetic distance and anatomical divergence. In other words, the use of such breeds for calibrating regression functions is preferable because their phylogenetic characters are more likely to be like those of their ancestors.

Aiming to estimate cattle body size based on skull remains, this study is designed to estimate body measurements (withers height and body length) in domestic cattle from simple cephalic measurements. For this purpose, a wide sample of Sanmartinero creole cattle was studied. This unselected native breed was chosen because it is presumed to have biotypes closer to ancient animals than those of highly selected modern breeds. Given that finding complete or nearly complete skulls in archaeological assemblages is not always possible, our variables were based on fragmentary parts of the neurocranium (length and breadth) and viscerocranium (length and breadth) as external independent size estimators. Considering these, we aim to generate linear equations for estimating body length and withers height from simple cranial skeletal features. By measuring variables directly from live animals in a large sample of an ancient cattle breed, our goal is to address some of the shortcomings associated with cattle body measurements and provide simple reliable means for estimating cattle height and length.

Materials and methods

The Sanmartinero breed

The creole Sanmartinero breed is a yellow-red or chestnut, well-built beef type (Porter 1991) of the eastern lowlands in the Colombian Orinoquia region (Martínez Correal *et al.* 2020) (Fig. 15.2). Its origins can be traced to the zootechnical work by the Jesuits, in the 16th century (Martínez Correal and Chaves Martínez 2001), with cattle introduced by the Spanish (Porter 1991) from Extremadura (Martínez *et al.* 2009). The breed currently has an average body mass of 232 and 215 kg, and a withers height of 135 and 130 cm for males and females respectively at 16 months of age (Martínez *et al.* 2001; 2009).

Although Creole breeds mainly derived from the Iberian Peninsula, since their importation to the Americas they have undergone a long period of selective adaptation to a diverse set of environmental conditions. Despite various threats and admixture with breeds from other European countries and from India, some Creole populations are still maintained nowadays, especially in marginal regions.

The Sanmartinero indigenous bovine breed is one of the seven indigenous breeds present in Colombia and is at risk of disappearing given the low number of individuals in the country due to the preferential use of improved exotic breeds. Since its introduction into Latin America, Sanmartinero cattle have experienced a prolonged process of natural selection that has allowed the breed to develop adaptive characteristics to live in extreme climatic environments (high temperature and relative humidity) and to feed on fibrous forage in the ecosystem where it lives (Salamanca-Carreño *et al.* 2024).

Figure 15.2. A Sanmartinero cow with her calf (photo by A. Salamanca-Carreño).

Among the remaining pure creole cattle in Americas, Sanmartinero cattle retain some primeval characters, which have escaped 'improvement' by selective crossing with other breeds. It is believed that Creole breeds are more closely related to medieval Iberian populations than any other modern Iberian breed. Some isolated populations of continental historical breeds in Iberia peninsula – such as Cachena and Barrosa, Vianese and Minhotan – could still bear similarities to the physical traits of ancient populations (Montserrat *et al.* 2001; Rivero Martínez 2015). However, it is very difficult to obtain measurements from them or from a representative sample size. On the other hand, ancient but not extinct taurine breeds such as Chianina from Italy, Pinzgau from Austria, and Chillingham, Shorthorn and Sussex from England are from non-Iberian countries so they could present a different conformation from Iberian ones due to their distant areas of origin. Moreover, these breeds have not come down through time intact, due to genetic manipulation and their productivity and efficiency yields. Therefore, we preferred to sample the creole Sanmartinero breed as it can give a more accurate insight into the appearance of an extinct (Iberian) cattle population.

Methods

We sampled 79 living female cattle of the Sanmartinero breed, aged between 2 months and 10 years. Following standard morphometric methods (Sañudo 2009), a set of six linear head variables, namely breadth of head (HdB), length of head (HdL), breadth of face (FaB), and length of face (FaL), breadth of skull (SkB), and length of skull (SkL), were obtained for the skull of each animal (Fig. 15.3). As body traits, we obtained body length (BL), which is the distance from the *pars cranialis tuberculi minoris* to the *tuber ischiadicum*; and withers height (WH), which is the distance from the ground to the highest point of the cervical vertebrae (Fig. 15.4). Although these measurements were obtained *in vivo*, the reference points were osseous (Sañudo 2009). This ensured a minimal error due to soft head tissues, not particularly thick in this part of the bovine anatomy (Barone 1986). As cattle display a marked degree of sexual dimorphism with males being considerably larger than females (Davis and Sendim 2020), males were excluded from this study. A zoometric stick was used to obtain the measurements of length and width. To facilitate the obtention of measurements, animals were immobilised in a chute.

Non-parametric Spearman's correlation coefficients were determined and Michaelis-Menten logistic regressions for BL and WH were established. To avoid problems associated with non-normal distributions, the data were logarithmically transformed (base 10). Bivariate regressions by means of an Ordinary Least Squares (OLS) of

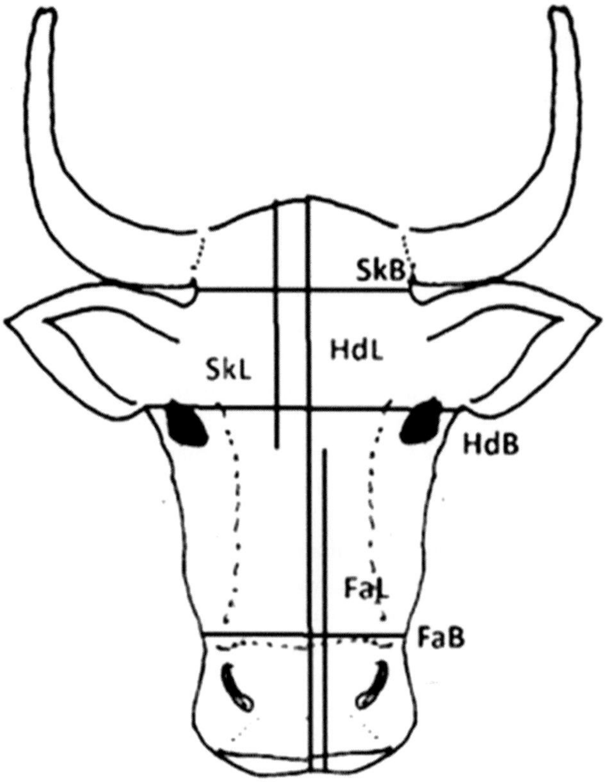

Figure 15.3. Lineal measurements on head: breadth of head (HdB, distance between both processus zygomaticus ossis frontalis); length of head (HdL, mid-sagittal distance from os parietale to the most rostral point of corpus ossis incisivi); breadth of face (FaB, distance between both tuber faciale); and length of face (FaL, mid-sagittal distance from a virtual line between the rostral point of both orbita to the most rostral point of corpus ossis incisivi); breadth of skull (SkB, minimum breadth between lateral walls of squama frontalis); and length of skull (SkL, mid-sagittal distance from os parietale to the intersection with a virtual line between rostral point of both orbita).

head measurements to BL and WH were then established. There is some debate surrounding the most appropriate choice for predicting an unknown quantity using regression (Elliott *et al.* 2016). We have chosen the OLS equation to define a line of best fit as it requires the assumption that X is measured without error, but it is asymmetric, so that the slope and resulting interpretation of the data are changed when the variables assigned to X and Y are reversed. Moreover, it has the advantage of having correction factors available to account for biases inherent in the technique (Elliott *et al.* 2016). In fact, most existing equations for estimating body mass in humans have employed OLS and inverse calibration on the grounds that it minimises the estimation error of the dependent variable (in our case, BL and WH) (Elliott *et al.* 2016). The resulting equations took the form $y=a+bx$, where 'y' is the estimated body

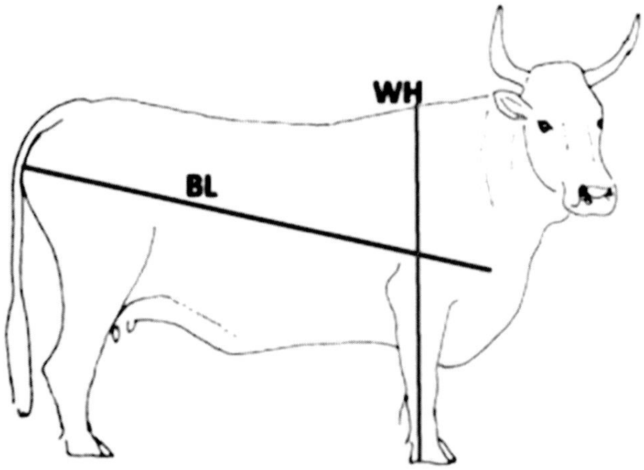

Figure 15.4. Lineal measurements on body: body length (BL, distance from pars cranialis tuberculi minoris to tuber ischiadicum) and withers height (BH, distance from ground to highest point of cervical vertebrae).

measurement, 'x' is the cephalic measurement, 'a' is the intercept of the regression line, and 'b' is its slope. BL and WH can thus be estimated as 'y' by inserting the target individual's value as 'x' for the skeletal variable into the equation.

Once the regression equations were derived, individual body measurements were entered into each corresponding formula and the estimated BL and WH ('Estimated Body Measurement' EBM) were then compared to the known BL and WH ('Known Body Measurement' KBM) for the individuals. Percentage differences (%PD) were calculated as percent error using the formula %PD=[(KBM-EBM/KBM)×100] (Elliott *et al.* 2016). %PD provides a directional difference between the known and the estimated BM. Positive %PDs indicate that known KBM>EBM (the equation underestimates BM), while negative %PDs indicate that KBM<EBM (the equation overestimates BM). To corroborate the fit of EBM to KBM, analysis of covariance (ANCOVA) tests was conducted for each head measurement.

To evaluate the predictive competence of given equations, we also used the acceptance criteria employed by Elliott *et al.* in human bones (Elliott *et al.* 2016). Specifically, the equations are considered valid when absolute percent errors fell within 20% of the known measurement. Therefore, we applied a '20% absolute error' criterion as a limit of acceptance for consistency and because it provides a relatively lenient baseline to assess each equation's competence.

Analyses were carried out using the PAST v. 4.17c package (Hammer *et al.* 2001). For all tests, statistical significance was demarcated at the 95% level.

Table 15.1. Main statistical descriptors for head measurements obtained in vivo in 79 cows of the Sanmartinero breed

	Minimal value	*Maximal value*	*Average value*	*Standard deviation*	*Coefficient of Variation (%)*
FaB	15.0	46.0	20.2	3.6	17.7
FaL	18.0	36.0	28.0	4.1	14.7
HdB	14.0	27.0	21.5	2.5	11.5
HdL	23.0	56.0	48.9	5.4	11.1
SkB	7.0	21.0	11.7	2.7	23.3
SkL	13.0	29.0	21.2	2.3	10.9
BL	95.0	162.0	138.0	14.6	10.6
WH	82.0	136.0	115.3	8.9	7.7

Breadth (HdB) and length (HdL) of head, breadth (FaB) and length (FaL) of face, breadth (SkB) and length (SkL) of skull, withers height (WH) and body length (BL) of Sanmartinero cows (n=79 females). Measurements in cm.

Table 15.2. Sexual dimorphism of Sanmartinero cattle (79 females and 21 males)

Measurement	*Male-female % difference*
FaB	8.3
FaL	0.0
HdB	2.2
HdL	-2.5
SkB	-4.3
SkL	-6.8
BL	-2.3
WH	-0.4

The differences between the average for each sex measurement is expressed as a percentage. Negative values showed a female-biased difference. See Table 15.1 for acronyms. Only head breadth (HdB) and facial breadth (FaB) showed substantial differences, being larger among bulls

Results

The main statistical descriptors for head measurements are presented in Table 15.1 and the differences between the average of females and males (not considered for the rest of the analysis), expressed as a percentage, are given in Table 15.2. Values were primarily female-biased, perhaps due to a gynomimicry phenomenon. Coefficients of variation were rather low except for SkB. As regards inter-sex differences, bulls appear wider in their face while cows have longer heads, with more marked foreheads.

To consider if head values were representative, it was essential to contrast that the values came from a normal *N*. For this purpose, the application of the Shapiro-Wilk test indicated that the cranial values did not conform to a normal ($\alpha<0.05$), except for HdB ($p=0.119$) allowing the null hypothesis (*i.e.* the distribution of the sample comes from a normally distributed population) to be rejected for all cranial variables. Moreover, logistic regression using the Michaelis-Menten model reflected clear sigmoidal increases [152.7x/(0.326+x)] and [122.9x/(0.199+x)] for BL and WH respectively. There was an initial rise in the growth velocity, which is almost linear, but as the age increased, there were diminishing returns with the eventual stop, around the age of 3 for BL and 2 years for WH. Therefore, only non-parametric tests were subsequently conducted.

Except HdB, all head values were correlated with BL and WH according to Spearman's correlation values (Table 15.3). In this way, the estimation of a simple lineal formula $y=a+bx$ for log-transformed values (except HdB) was performed. Predictive equations are presented in Table 15.4.

%PD was positive for all equations, meaning that there was an underestimation of BL and WH (Table 15.4). ANCOVA results are also presented in Table 15.4. In all cases, there were no slope differences between EBM and KBM. The '20% absolute errors' were null or very low for each variable (maximum 1.2% for FaB), meaning that there was a good limit of acceptance for all equations.

Discussion and conclusions

Estimating body measurements from skeletal remains is a crucial step in zooarchaeological practice. Body measurements, including withers height and body length, are essential tools for characterising archaeological animal populations. These measurements are particularly valuable for assessing the visual morphology of animals, as well as for estimating body mass, meat yields, or even sexual dimorphism within a given population (Kyselý 2016; Kirillova *et al.* 2019). Although various formulae based on post-cranial elements can estimate body size and mass in archaeological animals, these methods often yield inconsistent results when applied to archaeological bones (Von den Driesch and Boessneck 1974; Scott 1990; Wimberly 2023). Discrepancies can also arise due to the

Table 15.3. Spearman's correlation coefficients for each of the morphometric variables studied. P-values at the top. Significative coefficients appear in bold

	FaB	*FaL*	*HdB*	*HdL*	*SkB*	*SkL*	*BL*	*WH*
FaB		0.000	0.009	0.000	0.000	0.003	0.000	0.000
FaL	**0.449**		0.601	0.000	0.005	0.279	0.000	0.000
HdB	**0.290**	-0.060		0.104	0.826	0.000	0.124	0.075
HdL	**0.526**	**0.819**	0.184		0.001	0.003	0.000	0.000
SkB	**0.574**	**0.310**	0.025	**0.382**		0.079	0.000	0.000
SkL	**0.326**	-0.123	**0.640**	**0.335**	0.199		0.001	0.000
BL	**0.649**	**0.544**	0.174	**0.605**	**0.601**	**0.352**		0.000
WH	**0.641**	**0.441**	0.201	**0.549**	**0.594**	**0.391**	**0.812**	

See Table 15.1 for acronyms

Table 15.4. Simple lineal formulae y=a+bx to the estimation of body length (BL) and withers height (BH) from simple head measurements (log-transformed values) using the Ordinary Least Squares Equation Y=a+bX, where Y is the estimated body measurement (BL or BH), X is the skeletal cranial measurement, a is the intercept of the regression line, and b is its slope

	BL	R	*%PD*	p	*WH*	R	*%PD*	p
FaB	BL=1.56+0.44FaB	0.558	0.2	0.999	WH=1.67+0.20FaB	0.521	0.6	0.929
FaL	BL=1.68+0.31FaL	0.443	0.4	0.998	WH=1.79+0.18FaL	0.365	0.5	0.948
HdL	BL=1.25+0.52HdL	0.613	0.5	0.957	WH=1.49+0.33HdL	0.542	0.7	0.962
SkB	BL=1.85+0.26SkB	0.525	0.5	0.910	WH=1.83+0.20SkB	0.577	0.9	0.779
SkL	BL=1.46+0.51SkL	0.529	0.1	0.990	WH=1.58+0.35SkL	0.516	0.8	0.901

Percentage differences (%PD) were calculated as % error using the formula %PD=[(known body measurement – estimated body measurement/known body measurement)*100]. *P*-values correspond to ANCOVA tests. See Table 15.1 for acronyms

ratio of long bones, potentially affecting the accuracy of the body measurements (Johnstone 2004, 151–162, 248–287).

In the study of archaeological cattle, body measurements are generally estimated using formulae based primarily on long bones (Boessneck 1956; Matolcsi 1970; Von den Driesch and Boessneck 1974; Kyselý 2016; Badenhorst and Magoma 2019). However, complete cattle long bones are seldom found in archaeological assemblages, which are typically composed of fragmented bone parts. Similarly, finding complete or nearly complete cattle skulls is also rare in archaeological contexts. To date, no reliable formula has been produced for estimating cattle body measurements using head values. Therefore, the formula derived from Sanmartinero creole cattle offers an important alternative for estimating withers heights and body length in archaeological cattle populations. This is mainly because this study uses external independent size estimators based on fragmentary parts of the neurocranium (length and breadth) and viscerocranium (length and breadth), providing a valuable tool for zooarchaeological analysis.

In archaeological contexts, methodologies for assessing body measurements are distinctive to each taxon (Scott 1990, 302–3; Kyselý 2016; Wimberly 2023). The validity of these methodologies hinges on employing accurate methods to estimate body conformation. For the skeletal remains of domestic animals, constructing their taphonomic history is crucial, given their potential to provide insights into various cultural, ecological and behavioural phenomena of past human groups. Therefore, we posit that the variables outlined in this study will be pivotal for conducting subsequent research in what can be termed as 'animal palaeoethnology'.

The height at withers and body length of an individual are influenced by a range of internal factors such as genetics, population origin, sex, age and constitution, as well as external factors including nutrition, climate, management practices and environment. Therefore, studies of body conformation in both modern and ancient populations should take into account these diverse factors. In studies of living cattle, accurate formulae exist to estimate body weight based on measurements of height at withers and body length (Sieber *et al.* 1988; Vanvanhossou *et al.* 2018).

In our study, all head measurements except head breadth provided reliable estimations of withers height and body length. This suggests that linear equations can effectively capture variations in body height and length within archaeological domestic cattle populations, despite a slight tendency towards underestimation of actual measurements.

When examining statistical differences within the same method based on the cephalic references used, no significant intra-method differences were observed. Any slight underestimation is likely to be attributable to adjustments that would typically account for soft tissues of the living cows involved although statistically these differences are insignificant with the equations formulated.

It is a well-established fact that cranial elements are less frequent and have lower preservation rates compared to post-cranial elements in archaeological assemblages. However, despite their lower frequency, cranial elements, including those from cattle, are commonly found on archaeological sites. Therefore, based on fragmentary parts from the neurocranium and viscerocranium, our formula can reliably estimate cattle body shape. Moreover, the skull contains abundant qualitative information of ethological significance. This includes qualitative aspects such as head profile, which correlates with alloidism – a fundamental principle in animal ethnology. This information can also extend to horn design, rump inclination, tail position, and even chromatic aspects, collectively allowing for a phenotypical description of the populations studied.

However, several considerations must be considered when estimating the withers height and body length of domestic cattle from head values. One critical factor to consider is age. Both withers height (WH) and body length (BL) increase with age, typically until around 2 years for withers height and 3 years for body length in modern breeds. Since ancient populations may not have lived as long or experienced the same age-related growth patterns as modern breeds, it may not be necessary to include older cows in reference samples when applying these estimations to archaeological groups. Additionally, determining the sex of immature animals through the skull can be challenging. Moreover, skulls may be missing or poorly preserved, making sexual identification impossible. This limitation needs to be acknowledged when using skull-based estimates. Furthermore, it is essential to note that estimated conformation based on the skull cannot fully reflect the nutrition and health conditions under which an animal population developed. The size and shape of skulls are primarily genetically determined and show limited variation due to external factors. Therefore, while skull-based estimations provide valuable insights into body size and morphology, they do not capture the complete environmental and health history of the animals.

In conclusion, our study indicates that, except for head breadth, linear measurements taken from the face and skull provide reliable estimates for body length and withers height in cows. The simple equations we developed have demonstrated sufficient accuracy, fulfilling our objective of enhancing body mass estimation from cranial remains. The methodology applied to the Sanmartinero cattle breed shows promising potential for deriving body size information from archaeological cattle, even under non-ideal conditions where specimens are often scarce or fragmented and only partial skull remains are preserved in faunal assemblages. Optimal outcomes rely on the well-preserved status of both the viscerocranium and neurocranium.

Acknowledgements

The authors gratefully acknowledge the financial support provided by the Research Committee-CONADI (ID 2378-I) of the Cooperative University of Colombia for this study. Additionally, we extend our appreciation to all the research assistants for their valuable contributions to specimen measurements. We wish to express our gratitude to L. Lloveras and J. Nadal (University of Barcelona) for allowing us to cite the unpublished cattle metric results from the zooarchaeological study of Santa Maria de Pedralbes. This work is supported by research project PID2020-118194RJ-100 funded by AEI/MINECO.

References

Audoin-Rouzeau, F. (1986) *Ossements animaux du Moyen-Age au monastère de La Charité-sur-Loire*. Paris, Publications de la Sorbonne.

Badenhorst, S. and Magoma, M. (2019) The size of indigenous Venda cattle during the Late Iron Age in South Africa. *Journal of Archaeological Science: Reports* 24, 231–5. https://doi.org/10.1016/j.jasrep.2019.01.013

Barone, R. (1986) *Anatomie comparée des mammifères domestiques*. Paris, Vigot Editeurs.

Bene, S., Nagy, B., Nagy, L., Kiss, B., Polgár, J.P and Szabó, F. (2007) Comparison of body measurements of beef cows of different breeds. *Archives Animal Breeding* 50(4), 363–73. https://doi.org/10.5194/aab-50-363-2007

Boessneck, J. (1956). Ein Beitrag zur Errechnung der Widerristhöhe nach Metapodienmaßen bei Rindern. *Zeitschrift Für Tierzüchtung Und Züchtungsbiologie* 68(1), 75–90. https://doi.org/10.1111/j.1439-0388.1956.tb00619.x

Bover, P., Alcover, J.A., Michaux, J.J., Hautier, L., and Hutterer, R. (2010) Body shape and life style of the extinct Balearic dormouse Hypnomys (Rodentia, Gliridae): New evidence from the study of associated skeletons. *PLoS ONE* 5(12). https://doi.org/10.1371/journal.pone.0015817

Davis, S.J.M. and Sendim, A. (2020) Measurements of bones of seven female Barrosãs and one male Barrosão (Bos taurus L. 1758): A baseline for zooarchaeologists with notes on the evolution of Portuguese aurochsen and cattle. *Saguntum* 21, 61–86.

Elliott, M., Kurki, H., Weston, D.A. and Collard, M. (2016) Estimating body mass from skeletal material: new predictive equations and methodological insights from analyses of a known-mass sample of humans. *Archaeological and Anthropological Sciences* 8(4), 731–50. https://doi.org/10.1007/s12520-015-0252-5

Hammer, Ø., Harper, D.A.T. and Ryan, P.D. (2001) PAST v. 2.17c. *Palaeontologia Electronica* 4(1), 1–229.

Johnstone, C.J. (2004) *A Biometric Study of Equids in the Roman World*. PhD thesis, Department of Archaeology, University of York.

Kirillova, I.V., Shidlovskiy, F.K., and Zinoviev, A.V. (2019) Metapodials of ancient bison (Bison priscus) of northeast Russia: 'Stress markers,' sex and withers height. *Integrative Zoology* 14(3), 270–9. https://doi.org/10.1111/1749-4877.12356

Kyselý, R. (2016). The size of domestic cattle, sheep, goats and pigs in the Czech Neolithic and Eneolithic Periods. Temporal variations and their causes. *Archaeofauna* 25, 33–78. https://doi.org/10.15366/archaeofauna2016.25.003

Martínez Correal, G., and Chaves Martínez, G. (2001). *Ganado Criollo Sanmartinero: alternativa sustentable genética para la producción bovina en la Orinoquia.* Instituto Colombiano Agropecuario, ICA.

Martínez Correal, G., Flórez Díaz, H. and Martínez-Villate, G.C. (2020) El ganado criollo colombiano Sanmartinero, su conservación y aporte sostenible a la producción bovina en la Orinoquia colombiana. *Archivos Latinoamericanos de Producción Animal* 28(3–4), 69–86. https://doi.org/10.53588/alpa283402

Martínez, R., Onofre, G.and Polanco, N. (2009) Parámetros genéticos y tendencias para características de crecimiento en el ganado criollo sanmartinero en los Llanos Orientales de Colombia. *Ciencia & Tecnología Agropecuaria* 10(2), 196–204. https://doi.org/10.21930/rcta.vol10_num2_art:142

Matolcsi, J. (1970) Historische Erforschung der Körpergröße des Rindes auf Grund von ungarischem Knochenmaterial. *Zeitschrift Für Tierzüchtung Und Züchtungsbiologie* 87(1–4), 89–137. https://doi.org/10.1111/j.1439-0388.1970.tb01330.x

Montserrat, L., Sanchez, I., and Iglesias, A. (2001) Long horned cattle in the Iberian Peninsula. *The Origin of Hungarian Grey Cattle*, 99–106.

Porter, V. (1991) *Cattle. A Handbook to the Breeds of the World.* London, Christopher Helm Ltd.

Rivero Martínez, C.J. (2015) *Aportaciones a la caracterización de las razas bovinas autóctonas de Galicia en peligro de extinción.* Universidad de Santiago de Compostela.

Salamanca-Carreño, A., Parés-Casanova, P.M., Vélez-Terranova, M., Martínez-Correal, G. and Rangel-Pachón, D.E. (2024). Early Cannon Development in Females of the 'Sanmartinero' Creole Bovine Breed. *Animals* 14(4), 1–11. https://doi.org/10.3390/ani14040527

Sañudo, C. (2009) *Valoración morfológica de los animales domésticos.* Madrid, Ministerio de Medio Ambiente y Medio Rural y Marino.

Scott, K.M. (1990) Postcranial dimensions of ungulates as predictors of body mass. In J. Damuth and B.J. MacFadden (eds) *Body Size in Mammalian Paleobiology: Estimation and Biological Implications*, 301–35. Cambridge, Cambridge University Press.

Sieber, M., Freeman, A.E. and Kelley, D.H. (1988) Relationships between body measurements, body weight, and productivity in Holstein dairy cows. *Journal of Dairy Science* 71(12), 3437–45. https://doi.org/10.3168/jds.S0022-0302(88)79949-X

Silva-Jarquin, J.C., Román-Ponce, S.I., Durán-Aguilar, M., Vera-Ávila, H. R., Cambrón-Sandoval, V.H., and Andrade-Montemayor, H.M. (2019) Morphostructural characterization of the black creole goat raised in central Mexico, a currently threatened zoogenetic resource. *Animals* 9(7), 459. https://doi.org/10.3390/ani9070459

Vanvanhossou, S.F.U., Diogo, R.V.C. and Dossa, L.H. (2018) Estimation of live bodyweight from linear body measurements and body condition score in the West African Savannah Shorthorn cattle in North-West Benin. *Cogent Food & Agriculture* 4(1), 1549767. https://doi.org/10.1080/23311932.2018.1549767

Von den Driesch, A., and Boessneck, J. (1974) Kritische Anmerkungen zur Widerristhöhenberechnung aus Längenmassen vor-und frühgeschichtlicher Tierknochen. *Säugetierkundliche Mitteilung* 22, 325–48.

Wimberly, A.N. (2023) Predicting body mass in Ruminantia using postcranial measurements. *Journal of Morphology* 284(10), e21636.